MATHEMATICS RESEARCH DEVELOPMENTS

# NEW TRENDS IN FRACTIONAL PROGRAMMING

# MATHEMATICS RESEARCH DEVELOPMENTS

Additional books and e-books in this series can be found on Nova's website under the Series tab.

MATHEMATICS RESEARCH DEVELOPMENTS

# NEW TRENDS IN FRACTIONAL PROGRAMMING

RAM U. VERMA

nova science publishers
New York

For permission to use material from this book please contact us:
Telephone 631-231-7269; Fax 631-231-8175
Web Site: http://www.novapublishers.com



Additional color graphics may be available in the e-book version of this book.

**Library of Congress Cataloging-in-Publication Data**

Names: Verma, Ram U., author.
Title: New trends in fractional programming / Ram U. Verma (International Publications USA, Mathematical Sciences Division, Denton, TX, USA).
Description: Hauppauge, New York: Nova Science Publishers, Inc., 2019. |
Series: Mathematics research developments | Includes bibliographical references and index.
Identifiers: LCCN 2019015231 (print) | LCCN 2019017489 (ebook) | ISBN 9781536153729 () | ISBN 9781536153712 (hardcover)
Subjects: LCSH: Fractional programming. | Fractional calculus.
Classification: LCC QA402.7 (ebook) | LCC QA402.7 .V47 2019 (print) | DDC 515/.83--dc23
LC record available at https://lccn.loc.gov/2019015231

*Published by Nova Science Publishers, Inc. † New York*

# Contents

# Preface

This monograph is aimed at presenting a smooth and unified generalized fractional programming problems, especially advanced duality models for discrete minmax fractional programming problems. Under the current interdisciplinary computer - oriented research environment, these programs are among the most rapidly expanding research areas in terms of its multi - faceted applications-empowerment for real-world problems ranging from the robotics to money market portfolio management. The other more significant aspect of this monograph is to consider minimax fractional integral type problems using higher order sonvexity and sounivexity notions for developing different type duality models in terms of weak, strong and strictly converse duality theorems, which can be handled by transforming them into generalized fractional programming problems. The fractional integral type programming is one of the most application - enhanced fast-expanding areas of optimization which features in several types of real - world problems. It can be applied to different branches of engineering (including multitime multiobjective mechanical engineering problems) as well as to economics to minimize a ratio of functions between given periods of time. Furthermore, it can be utilized as a resource in order to measure the efficiency or productivity of a system. In such type of problems the objective function is given as a ratio of functions. For example, we consider a problem that deals with minimizing a maximum of several time dependent ratios involving integral expressions.

We envision that this monograph is a unique presentation for interdisciplinary research for the worldwide scientific community (including graduate students, faculty, general readers, and research scientists). Furthermore, some of the new concepts can be applied to duality theorems based on using a new class of multitime multiobjective variational problems as well.

As our problem involves a state function $x(t)$ in the integrand of the integrals, the obtained results also relate to optimal control problems.

Consider the following an advanced minmax fractional variational problem:

$$(\mathrm{P}) \qquad \min_{x} \left\{ \sum_{1 \le i \le p} \max \frac{\int_a^b f^i(t,x,\dot{x})dt}{\int_a^b g^i(t,x,\dot{x})dt} \right\}$$

$$\text{subject to } x \in PS(T,R^n), x(a) = \alpha, x(b) = \beta,$$

$$\int_a^b h^j(t,x,\dot{x})dt \le 0, j \in \underline{m} \equiv \{1,2,\ldots,m\},$$

where functions $f^i, g^i, i \in \underline{p}$ and $h^j, j \in \underline{m}$ are continuous in $t$ and have continuous partial derivatives with respect to $x$ and $\dot{x}$; $T = [a,b]$ denotes the time space, and $PS(T,R^n)$ stands for the state space of all piecewise smooth functions $x : T \rightarrow R^n$ with norm defined by

$$\|x\| = \|x\|_\infty + \|Dx\|_\infty$$

and $D$ is the differential operator on $PS(T,R^n)$ defined by

$$y = Dx \text{ if and only if } x(t) = x(a) + \int_a^b y(s)ds.$$

Thus $D = \frac{d}{dt}$ except at the point of discontinuity.

$$\int_a^b g^i(t,x,\dot{x}) > 0, \int_a^b f^i(t,x,\dot{x}) \geq 0 \text{ for each } i \in \underline{p}.$$

# Chapter 1

# Advanced Parameter-Free Duality Models

A class of second order parameter-free duality models are constructed for semiinfinite (with infinitely many inequality and equality constraints) discrete minmax fractional programming problems, and then using various generalized second-order ($\phi$, $\eta$, $\zeta$, $\tau$, $\rho$, $\theta$, $\tilde{m}$)- sonvexity frameworks, several weak, strong, and strictly converse duality theorems are established. The semiinfinite fractional programming is a rapidly fast-expanding research field, and whose real-world applications range from the robotics to money market portfolio management, while duality models and duality theorems developed in this chapter encompass most of the results on the generalized fractional programming problems (with finite number of constraints), not just for parameter-free duality results but beyond in the literature. To the best of our estimates, the obtained parameter-free duality results are innovative, and more importantly are highly significant to the context of the semiinfinite aspects of the fractional programming in terms of interdisciplinary applications worldwide.

## 1. Introduction

We intend in this chapter to establish a certain class of second-order parameter-free duality models and corresponding duality results to the context of the following semiinfinite discrete minmax fractional programming problem:

$$(P) \qquad \text{Minimize} \max_{1 \leq i \leq p} \frac{f_i(x)}{g_i(x)}$$

subject to

$$G_j(x,t) \leq 0 \text{ for all } t \in T_j, \ j \in \underline{q},$$

$$H_k(x,s) = 0 \text{ for all } s \in S_k, \ k \in \underline{r},$$

$$x \in X,$$

where $p$, $q$, and $r$ are positive integers, $X$ is a nonempty open convex subset of $\mathbb{R}^n$ ($n$-dimensional Euclidean space), for each $j \in \underline{q} \equiv \{1,2,\ldots,q\}$ and $k \in \underline{r}$, $T_j$ and $S_k$ are compact subsets of complete metric spaces, for each $i \in \underline{p}$, $f_i$ and $g_i$ are real-valued functions defined on $X$, for each $j \in \underline{q}$, $z \to G_j(z,t)$ is a real-valued function defined on $X$ for all $t \in T_j$, for each $k \in \underline{r}$, $z \to H_k(z,s)$ is a real-valued function defined on $X$ for all $s \in S_k$, for each $j \in \underline{q}$ and $k \in \underline{r}$, $t \to G_j(x,t)$ and $s \to H_k(x,s)$ are continuous real-valued functions defined, respectively, on $T_j$ and $S_k$ for all $x \in X$, and for each $i \in \underline{p}$, $g_i(x) > 0$ for all $x$ satisfying the constraints of $(P)$.

We plan further to investigate the second-order parameter-free duality aspects of our principal problem $(P)$ in semiinfinite discrete minmax fractional programming setting (with a finite number of variables and infinitely many constraints). Problems of this nature have been applied for handling modeling and analysis of a broad spectrum of theoretical as well as concrete real-world problems. Furthermore, the scope of the semiinfinite fractional programming expands well beyond handling problems from game theory, engineering design, boundary value problems, geometry, random graphs, reliability testing, optimal control problems, robotics and continuum mechanics, and others. The second-order duality results will be based on the properties of some new classes of generalized second-order $(\phi,\eta,\zeta,\tau,\rho,\theta,\tilde{m})$- sonvexity frameworks. Based on the generalized $(\phi,\eta,\zeta,\tau,\rho,\theta,\tilde{m})$-sonvexity, we will establish several second-order parameter - free duality models for a class of semiinfinite discrete minmax fractional programming problems, and related results that encompass most of the investigations available in the current literature. A new class of second order parameter-free duality models are constructed for semiinfinite (with infinitely many inequality and equality constraints) discrete minmax fractional programming problems, and then using various generalized second-order $(\phi,\eta,\zeta,\tau,\rho,\theta,\tilde{m})$- sonvexity frameworks, several weak, strong, and strictly converse duality theorems are established. Semiinfinite fractional programming is a rapidly fast-expanding research field, and whose real-world applications range from the robotics to money market portfolio management, while duality models and duality theorems developed in this chapter encompass most of the results on the generalized fractional programming problems (with finite number of constraints), not just for parameter-free duality results but well beyond in the literature. Furthermore, the presentation approach adopted in this chapter as well as the main duality results explored here will prove useful investigating other related categories of nonlinear programming problems and utilizing similar generalized convexity concepts. It seems that employing similar techniques, one can investigate the sufficient optimality and other duality aspects of the semiinfinite minmax fractional programming problems.

## 2. Preliminaries

In this section we recall the definitions of certain classes of generalized convex functions, introduced recently in [38], which will be needed in the sequel. Let $x^* \in X$ and assume that the function $f : X \to \mathbb{R}$ is twice differentiable at $x^*$. A second-order invex function is referred to as "sonvex function" in the literature.

**Definition 1.1.** The function $f$ is said to be *(strictly)* $(\phi,\eta,\zeta,\tau,\rho,\theta,\tilde{m})$*-sonvex* at $x^*$ if there exist functions $\phi : \mathbb{R} \to \mathbb{R}$, $\eta,\zeta,\tau : X \times X \to \mathbb{R}^n$, $\rho : X \times X \to \mathbb{R}$, and $\theta : X \times X \to \mathbb{R}^n$, and a positive integer $\tilde{m}$ such that for each $x \in X$ $(x \neq x^*)$ and $z \in \mathbb{R}^n$,

$$\begin{aligned} & \phi\big(f(x) - f(x^*)\big)(>) \geq \langle \nabla f(x^*), \eta(x,x^*)\rangle + \langle \zeta(x,x^*), \nabla^2 f(x^*)z\rangle \\ - & \frac{1}{2}\langle \nabla^2 f(x^*)z, \tau(x,x^*)\rangle \\ + & \rho(x,x^*)\|\theta(x,x^*)\|^{\tilde{m}}. \end{aligned}$$

The function $f$ is said to be (strictly) $(\phi,\eta,\zeta,\tau,\rho,\theta,\tilde{m})$-sonvex on $X$ if it is (strictly) $(\phi,\eta,\zeta,\tau,\rho,\theta,\tilde{m})$-sonvex at each $x^* \in X$.

**Definition 1.2.** The function $f$ is said to be *(strictly)* $(\phi,\eta,\zeta,\tau,\rho,\theta,\tilde{m})$*-pseudosonvex* at $x^*$ if there exist functions $\phi : \mathbb{R} \to \mathbb{R}$, $\eta,\zeta,\tau : X \times X \to \mathbb{R}^n$, $\rho : X \times X \to \mathbb{R}$, and $\theta : X \times X \to \mathbb{R}^n$, and a positive integer $\tilde{m}$ such that for each $x \in X$ $(x \neq x^*)$ and $z \in \mathbb{R}^n$,

$$\begin{aligned} & \langle \nabla f(x^*), \eta(x,x^*)\rangle + \langle \zeta(x,x^*), \nabla^2 f(x^*)z\rangle \\ - & \frac{1}{2}\langle \nabla^2 f(x^*)z, \tau(x,x^*)\rangle \\ \geq & -\rho(x,x^*)\|\theta(x,x^*)\|^{\tilde{m}} \\ \Rightarrow & \ \phi\big(f(x) - f(x^*)\big)(>) \geq 0, \end{aligned}$$

equivalently,

$$\begin{aligned} & \phi\big(f(x) - f(x^*)\big)(\leq) < 0 \Rightarrow \\ & \langle \nabla f(x^*), \eta(x,x^*)\rangle + \langle \zeta(x,x^*), \nabla^2 f(x^*)z\rangle \\ - & \frac{1}{2}\langle \nabla^2 f(x^*)z, \tau(x,x^*)\rangle \\ < & -\rho(x,x^*)\|\theta(x,x^*)\|^{\tilde{m}}. \end{aligned}$$

**Definition 1.3.** The function $f$ is said to be *(prestrictly)* $(\phi,\eta,\zeta,\tau,\rho,\theta,\tilde{m})$*-quasisonvex* at $x^*$ if there exist functions $\phi : \mathbb{R} \to \mathbb{R}$, $\eta,\zeta,\tau : X \times X \to \mathbb{R}^n$, $\rho : X \times X \to \mathbb{R}$, and $\theta : X \times X \to \mathbb{R}^n$, and a positive integer $\tilde{m}$ such that for each $x \in X$ and $z \in \mathbb{R}^n$,

$$\begin{aligned} & \phi\big(f(x) - f(x^*)\big)(<) \leq 0 \ \Rightarrow \\ & \langle \nabla f(x^*), \eta(x,x^*)\rangle + \langle \zeta(x,x^*), \nabla^2 f(x^*)z\rangle \\ - & \frac{1}{2}\langle \nabla^2 f(x^*)z, \tau(x,x^*)\rangle \\ \leq & -\rho(x,x^*)\|\theta(x,x^*)\|^{\tilde{m}}, \end{aligned}$$

equivalently

$$\begin{aligned} & \langle \nabla f(x^*), \eta(x,x^*)\rangle + \langle \zeta(x,x^*), \nabla^2 f(x^*)z\rangle \\ - & \frac{1}{2}\langle \nabla^2 f(x^*)z, \tau(x,x^*)\rangle \\ > & -\rho(x,x^*)\|\theta(x,x^*)\|^{\tilde{m}} \\ & \Rightarrow \phi\big(f(x) - f(x^*)\big)(\geq) > 0. \end{aligned}$$

Now we observe some specialized forms of above definitions:

**Definition 1.4.** The function $f$ is said to be *(strictly)* $(\phi,\eta,\rho,\theta,\tilde{m})$*-sonvex* at $x^*$ if there exist functions $\phi:\mathbb{R}\to\mathbb{R}$, $\eta: X\times X\to\mathbb{R}^n$, $\rho: X\times X\to\mathbb{R}$, and $\theta: X\times X\to\mathbb{R}^n$, and a positive integer $\tilde{m}$ such that for each $x\in X$ $(x\neq x^*)$ and $z\in\mathbb{R}^n$,

$$\phi\big(f(x)-f(x^*)\big)(>)\geq\langle\nabla f(x^*),\eta(x,x^*)\rangle+\frac{1}{2}\langle z,\nabla^2 f(x^*)z\rangle+\rho(x,x^*)\|\theta(x,x^*)\|^{\tilde{m}}.$$

The function $f$ is said to be (strictly) $(\phi,\eta,\rho,\theta,\tilde{m})$-sonvex on $X$ if it is (strictly) $(\phi,\eta,\rho,\theta,\tilde{m})$-sonvex at each $x^*\in X$.

**Definition 1.5.** The function $f$ is said to be *(strictly)* $(\phi,\ \eta,\ \rho,\ \theta,\ \tilde{m})$ *-pseudosonvex* at $x^*$ if there exist functions $\phi:\mathbb{R}\to\mathbb{R}$, $\eta: X\times X\to\mathbb{R}^n$, $\rho: X\times X\to\mathbb{R}$, and $\theta: X\times X\to\mathbb{R}^n$, and a positive integer $\tilde{m}$ such that for each $x\in X$ $(x\neq x^*)$ and $z\in\mathbb{R}^n$,

$$\begin{aligned}&\langle\nabla f(x^*),\eta(x,x^*)\rangle+\frac{1}{2}\langle z,\nabla^2 f(x^*)z\rangle\\&\geq-\rho(x,x^*)\|\theta(x,x^*)\|^{\tilde{m}}\\&\Rightarrow\ \phi\big(f(x)-f(x^*)\big)(>)\geq 0,\end{aligned}$$

equivalently,

$$\begin{aligned}&[\phi\big(f(x)-f(x^*)\big)(\leq)<0\\&\Rightarrow\langle\nabla f(x^*),\eta(x,x^*)\rangle+\frac{1}{2}\langle z,\nabla^2 f(x^*)z\rangle\\&<-\rho(x,x^*)\|\theta(x,x^*)\|^{\tilde{m}}.\end{aligned}$$

**Definition 1.6.** The function $f$ is said to be *(prestrictly)* $(\phi,\ \eta,\ \rho,\ \theta,\ \tilde{m})$ *-quasisonvex* at $x^*$ if there exist functions $\phi:\mathbb{R}\to\mathbb{R}$, $\eta: X\times X\to\mathbb{R}^n$, $\rho: X\times X\to\mathbb{R}$, and $\theta: X\times X\to\mathbb{R}^n$, and a positive integer $\tilde{m}$ such that for each $x\in X$ and $z\in\mathbb{R}^n$,

$$\begin{aligned}&\phi\big(f(x)-f(x^*)\big)(<)\leq 0\\&\Rightarrow\ \langle\nabla f(x^*),\eta(x,x^*)\rangle+\frac{1}{2}\langle z,\nabla^2 f(x^*)z\rangle\\&\leq-\rho(x,x^*)\|\theta(x,x^*)\|^{\tilde{m}},\end{aligned}$$

equivalently

$$\begin{aligned}&\langle\nabla f(x^*),\eta(x,x^*)\rangle+\frac{1}{2}\langle z,\nabla^2 f(x^*)z\rangle\\&>-\rho(x,x^*)\|\theta(x,x^*)\|^{\tilde{m}}\\&\Rightarrow\phi\big(f(x)-f(x^*)\big)(\geq)>0.\end{aligned}$$

**Example 1.7.** The function $f$ is said to be *(strictly)* $(\phi,\eta,\rho,\theta,\tilde{m})$*-sonvex* at $x^*$ if there exist functions $\phi:\mathbb{R}\to\mathbb{R}$, $\eta: X\times X\to\mathbb{R}^n$, $\rho: X\times X\to\mathbb{R}$, and $\theta: X\times X\to\mathbb{R}^n$, and a positive integer $\tilde{m}$ such that for each $x \in X$ $(x \neq x^*)$ and $z \in \mathbb{R}^n$,

$$\begin{aligned} &\phi\big(f(x) - f(x^*)\big)(>) \geq \langle \nabla f(x^*) + \nabla^2 f(x^*)z, \eta(x,x^*)\rangle \\ - \; &\frac{1}{2}\langle z, \nabla^2 f(x^*)z\rangle + \rho(x,x^*)\|\theta(x,x^*)\|^{\tilde{m}}. \end{aligned}$$

Next, we present some specializations of Definitions 1.1 - 1.3, which are readily available in the literature.

**Definition 1.8.** The function $f$ is said to be *(strictly)* $(\phi, \eta, \rho, \theta)$*-invex* at $x^*$ if there exist functions $\phi : \mathbb{R} \to \mathbb{R}$, $\eta : X \times X \to \mathbb{R}^n$, $\rho : X \times X \to \mathbb{R}$, and $\theta : X \times X \to \mathbb{R}^n$, such that for each $x \in X$ $(x \neq x^*)$ and $z \in \mathbb{R}^n$,

$$\phi\big(f(x) - f(x^*)\big)(>) \geq \langle \nabla f(x^*), \eta(x,x^*)\rangle + \rho(x,x^*)\|\theta(x,x^*)\|^2.$$

**Definition 1.9.** The function $f$ is said to be *(strictly)* $(\phi, \eta, \rho, \theta)$*-pseudoinvex* at $x^*$ if there exist functions $\phi : \mathbb{R} \to \mathbb{R}$, $\eta : X \times X \to \mathbb{R}^n$, $\rho : X \times X \to \mathbb{R}$, and $\theta : X \times X \to \mathbb{R}^n$, such that for each $x \in X$ $(x \neq x^*)$ and $z \in \mathbb{R}^n$,

$$\langle \nabla f(x^*), \eta(x,x^*)\rangle \geq -\rho(x,x^*)\|\theta(x,x^*)\|^2 \;\Rightarrow\; \phi\big(f(x) - f(x^*)\big)(>) \geq 0.$$

**Definition 1.10.** The function $f$ is said to be *(prestrictly)* $(\phi, \eta, \rho, \theta)$*-quasiinvex* at $x^*$ if there exist functions $\phi : \mathbb{R} \to \mathbb{R}$, $\eta : X \times X \to \mathbb{R}^n$, $\rho : X \times X \to \mathbb{R}$, and $\theta : X \times X \to \mathbb{R}^n$, such that for each $x \in X$ and $z \in \mathbb{R}^n$,

$$\phi\big(f(x) - f(x^*)\big)(<) \leq 0 \;\Rightarrow\; \langle \nabla f(x^*), \eta(x,x^*)\rangle \leq -\rho(x,x^*)\|\theta(x,x^*)\|^2.$$

We conclude this section by recalling a result [37] on the set of second-order parameter-free necessary optimality conditions for $(P)$, which will be needed for proving strong and strict converse duality theorems.

**Theorem 1.11.** Let $x^* \in \mathbb{F}$ and $\lambda^* = \max_{1 \leq i \leq p} f_i(x^*)/g_i(x^*)$, for each $i \in \underline{p}$, let $f_i$ and $g_i$ be twice continuously differentiable at $x^*$, for each $j \in \underline{q}$, let the function $z \to G_j(z,t)$ be twice continuously differentiable at $x^*$ for all $t \in T_j$, and for each $k \in \underline{r}$, let the function $z \to H_k(z,s)$ be twice continuously differentiable at $x^*$ for all $s \in S_k$. If $x^*$ is an optimal solution of (P), if the second order generalized Abadie constraint qualification holds at $x^*$, and if for any critical direction y, the set cone

$$\begin{aligned} &\{\Big(\nabla G_j(x^*,t), \langle y, \nabla^2 G_j(x^*,t)y\rangle\Big) : t \in \hat{T}_j(x^*), j \in \underline{q}\} \\ + \; &span\{\Big(\nabla H_k(x^*,s), \langle y, \nabla^2 H_k(x^*,s)y\rangle\Big) : s \in S_k, k \in \underline{r}\}, \\ &where\ \hat{T}_j(x^*) \equiv \{t \in T_j \,:\, G_j(x^*,t) = 0\}, \end{aligned}$$

is closed, then there exist $u^* \in U \equiv \{u \in \mathbb{R}^p : u \geq 0, \sum_{i=1}^{p} u_i = 1\}$ and integers $\nu_0^*$ and $\nu^*$, with $0 \leq \nu_0^* \leq \nu^* \leq n+1$, such that there exist $\nu_0^*$ indices $j_m$, with $1 \leq j_m \leq q$, together with $\nu_0^*$ points $t^m \in \hat{T}_{j_m}(x^*)$, $m \in \underline{\nu_0^*}$, $\nu^* - \nu_0^*$ indices $k_m$, with $1 \leq k_m \leq r$, together with

$\nu^* - \nu_0^*$ points $s^m \in S_{k_m}$ for $m \in \underline{\nu^*} \backslash \underline{\nu_0^*}$, and $\nu^*$ real numbers $v_m^*$, with $v_m^* > 0$ for $m \in \underline{\nu_0^*}$, and $w_m^* \in \mathbb{R}$ for $m \in \underline{\nu^*} \setminus \underline{\nu_0}$ with the property that

$$\sum_{i=1}^{p} u_i^*[D(x^*,u^*)\nabla f_i(x^*) - N(x^*,u^*)\nabla g_i(x^*)] + \sum_{m=1}^{\nu_0^*} v_m^*[\nabla G_{j_m}(x^*,t^m)$$
$$+ \sum_{m=\nu_0^*+1}^{\nu^*} w_m^* \nabla H_k(x^*,s^m) = 0, \tag{1.1}$$

$$\langle z^*, \Big[\sum_{i=1}^{p} u_i^*[D(x^*,u^*)\nabla^2 f_i(x^*) - N(x^*,u^*)\nabla^2 g_i(x^*)]$$
$$+ \sum_{m=1}^{\nu_0^*} v_m^* \nabla^2 G_{j_m}(x^*,t^m)$$
$$+ \sum_{m=\nu_0^*+1}^{\nu^*} w_m^* \nabla^2 H_k(x^*,s^m)\Big] z^*\rangle \geq 0, \tag{1.2}$$

$$u_i^*[D(x^*,u^*)f_i(x^*) - N(x^*,u^*)g_i(x^*)] = 0, \ \ i \in \underline{p}, \tag{1.3}$$

$$\max_{1 \leq i \leq p} \frac{f_i(x^*)}{g_i(x^*)} = \frac{N(x^*,u^*)}{D(x^*,u^*)},$$

$$v_j^* G_j(x^*) = 0, \ \ j \in \underline{q},$$

where $C(x^*)$ is the set of all critical directions of (P) at $x^*$, that is,

$$C(x^*) = \{z^* \in \mathbb{R}^n : \langle \nabla f_i(x^*) - \lambda \nabla g_i(x^*), z^* \rangle = 0, \ \ i \in A(x^*),$$
$$\langle \nabla G_j(x^*), z^* \rangle \leq 0, \ \ j \in B(x^*),$$
$$\langle \nabla H_k(x^*), z^* \rangle = 0, \ \ k \in \underline{r}\},$$

$A(x^*) = \{j \in \underline{p} : f_j(x^*)/g_j(x^*) = \max_{1 \leq i \leq p} f_i(x^*)/g_i(x^*)\}$, $B(x^*) = \{j \in \underline{q} : G_j(x^*) = 0\}$, $N(x^*,u^*) = \sum_{i=1}^{p} u_i^* f_i(x^*)$, *and* $D(x^*,u^*) = \sum_{i=1}^{p} u_i^* g_i(x^*)$.

We shall assume that the functions $f_i$, $g_i$, $i \in \underline{p}$, $G_j$, $j \in \underline{q}$, and $H_k$, $k \in \underline{r}$, are twice continuously differentiable on the open set $X$.

## 3. Duality Model I

In this section, we consider two second-order parameter-free duality models with relatively modest constraint structures and prove weak, strong, and strict converse duality theorems. Consider the following two problems:

$(DI)$ Maximize $\psi_I(y,z,u,v,v) = \dfrac{\sum_{i=1}^p u_i f_i(y)}{\sum_{i=1}^p u_i g_i(y)}$

subject to

$$\sum_{i=1}^{p} u_i[D(y,u)\nabla f_i(y) - N(y,u)\nabla g_i(y)] + \sum_{m=1}^{v_0} v_m \nabla G_{j_m}(y,t^m)$$

$$+ \sum_{m=v_0+1}^{v} v_m \nabla H_{k_m}(y,s^m) = 0, \tag{1.4}$$

$$\Big\langle z, \Big\{ \sum_{i=1}^{p} u_i[D(y,u)\nabla^2 f_i(y) - N(y,u)\nabla^2 g_i(y)] + \sum_{m=1}^{v_0} v_m \nabla^2 G_{j_m}(y,t^m)$$

$$+ \sum_{m=v_0+1}^{v} w_m \nabla^2 H_{k_m}(y,s^m) \Big\} z \Big\rangle \geq 0, \tag{1.5}$$

$$\sum_{m=1}^{v_0} v_m G_{j_m}(y,t^m) + \sum_{m=v_0+1}^{v} w_m H_{k_m}(y,s^m) \geq 0, \tag{1.6}$$

$$y \in X,\ z \in C(y),\ u \in U,\ v \in \mathbb{R}_+^q,\ w \in \mathbb{R}^r; \tag{1.7}$$

$(\tilde{D}I)$ Maximize $\psi_I(y,z,u,v,w) = \dfrac{\sum_{i=1}^p u_i f_i(y)}{\sum_{i=1}^p u_i g_i(y)}$

subject to (5) - (7) and

$$\Big\langle \sum_{i=1}^{p} u_i[D(y,u)\nabla f_i(y) - N(y,u)\nabla g_i(y)]$$

$$+ \sum_{m=1}^{v_0} v_m \nabla G_{j_m}(y,t^m)$$

$$+ \sum_{m=v_0+1}^{v} w_m \nabla H_{k_m}(y,s^m), \eta(x,y) \Big\rangle \geq 0 \text{ for all } x \in \mathbb{F}, \tag{1.8}$$

where $\eta$ is a function from $X \times X$ to $\mathbb{R}^n$.

Comparing $(DI)$ and $(\tilde{D}I)$, we see that $(\tilde{D}I)$ is relatively more general than $(DI)$ in the sense that any feasible solution of $(DI)$ is also feasible for $(\tilde{D}I)$, but the converse is not necessarily true. Furthermore, we observe that (4) is a system of $n$ equations, whereas (8) is a single inequality. Clearly, from a computational point of view, $(DI)$ is preferable to $(\tilde{D}I)$ because of the dependence of (8) on the feasible set of $(P)$.

Despite these apparent differences, it turns out that the statements and proofs of all the duality theorems for $(P)-(DI)$ and $(P)-(\tilde{D}I)$ are almost identical and, therefore, we shall consider only the pair $(P)-(DI)$.

For the sake of the convenience, we shall use the following list of symbols in the statements and proofs of our duality theorems:

$$D(x,u) = \sum_{i=1}^{p} u_i g_i(x)$$

$$N(x,u) = \sum_{i=1}^{p} u_i f_i(x)$$

$$\mathscr{C}(x,v) = \sum_{m=1}^{v_0} v_m G_{j_m}(x,t^m),$$

$$\mathscr{D}(x,w) = \sum_{m=v_0+1}^{v} w_m H_{k_m}(x,s^m),$$

$$\mathscr{E}_i(x,y,u) = D(y,u)f_i(x) - N(y,u)g_i(x), \quad i \in \underline{p}$$

$$\mathscr{E}(x,y,u) = \sum_{i=1}^{p} u_i[D(y,u)f_i(x) - N(y,u)g_i(x)],$$

$$\mathscr{G}(x,v,w) = \sum_{m=1}^{v_0} v_m G_{j_m}(x,t^m) + \sum_{m=v_0+1}^{v} w_m H_{k_m}(x,s^m),$$

$$I_+(u) = \{i \in \underline{p} : u_i > 0\}, \quad J_+(v) = \{m \in \underline{v_0} : v_m > 0\},$$

$$K_*(w) = \{m \in \underline{v} \setminus \underline{v_0} : w_m \neq 0\}.$$

In the proofs of our duality theorems, we shall make frequent use of the following auxiliary result which provides an alternative expression for the objective function of $(P)$.

**Lemma 1.12.** [3] For each $x \in X$,

$$\varphi(x) \equiv \max_{1 \le i \le p} \frac{f_i(x)}{g_i(x)} = \max_{u \in U} \frac{\sum_{i=1}^{p} u_i f_i(x)}{\sum_{i=1}^{p} u_i g_i(x)}.$$

The next two theorems show that $(DI)$ is a dual problem for $(P)$.

**Theorem 1.13.** (Weak Duality) Let $x$ and $\mathscr{S} \equiv (y,z,u,v,w)$ be arbitrary feasible solutions of (P) and (DI), respectively, and assume that $N(y,u) \ge 0$, $D(y,u) > 0$, and either one of the following two sets of hypotheses is satisfied:

(a) (i) for each $i \in I_+ \equiv I_+(u)$, $f_i$ is $(\phi,\eta,\zeta,\tau,\bar{\rho}_i,\theta,\tilde{m})$-sonvex and $-g_i$ is $(\phi,\eta,\tilde{\rho}_i,\theta,\tilde{m})$-sonvex at $y$;

(ii) for each $m \in J_+ \equiv J_+(v)$, $G_{j_m}$ is $(\phi,\eta,\zeta,\tau,\hat{\rho}_m,\theta,\tilde{m})$-sonvex at $y$;

(iii) for each $m \in K_* \equiv K_*(w)$, $H_{k_m}$ is $(\phi,\eta,\zeta,\tau,\breve{\rho}_m,\theta,\tilde{m})$-sonvex at $y$;

(iv) $\phi$ is superlinear and $\phi(a) \geq 0 \Rightarrow a \geq 0$;

(v) $\rho^*(x,y) + \sum_{m\in J_+} v_m\hat{\rho}_m(x,y) + \sum_{m=v_0+1}^{v} w_m\breve{\rho}_m(x,y) \geq 0$, where $\rho^*(x,y) = \sum_{i\in I_+} u_i[D(y,u)\bar{\rho}_i(x,y) + N(y,u)\tilde{\rho}_i(x,y)]$;

(b) the Lagrangian-type function

$$\xi \to L(\xi,y,u,v,w) = \sum_{i=1}^{p} u_i[D(y,u)f_i(\xi) - N(y,u)g_i(\xi)]$$

$$+ \sum_{m=1}^{v_0} v_m G_{j_m}(\xi,t^m)$$

$$+ \sum_{m=v_0+1}^{v} w_m H_k(\xi,s^m)$$

is $(\phi,\eta,\zeta,\tau,\rho,\theta,\tilde{m})$-pseudosonvex at $y$, $\rho(x,y) \geq 0$, and $\phi(a) \geq 0 \Rightarrow a \geq 0$.

Then $\varphi(x) \geq \psi_I(y,z,u,v,w)$.

*Proof.* Using the hypotheses specified in (i) - (iii) for (a), we have

$$\phi\big(f_i(x) - f_i(y)\big) \geq \langle \nabla f_i(y), \eta(x,y)\rangle + \langle \zeta(x,x^*), \nabla^2 f_i(y)z\rangle$$

$$- \frac{1}{2}\langle \tau(x,x^*)z, \nabla^2 f_i(y)z\rangle$$

$$+ \bar{\rho}_i(x,y)\|\theta(x,y)\|^{\tilde{m}}, \ i \in I_+, \tag{1.9}$$

$$\phi\big(-g_i(x)+g_i(y)\big) \geq \langle -\nabla g_i(y), \eta(x,y)\rangle$$

$$+ \quad \langle \zeta(x,x^*), \nabla^2 g_i(y)z\rangle$$

$$- \quad \frac{1}{2}\langle \tau(x,x^*), \nabla^2 g_i(y)z\rangle$$

$$+ \quad \tilde{\rho}_i(x,y)\|\theta(x,y)\|^{\tilde{m}}, \; i \in I_+, \tag{1.10}$$

$$\phi\big(G_{j_m}(x,t^m) - G_{j_m}(y,t^m)\big)$$

$$\geq \langle \nabla G_{j_m}(y,t^m), \eta(x,y)\rangle$$

$$+ \quad \langle \zeta(x,x^*), \nabla^2 G_{j_m}(y,t^m)z\rangle$$

$$- \quad \frac{1}{2}\langle \tau, \nabla^2 G_{j_m}(y,t^m)z\rangle$$

$$+ \quad \hat{\rho}_m(x,y)\|\theta(x,y)\|^{\tilde{m}}, \tag{1.11}$$

$$\phi\big(H_{k_m}(x,s^m) - H_{k_m}(y,s^m)\big)$$

$$\geq \langle \nabla H_{k_m}(y,s^m), \eta(x,y)\rangle$$

$$+ \quad \langle \zeta(x,x^*), \nabla^2 H_{k_m}(y,s^m)z\rangle$$

$$- \quad \frac{1}{2}\langle \tau(x,x^*), \nabla^2 H_{k_m}(y,s^m)z\rangle$$

$$+ \quad \breve{\rho}_m(x,y)\|\theta(x,y)\|^{\tilde{m}}. \tag{1.12}$$

Now, multiplying (9) by $u_i D(y,u)$ and then summing over $i \in I_+$, (10) by $u_i N(y,u)$ and then summing $i \in I_+$, (11) by $v_m$ and then summing over $m \in \underline{v_0}$, (12) by $w_m$ and then summing (12) over $m \in \underline{v} \setminus \underline{v_0}$, adding the resulting inequalities, and applying the superlinearity of $\phi$, we obtain

$$\phi\Big(\sum_{i=1}^{p} u_i[D(y,u)f_i(x)-N(y,u)g_i(x)]+\sum_{m=1}^{\nu_0} v_m G_{j_m}(x,t^m)$$
$$+\sum_{m=\nu_0+1}^{\nu} w_m H_{k_m}(x,s^m)$$
$$-\Big\{\sum_{i=1}^{p} u_i[D(y,u)f_i(y)-N(y,u)g_i(y)]+\sum_{m=1}^{\nu_0} v_m G_{j_m}(y,t^m)$$
$$+\sum_{m=\nu_0+1}^{\nu} w_m H_{k_m}(y,s^m)\Big\}\Big) \geq \Big\langle \sum_{i=1}^{p} u_i[D(y,u)\nabla f_i(y)-N(y,u)\nabla g_i(y)]$$
$$+\sum_{m=1}^{\nu_0} v_m \nabla G_{j_m}(y,t^m)+\sum_{m=\nu_0+1}^{\nu} w_m \nabla H_{k_m}(y,s^m), \eta(x,y)\Big\rangle$$
$$+\Big\langle \zeta(x,y), \Big\{\sum_{i=1}^{p} u_i[D(y,u)\nabla^2 f_i(y)-N(y,u)\nabla^2 g_i(y)]$$
$$+\sum_{m=1}^{\nu_0} v_m \nabla^2 G_{j_m}(y,t^m)+\sum_{m=\nu_0+1}^{\nu} w_m \nabla^2 H_{k_m}(y,s^m)\Big\}z\Big\rangle$$
$$-\frac{1}{2}\Big\langle \tau(x,y), \Big\{\sum_{i=1}^{p} u_i[D(y,u)\nabla^2 f_i(y)-N(y,u)\nabla^2 g_i(y)]$$
$$+\sum_{m=1}^{\nu_0} v_m \nabla^2 G_{j_m}(y,t^m)+\sum_{m=\nu_0+1}^{\nu} w_m \nabla^2 H_{k_m}(y,s^m)\Big\}z\Big\rangle$$
$$+\Big\{\sum_{i=1}^{p} u_i[D(y,u)\bar{\rho}_i(x,y)+N(y,u)\tilde{\rho}_i(x,y)]$$
$$+\sum_{m=1}^{\nu_0} v_m \hat{\rho}_m(x,y)+\sum_{m=\nu_0+1}^{\nu} w_m \breve{\rho}_m(x,y)\Big\}\|\theta(x,y)\|^{\tilde{m}}.$$

Because of the dual feasibility of $\mathscr{S}$, (4), (5), and (v), the above inequality becomes

$$\phi\Big(\sum_{i=1}^{p} u_i[D(y,u)f_i(x)-N(y,u)g_i(x)]+\sum_{m=1}^{\nu_0} v_m G_{j_m}(x,t^m)$$
$$+\sum_{m=\nu_0+1}^{\nu} w_m H_{k_m}(x,s^m)$$
$$-\Big\{\sum_{i=1}^{p} u_i[D(y,u)f_i(y)-N(y,u)g_i(y)]+\sum_{m=1}^{\nu_0} v_m G_{j_m}(y,t^m)$$
$$+\sum_{m=\nu_0+1}^{\nu} w_m H_{k_m}(y,s^m)\Big\}\Big) \geq 0.$$

But $\phi(a) \geq 0 \Rightarrow a \geq 0$, and hence

$$\sum_{i=1}^{p} u_i[D(y,u)f_i(x) - N(y,u)g_i(x)] + \sum_{m=1}^{v_0} v_j G_j(x,t^m) + \sum_{m=v_0+1}^{v} w_m H_{k_m}(x,s^m) - \Big\{\sum_{i=1}^{p} u_i[D(y,u)f_i(y) - N(y,u)g_i(y)] + \sum_{j=1}^{q} v_j G_j(y) + \sum_{k=1}^{r} w_k H_k(y)\Big\} \geq 0.$$

In view of the definitions of $N(y,u)$ and $D(y,u)$, (6), and the primal feasibility of $x$, this inequality further reduces to

$$\sum_{i=1}^{p} u_i[D(y,u)f_i(x) - N(y,u)g_i(x)] \geq 0. \tag{1.13}$$

Now using (13) and Lemma 1.12, we obtain the weak duality inequality as follows:

$$\begin{aligned} \varphi(x) &= \max_{a\in U} \frac{\sum_{i=1}^{p} a_i f_i(x)}{\sum_{i=1}^{p} a_i g_i(x)} \\ &\geq \max_{1\leq i\leq p} \frac{\sum_{i=1}^{p} u_i f_i(x)}{\sum_{i=1}^{p} u_i g_i(x)} \\ &\geq \frac{N(y,u)}{D(y,u)} = \psi_I(y,z,u,v,w). \end{aligned}$$

(b) : Using the dual feasibility of $(y,z,u,v,w)$, nonnegativity of $\rho(x,y)$, (4), and (5), we obtain the following inequality:

$$\langle \nabla L(y,y,u,v,w), \eta(x,y)\rangle + \langle \zeta(x,y), \nabla^? L(y,y,u,v,w)z\rangle$$

$$-\frac{1}{2}\langle \tau(x,y), \nabla^2 L(y,y,u,v,w)z\rangle$$

$$\geq 0 \geq -\rho(x,y)\|\theta(x,y)\|^{\tilde{m}},$$

which in view of our $(\phi,\eta,\zeta,\tau,\rho,\theta,m)$-pseudosonvexity assumption implies that

$$\phi\big(L(x,y,u,v,w) - L(y,y,u,v,w)\big) \geq 0.$$

But $\phi(a) \geq 0 \Rightarrow a \geq 0$ and hence we have

$$L(x,y,u,v,w) - L(y,y,u,v,w) \geq 0.$$

Because $\sum_{i=1}^{p} u_i[D(y,u)f_i(y) - N(y,u)g_i(y)] = 0$, $x \in \mathbb{F}$, $v \geq 0$, and (6) holds, the above inequality reduces to

$$\sum_{i=1}^{p} u_i[D(y,u)f_i(x) - N(y,u)g_i(x)] \geq 0,$$

which is precisely (13). As seen in the proof of part (a), this inequality leads to the desired conclusion that $\varphi(x) \geq \psi_I(y,z,u,v,w)$.

□

**Corollary 1.14.** (Weak Duality) Let $x$ and $\mathscr{S} \equiv (y,z,u,v,w)$ be arbitrary feasible solutions of (P) and (DI), respectively, and assume that $N(y,u) \geq 0$, $D(y,u) > 0$, and either one of the following two sets of hypotheses is satisfied:

(a) (i) for each $i \in I_+ \equiv I_+(u)$, $f_i$ is $(\phi,\eta,\bar{\rho}_i,\theta,\tilde{m})$-sonvex and $-g_i$ is $(\phi,\eta,\tilde{\rho}_i,\theta,\tilde{m})$-sonvex at $y$;

(ii) for each $m \in J_+ \equiv J_+(v)$, $G_{j_m}$ is $(\phi,\eta,\hat{\rho}_m,\theta,\tilde{m})$-sonvex at $y$;

(iii) for each $m \in K_* \equiv K_*(w)$, $H_{k_m}$ is $(\phi,\eta,\breve{\rho}_m,\theta,\tilde{m})$-sonvex at $y$;

(iv) $\phi$ is superlinear and $\phi(a) \geq 0 \Rightarrow a \geq 0$;

(v) $\rho^*(x,y) + \sum_{m\in J_+} v_m\hat{\rho}_m(x,y) + \sum_{m=v_0+1}^{v} w_m\breve{\rho}_m(x,y) \geq 0$, where $\rho^*(x,y) = \sum_{i\in I_+} u_i[D(y,u)\bar{\rho}_i(x,y) + N(y,u)\tilde{\rho}_i(x,y)]$;

(b) the Lagrangian-type function

$$\xi \to L(\xi,y,u,v,w) = \sum_{i=1}^{p} u_i[D(y,u)f_i(\xi) - N(y,u)g_i(\xi)]$$

$$+ \sum_{m=1}^{v_0} v_m G_{j_m}(\xi,t^m)$$

$$+ \sum_{m=v_0+1}^{v} w_m H_k(\xi,s^m)$$

is $(\phi,\eta,\rho,\theta,\tilde{m})$-pseudosonvex at $y$, $\rho(x,y) \geq 0$, and $\phi(a) \geq 0 \Rightarrow a \geq 0$.

Then $\varphi(x) \geq \psi_I(y,z,u,v,w)$.

*Proof.* Using the hypotheses specified in (i) - (iii), we have

$$\phi\big(f_i(x) - f_i(y)\big) \geq \langle \nabla f_i(y), \eta(x,y)\rangle$$

$$+ \quad \langle z, \nabla^2 f_i(y)z\rangle - \frac{1}{2}\langle z, \nabla^2 f_i(y)z\rangle$$

$$+ \quad \bar{\rho}_i(x,y)\|\theta(x,y)\|^{\tilde{m}}, \ i \in I_+, \tag{1.14}$$

$$\phi\big(-g_i(x) + g_i(y)\big) \geq \langle -\nabla g_i(y), \eta(x,y)\rangle$$

$$- \quad [\langle z, \nabla^2 g_i(y)z\rangle - \frac{1}{2}\langle z, \nabla^2 g_i(y)z\rangle]$$

$$+ \quad \tilde{\rho}_i(x,y)\|\theta(x,y)\|^{\tilde{m}}, \ i \in I_+, \tag{1.15}$$

$$\begin{aligned} & \phi\big(G_{j_m}(x,t^m) - G_{j_m}(y,t^m)\big) \\ \geq \ & \langle \nabla G_{j_m}(y,t^m), \eta(x,y)\rangle \\ + \ & \langle z, \nabla^2 G_{j_m}(y,t^m)z\rangle \\ - \ & \frac{1}{2}\langle z, \nabla^2 G_{j_m}(y,t^m)z\rangle \\ + \ & \hat{\rho}_m(x,y)\|\theta(x,y)\|^{\tilde{m}}, \end{aligned} \tag{1.16}$$

$$\begin{aligned} & \phi\big(H_{k_m}(x,s^m) - H_{k_m}(y,s^m)\big) \\ \geq \ & \langle \nabla H_{k_m}(y,s^m), \eta(x,y)\rangle \\ + \ & \langle z, \nabla^2 H_{k_m}(y,s^m)z\rangle \\ - \ & \frac{1}{2}\langle z, \nabla^2 H_{k_m}(y,s^m)z\rangle \\ + \ & \check{\rho}_m(x,y)\|\theta(x,y)\|^{\tilde{m}}. \end{aligned} \tag{1.17}$$

Now, multiplying (14) by $u_iD(y,u)$ and then summing over $i \in I_+$, (15) by $u_iN(y,u)$ and then summing $i \in I_+$, (16) by $v_m$ and then summing over $m \in \underline{v_0}$, (17) by $w_m$ and then summing over $m \in \underline{v} \setminus \underline{v_0}$, adding the resulting inequalities, and applying the superlinearity of $\phi$, we obtain

$$\phi\Big(\sum_{i=1}^{p} u_i[D(y,u)f_i(x) - N(y,u)g_i(x)] + \sum_{m=1}^{v_0} v_m G_{j_m}(x,t^m) + \sum_{m=v_0+1}^{v} w_m H_{k_m}(x,s^m)$$

$$- \Big\{\sum_{i=1}^{p} u_i[D(y,u)f_i(y) - N(y,u)g_i(y)] + \sum_{m=1}^{v_0} v_m G_{j_m}(y,t^m) + \sum_{m=v_0+1}^{v} w_m H_{k_m}(y,s^m)\Big\}\Big) \geq \Big\langle \sum_{i=1}^{p} u_i[D(y,u)\nabla f_i(y) - N(y,u)\nabla g_i(y)]$$

$$+ \sum_{m=1}^{v_0} v_m \nabla G_{j_m}(y,t^m) + \sum_{m=v_0+1}^{v} w_m \nabla H_{k_m}(y,s^m), \eta(x,y)\Big\rangle$$

$$+ \Big\langle z, \Big\{\sum_{i=1}^{p} u_i[D(y,u)\nabla^2 f_i(y) - N(y,u)\nabla^2 g_i(y)] + \sum_{m=1}^{v_0} v_m \nabla^2 G_{j_m}(y,t^m) + \sum_{m=v_0+1}^{v} w_m \nabla^2 H_{k_m}(y,s^m)\Big\} z\Big\rangle$$

$$- \frac{1}{2}\Big\langle z, \Big\{\sum_{i=1}^{p} u_i[D(y,u)\nabla^2 f_i(y) - N(y,u)\nabla^2 g_i(y)] + \sum_{m=1}^{v_0} v_m \nabla^2 G_{j_m}(y,t^m) + \sum_{m=v_0+1}^{v} w_m \nabla^2 H_{k_m}(y,s^m)\Big\} z\Big\rangle$$

$$+ \Big\{\sum_{i=1}^{p} u_i[D(y,u)\bar{\rho}_i(x,y) + N(y,u)\tilde{\rho}_i(x,y)] + \sum_{m=1}^{v_0} v_m \hat{\rho}_m(x,y) + \sum_{m=v_0+1}^{v} w_m \breve{\rho}_m(x,y)\Big\} \|\theta(x,y)\|^{\tilde{m}}.$$

Because of the dual feasibility of $\mathscr{S}$, (4), (5), and (v), the above inequality becomes

$$\phi\Big(\sum_{i=1}^{p} u_i[D(y,u)f_i(x)-N(y,u)g_i(x)]+\sum_{m=1}^{\nu_0} v_m G_{j_m}(x,t^m)$$

$$+\sum_{m=\nu_0+1}^{\nu} w_m H_{k_m}(x,s^m)$$

$$-\Big\{\sum_{i=1}^{p} u_i[D(y,u)f_i(y)-N(y,u)g_i(y)]+\sum_{m=1}^{\nu_0} v_m G_{j_m}(y,t^m)$$

$$+\sum_{m=\nu_0+1}^{\nu} w_m H_{k_m}(y,s^m)\Big\}\Big)\geq 0.$$

But $\phi(a)\geq 0 \Rightarrow a\geq 0$, and hence

$$\sum_{i=1}^{p} u_i[D(y,u)f_i(x)-N(y,u)g_i(x)]+\sum_{m=1}^{\nu_0} v_j G_j(x,t^m)$$

$$+\sum_{m=\nu_0+1}^{\nu} w_m H_{k_m}(x,s^m)$$

$$-\Big\{\sum_{i=1}^{p} u_i[D(y,u)f_i(y)-N(y,u)g_i(y)]+\sum_{j=1}^{q} v_j G_j(y)$$

$$+\sum_{k=1}^{r} w_k H_k(y)\Big\}\geq 0.$$

In view of the definitions of $N(y,u)$ and $D(y,u)$, (6), and the primal feasibility of $x$, this inequality further reduces to

$$\sum_{i=1}^{p} u_i[D(y,u)f_i(x)-N(y,u)g_i(x)]\geq 0. \tag{1.18}$$

Now using (18) and Lemma 1.12, we obtain the weak duality inequality as follows:

$$\begin{aligned}\varphi(x) &= \max_{a\in U}\frac{\sum_{i=1}^{p} a_i f_i(x)}{\sum_{i=1}^{p} a_i g_i(x)}\\ &\geq \max_{1\leq i\leq p}\frac{\sum_{i=1}^{p} u_i f_i(x)}{\sum_{i=1}^{p} u_i g_i(x)}\geq\frac{N(y,u)}{D(y,u)}\\ &= \psi_I(y,z,u,v,w).\end{aligned}$$

(b) : Using the dual feasibility of $(y,z,u,v,w)$, nonnegativity of $\rho(x,y)$, (4), and (5), we obtain the following inequality:

$$\langle \nabla L(y,y,u,v,w), \eta(x,y)\rangle + \langle z, \nabla^2 L(y,y,u,v,w)z\rangle$$
$$-\frac{1}{2}\langle z, \nabla^2 L(y,y,u,v,w)z\rangle \geq 0 \geq -\rho(x,y)\|\theta(x,y)\|^{\tilde{m}},$$

which in view of our $(\phi,\eta,\rho,\theta,m)$-pseudosonvexity assumption implies that

$$\phi\big(L(x,y,u,v,w) - L(y,y,u,v,w)\big) \geq 0.$$

But $\phi(a) \geq 0 \Rightarrow a \geq 0$ and hence we have

$$L(x,y,u,v,w) - L(y,y,u,v,w) \geq 0.$$

Because $\sum_{i=1}^{p} u_i[D(y,u)f_i(y) - N(y,u)g_i(y)] = 0$, $x \in \mathbb{F}$, $v \geq 0$, and (6) holds, the above inequality reduces to

$$\sum_{i=1}^{p} u_i[D(y,u)f_i(x) - N(y,u)g_i(x)] \geq 0,$$

which is precisely (13). As seen in the proof of part (a), this inequality leads to the desired conclusion that $\varphi(x) \geq \psi_I(y,z,u,v,w)$. □

**Theorem 1.15.** (Strong Duality) Let $x^*$ be a normal optimal solution of (P) and assume that either one of the two sets of conditions specified in Theorem 1.13 is satisfied for all feasible solutions of (DI). Then for each $z^* \in C(x^*)$, there exist $u^* \in U$, $v^* \in \mathbb{R}^q_+$, and $w^* \in \mathbb{R}^r$ such that $\mathscr{S}^* \equiv (x^*,z^*,u^*,v^*,w^*)$ is an optimal solution of (DI) and $\varphi(x^*) = N(x^*,u^*)/D(x^*,u^*) = \psi_I(x^*,z^*,u^*,v^*,w^*)$.

*Proof.* Since $x^*$ is a normal optimal solution of $(P)$, by Theorem 1.11, there exist $z^*$, $u^*$, $v^*$, and $w^*$, as specified above, such that $\mathscr{S}^*$ is a feasible solution of $(DI)$. If $\mathscr{S}^*$ were not optimal, then there would exist a feasible solution $(\tilde{y},\tilde{z},\tilde{u},\tilde{v},\tilde{w})$ of $(DI)$ such that $\psi_I(\tilde{y},\tilde{z},\tilde{u},\tilde{v},\tilde{w}) > \psi_I(x^*,z^*,u^*,v^*,w^*) = \varphi(x^*)$, contradicting Theorem 1.13. Therefore, $\mathscr{S}^*$ is an optimal solution of $(DI)$ and $\varphi(x^*) = \psi_I(x^*,z^*,u^*,v^*,w^*)$. □

We also have the following converse duality result for $(P)$ and $(DI)$.

**Theorem 1.16.** (Strict Converse Duality) Let $x^*$ be a normal optimal solution of (P), let $\tilde{\mathscr{S}} \equiv (\tilde{x},\tilde{z},\tilde{u},\tilde{v},\tilde{w})$ be an optimal solution of (DI), and assume that either one of the following two sets of hypotheses is satisfied:

(a) The assumptions of part (a) of Theorem 1.13 are satisfied for the feasible solution $\tilde{\mathscr{S}}$ of (DI), $\phi(a) > 0 \Rightarrow a > 0$, and $f_i$ is strictly $(\phi,\eta,\bar{\rho}_i,\theta,\tilde{m})$-sonvex at $\tilde{x}$ for at least one index $i \in I_+(\tilde{u})$, or $-g_i$ is strictly $(\phi,\eta,\tilde{\rho}_i,\theta,\tilde{m})$-sonvex at $\tilde{x}$ for at least one index $i \in I_+(\tilde{u})$ (and $N(\tilde{x},\tilde{u}) > 0$), or $G_j$ is strictly $(\phi,\eta,\hat{\rho}_j,\theta,\tilde{m})$-sonvex at $\tilde{x}$ for at least one index $j \in J_+(\tilde{v})$, or $\xi \to \mathscr{D}_k(\xi,\tilde{w})$ is strictly $(\phi,\eta,\breve{\rho}_k,\theta,\tilde{m})$-sonvex at $\tilde{x}$ for at least one index $k \in K_*(\tilde{w})$, or

$$\sum_{i\in I_+(\tilde{u})} \tilde{u}_i[D(x^*,\tilde{u})\bar{\rho}_i(x^*,\tilde{x})+N(x^*,\tilde{u})\tilde{\rho}_i(x^*,\tilde{x})]+\sum_{m=1}^{\nu_0}\tilde{v}_m\hat{\rho}_m(x^*,\tilde{x})$$
$$+\sum_{m=\nu_0+1}^{\nu}\tilde{w}_m\breve{\rho}_m(x^*,\tilde{x})>0.$$

(b) The Lagrangian-type function $\xi \to \Lambda(\xi,\tilde{x},\tilde{u},\tilde{v},\tilde{w})$ is strictly $(\phi,\eta,\rho,\theta,\tilde{m})$-pseudosonvex at $\tilde{x}$, $\rho(x^*,\tilde{x})\geq 0$, and $\phi(a)>0 \Rightarrow a>0$.

Then $\tilde{x}=x^*$, that is, $\tilde{x}$ is an optimal solution of (P), and $\varphi(x^*)=\psi_I(\tilde{x},\tilde{z},\tilde{u},\tilde{v},\tilde{w})$.

*Proof.* (a) : Since $x^*$ is a normal optimal solution of $(P)$, by Theorem 1.11, there exist $z^*\in C(x^*)$, $u^*\in U$, $v^*\in\mathbb{R}^q_+$, and $w^*\in\mathbb{R}^r$ such that $\mathscr{S}^*\equiv(x^*,z^*,u^*,v^*,w^*)$ is an optimal solution of $(DI)$ and $\varphi(x^*)=\psi_I(x^*,z^*,u^*,v^*,w^*)$. Suppose to the contrary that $\tilde{x}\neq x^*$. Now proceeding as in the proof of Theorem 1.13 (with $x$ replaced by $x^*$ and $\mathscr{S}$ by $\tilde{\mathscr{S}}$) and using any of the conditions set forth above, we arrive at the strict inequality

$$\sum_{i=1}^{p}\tilde{u}_i[D(\tilde{x},\tilde{u})f_i(x^*)-N(\tilde{x},\tilde{u})g_i(x^*)]>0.$$

Now using this inequality in conjunction with Lemma 1.12, as in the proof of part (a), we arrive at the strict inequality $\varphi(x^*)>N(\tilde{x},\tilde{u})/D(\tilde{x},\tilde{u})=\psi_I(\tilde{x},\tilde{z},\tilde{u},\tilde{v},\tilde{w})$ which contradicts the fact that $\varphi(x^*)=\psi_I(x^*,z^*,u^*,v^*,w^*)=\psi_I(\tilde{x},\tilde{z},\tilde{u},\tilde{v},\tilde{w})$.

(b) : The proof is similar to that of part (a). □

## 4. Duality Model II

In this section, we consider two duality models with special constraint structures that allow for a greater variety of generalized $(\phi,\eta,\rho,\theta,\tilde{m})$-sonvexity conditions under which duality can be established.

Consider the following two problems:

$$(DII) \quad \text{Maximize} \quad \psi_{II}(y,z,u,v,w)=\frac{\sum_{i=1}^{p}u_if_i(y)}{\sum_{i=1}^{p}u_ig_i(y)}$$

subject to

$$\sum_{i=1}^{p}u_iD(y,u)[\nabla f_i(y)-N(y,u)\nabla g_i(y)]+\sum_{m=1}^{\nu_0}v_m\nabla G_{j_m}(y,t^m)$$
$$+\sum_{m=\nu_0+1}^{\nu}w_m\nabla H_{k_m}(y,s^m)=0, \quad (1.19)$$

$$\Big\langle z, \Big\{ \sum_{i=1}^{p} u_i[D(y,u)\nabla^2 f_i(y) - N(y,u)\nabla^2 g_i(y)]$$

$$+ \sum_{m=1}^{\nu_0} v_m \nabla^2 G_{j_m}(y,t^m)$$

$$+ \sum_{m=\nu_0+1}^{\nu} w_m \nabla^2 H_{k_m}(y,s^m) \Big\} z \Big\rangle \geq 0, \quad (1.20)$$

$$v_m G_{j_m}(y,t^m) \geq 0, \ m \in \underline{\nu_0}, \quad (1.21)$$

$$w_m H_{k_m}(y,s^m) \geq 0, \ m \in \underline{\nu} \backslash \underline{\nu_0}; \quad (1.22)$$

$(\tilde{D}II)$ Maximize $\psi_{II}(y,z,u,v,w) = \dfrac{\sum_{i=1}^{p} u_i f_i(y)}{\sum_{i=1}^{p} u_i g_i(y)}$

subject to (7) and (20) - (22).

The remarks made earlier about the relationships between $(DI)$ and $(\tilde{D}I)$ are, of course, also hold for $(DII)$ and $(\tilde{D}II)$ as well.

The next two theorems show that $(DII)$ is a dual problem for $(P)$.

**Theorem 1.17.** (Weak Duality) Let $x$ and $w \equiv (y,\ z,\ u,\ v,\ \lambda,\ \nu,\ \nu_0,\ J_{\nu_0},\ K_{\nu\backslash\nu_0},\ \bar{t},\ \bar{s})$ be arbitrary feasible solutions of (P) and (DII), respectively, and assume that any one of the following five sets of hypotheses is satisfied:

(a) (i) for each $i \in \underline{p}$, $f_i$ is $(\bar{\phi},\eta,\bar{\rho}_i,\theta,\tilde{m})$-sonvex and $-g_i$ is $(\bar{\phi},\eta,\tilde{\rho}_i,\theta,\tilde{m})$-sonvex at $y$, $\bar{\phi}$ is superlinear, and $\bar{\phi}(a) \geq 0 \Rightarrow a \geq 0$;

(ii) the function $\xi \to G_{j_m}(\xi,t^m)$ is $(\hat{\phi}_m,\eta,\hat{\rho}_m,\theta,\tilde{m})$-quasisonvex at $y$, $\hat{\phi}_m$ is increasing, and $\hat{\phi}_m(0) = 0$ for each $m \in \underline{\nu_0}$;

(iii) the function $\xi \to w_m H_{k_m}(\xi,s^m)$ is $(\breve{\phi}_m,\eta,\breve{\rho}_m,\theta,\tilde{m})$-quasisonvex at $y$, $\breve{\phi}_m$ is increasing, and $\breve{\phi}_m(0) = 0$ for each $m \in \underline{\nu}\backslash\underline{\nu_0}$;

(iv) $\rho^*(x,y) + \sum_{m=1}^{\nu_0} v_m \hat{\rho}_m(x,y) + \sum_{m=\nu_0+1}^{\nu} w_m \breve{\rho}_m(x,y) \geq 0$, where

$$\rho^*(x,y) = \sum_{i=1}^{p} u_i[\bar{\rho}_i(x,y) + \lambda \tilde{\rho}_i(x,y)];$$

(b) (i) for each $i \in \underline{p}$, $f_i$ is $(\bar{\phi},\eta,\bar{\rho}_i,\theta,\tilde{m})$-sonvex and $-g_i$ is $(\bar{\phi},\eta,\tilde{\rho}_i,\theta,\tilde{m})$-sonvex at $y$, $\bar{\phi}$ is superlinear, and $\bar{\phi}(a) \geq 0 \Rightarrow a \geq 0$;

(ii) the function $\xi \to v_m G_{j_m}(\xi, t^m)$ is $(\hat{\phi}, \eta, \hat{\rho}, \theta, \tilde{m})$-quasisonvex at $y$, $\hat{\phi}$ is increasing, and $\hat{\phi}(0) = 0$;

(iii) the function $\xi \to w_m H_{k_m}(\xi, s^m)$ is $(\breve{\phi}_m, \eta, \breve{\rho}_m, \theta, \tilde{m})$-quasisonvex at $y$, $\breve{\phi}_m$ is increasing, and $\breve{\phi}_m(0) = 0$ for each $m \in \underline{v} \backslash \underline{v_0}$;

(iv) $\rho^*(x,y) + \hat{\rho}(x,y) + \sum_{m=v_0+1}^{v} w_m \breve{\rho}_m(x,y) \geq 0$;

(c) (i) for each $i \in \underline{p}$, $f_i$ is $(\bar{\phi}, \eta, \bar{\rho}_i, \theta, \tilde{m})$-sonvex and $-g_i$ is $(\bar{\phi}, \eta, \tilde{\rho}_i, \theta, \tilde{m})$-sonvex at $y$, $\bar{\phi}$ is superlinear, and $\bar{\phi}(a) \geq 0 \Rightarrow a \geq 0$;

(ii) the function $\xi \to G_{j_m}(\xi, t^m)$ is $(\hat{\phi}_m, \eta, \hat{\rho}_m, \theta, \tilde{m})$-quasisonvex at $y$, $\hat{\phi}_m$ is increasing, and $\hat{\phi}_m(0) = 0$ for each $m \in \underline{v_0}$;

(iii) the function $\xi \to w_m H_{k_m}(\xi, s^m)$ is $(\breve{\phi}, \eta, \breve{\rho}, \theta, \tilde{m})$-quasisonvex at $y$, $\breve{\phi}$ is increasing, and $\breve{\phi}(0) = 0$;

(iv) $\rho^*(x,y) + \sum_{m=1}^{v_0} v_m \hat{\rho}_m(x,y) + \breve{\rho}(x,y) \geq 0$;

(d) (i) for each $i \in \underline{p}$, $f_i$ is $(\bar{\phi}, \eta, \bar{\rho}_i, \theta, \tilde{m})$-sonvex and $-g_i$ is $(\bar{\phi}, \eta, \tilde{\rho}_i, \theta, \tilde{m})$-sonvex at $y$, $\bar{\phi}$ is superlinear, and $\bar{\phi}(a) \geq 0 \Rightarrow a \geq 0$;

(ii) the function $\xi \to \sum_{m=1}^{v_0} v_m G_{j_m}(\xi, t^m)$ is $(\hat{\phi}, \eta, \hat{\rho}, \theta, \tilde{m})$-quasisonvex at $y$, $\hat{\phi}$ is increasing, and $\hat{\phi}(0) = 0$;

(iii) the function $\xi \to \sum_{m=v_0+1}^{v} w_m H_{k_m}(\xi, s^m)$ is $(\breve{\phi},\eta,\breve{\rho},\theta,\tilde{m})$- quasisounivex at $y$, $\breve{\phi}$ is increasing, and $\breve{\phi}(0) = 0$;

(iv) $\rho^*(x,y) + \hat{\rho}(x,y) + \breve{\rho}(x,y) \geq 0$;

(e) (i) for each $i \in \underline{p}$, $f_i$ is $(\bar{\phi}, \eta, \bar{\rho}_i, \theta, \tilde{m})$-sonvex and $-g_i$ is $(\bar{\phi}, \eta, \tilde{\rho}_i, \theta, \tilde{m})$-sonvex at $y$, $\bar{\phi}$ is superlinear, and $\bar{\phi}(a) \geq 0 \Rightarrow a \geq 0$;

(ii) the function $\xi \to \sum_{m=1}^{v_0} v_m G_{j_m}(\xi, t^m) + \sum_{m=v_0+1}^{v} w_m H_{k_m}(\xi, s^m)$ is $(\hat{\phi}, \eta, \hat{\rho}, \theta, \tilde{m})$-quasisonvex at $y$, $\hat{\phi}$ is increasing, and $\hat{\phi}(0) = 0$;

(iii) $\rho^*(x,y) + \hat{\rho}(x,y) \geq 0$.

Then $\varphi(x) \geq \psi_{II}(y, z, u, v, w)$.

*Proof.* (a): Proceeding as in the proof of part (a) of Theorem 1.13, we see that our assumptions in (i) lead to the following inequality:

$$\phi\Big(\sum_{i=1}^{p}u_i\{[D(y,u)f_i(x)-N(y,u)g_i(x)]-[D(y,u)f_i(y)-D(y,u)g_i(y)]\}\Big)$$

$$\geq\langle\sum_{i=1}^{p}u_i\{D(y,u)\nabla f_i(y)-N(y,u)\nabla g_i(y),\eta(x,y)\rangle$$

$$+\frac{1}{2}\Big\langle z,\sum_{i=1}^{p}u_i[D(y,u)\nabla^2 f_i(y)-N(y,u)\nabla^2 g_i(y)]z\Big\rangle$$

$$+\sum_{i=1}^{p}u_i[D(y,u)\bar{\rho}_i(x,y)+N(y,u)\tilde{\rho}_i(x,y)]\|\theta(x,y)\|^{\tilde{m}}. \quad (1.23)$$

From the primal feasibility of $x$, and (21) holds, we find that $G_{j_m}(x,t^m)\leq 0\leq G_{j_m}(y,t^m)$ and using (ii), we have

$$\langle\nabla G_{j_m}(y,t^m),\eta(x,y)\rangle+\frac{1}{2}\langle z,\nabla^2 G_{j_m}(y,t^m)z\rangle\leq-\hat{\rho}_m(x,y)\|\theta(x,y)\|^{\tilde{m}}$$

for each $m\in\underline{\nu_0}$, and hence, we get

$$\langle\sum_{m=1}^{\nu_0}v_m[\nabla G_{j_m}(y,t^m),\eta(x,y)\rangle+\frac{1}{2}\langle z,\sum_{m=1}^{\nu_0}v_m[\nabla^2 G_{j_m}(y,t^m)z\rangle$$

$$\leq-\sum_{m=1}^{\nu_0}v_m\hat{\rho}_m(x,y)\|\theta(x,y)\|^{\tilde{m}}. \quad (1.24)$$

Similarly, we have

$$\langle\sum_{m=\nu_0+1}^{\nu}w_m[\nabla H_{k_m}(y,s^m),\eta(x,y)\rangle$$

$$+\frac{1}{2}\langle z,\sum_{m=\nu_0+1}^{\nu}w_m[\nabla^2 H_{j_m}(y,s^m)z\rangle$$

$$\leq-\sum_{m=\nu_0+1}^{\nu_0}w_m\breve{\rho}_m(x,y)\|\theta(x,y)\|^{\tilde{m}}. \quad (1.25)$$

Now using (19), (20) and (23) - (25), we have

$$\bar{\phi}\Big(\sum_{i=1}^{p}u_i[D(y,u)f_i(x)-N(y,u)g_i(x)]-\sum_{i=1}^{p}u_i[D(y,u)f_i(y)-N(y,u)g_i(y)]\Big)$$

$$\geq -\Big[\Big\langle \sum_{m=1}^{\nu_0} v_m\nabla G_{j_m}(y,t^m),\eta(x,y)\Big\rangle$$

$$+\frac{1}{2}\Big\langle z,\sum_{m=1}^{\nu_0} v_m\nabla^2 G_{j_m}(y,t^m)z\Big\rangle$$

$$+\Big\langle \sum_{m=\nu_0+1}^{\nu} w_m\nabla H_{k_m}(y,s^m),\eta(x,y)\Big\rangle+\frac{1}{2}\Big\langle z,\sum_{m=\nu_0+1}^{\nu} w_m\nabla^2 H_{k_m}(y,s^m)z\Big\rangle\Big]$$

$$+\sum_{i\in I_+}u_i[\bar{\rho}_i(x,y)+\lambda\tilde{\rho}_i(x,y)]\|\theta(x,y)\|^{\tilde{m}}$$

$$\geq\Big\{\sum_{i\in I_+}u_i[\bar{\rho}_i(x,y)+\lambda\tilde{\rho}_i(x,y)]+\sum_{m=1}^{\nu_0}v_m\hat{\rho}_m(x,y)$$

$$+\sum_{m=\nu_0+1}^{\nu}w_m\breve{\rho}_m(x,y)\Big\}\|\theta(x,y)\|^{\tilde{m}}$$

$\geq 0$ (by (iv)).

But $\bar{\phi}(a)\geq 0\Rightarrow a\geq 0$, and hence we have

$$\sum_{i=1}^{p}u_i[D(y,u)f_i(x)-N(y,u)g_i(x)]\geq\sum_{i=1}^{p}u_i[D(y,u)f_i(y)-N(y,u)g_i(y)]\geq 0,$$

where the second inequality follows from the dual feasibility of $w$ and (21). As shown in the proof of part (a) of Theorem 1.13, this inequality leads to the

$$\varphi(x)\geq\frac{N(y,u)}{D(y,u)}=\psi_{II}(y,z,u,v,w).$$

(b) - (e) : The proofs are similar to that of part (a).

□

**Theorem 1.18.** (Strong Duality) Let $x^*$ be a normal optimal solution of (P), and assume that any one of the five sets of conditions set forth in Theorem 1.17 is satisfied for all feasible solutions of (DII). Then for each $z^*\in C(x^*)$, there exist $u^*,v^*,\nu^*,\nu_0^*,J_{\nu_0^*},K_{\nu^*\setminus\nu_0^*},\bar{t}^*$, and $\bar{s}^*$ such that $(x^*,z^*,u^*,v^*,w^*,\nu^*,\nu_0^*,J_{\nu_0^*},K_{\nu^*\setminus\nu_0^*},\bar{t}^*,\bar{s}^*)$ is an optimal solution of (DII) and $\varphi(x^*)=\psi_{II}(x^*,z^*,u^*,v^*,w^*)$, where $C(x^*)$ is the set of all critical directions of $(P)$ at $x^*$.

*Proof.* The proof is similar to that of Theorem 1.15. □

**Theorem 1.19.** (Strict Converse Duality) Let $x^*$ be a normal optimal solution of (P), let $\tilde{S} \equiv (\tilde{x}, \tilde{z}, \tilde{u}, \tilde{v}, \tilde{w}, \tilde{\nu}, \tilde{\nu}_0, J_{\tilde{\nu}_0}, K_{\tilde{\nu}\backslash\tilde{\nu}_0}, \bar{\tilde{t}}, \bar{\tilde{s}})$ be an optimal solution of (DII), and assume that any one of the following five sets of conditions is satisfied:

(a) The assumptions specified in part (a) of Theorem 1.17 are satisfied for the feasible solution $\tilde{S}$ of (DII). Moreover, $\phi(a) > 0 \Rightarrow a > 0$, $f_i$ is strictly $(\phi, \eta, \bar{\rho}_i, \theta), \tilde{m})$-sonvex at $\tilde{x}$ for at least one $i \in \underline{p}$ with the corresponding component $\tilde{u}_i$ of $\tilde{u}$ positive, or $-g_i$ is strictly $(\phi, \eta, \tilde{\rho}_i, \theta, \tilde{m})$-sonvex at $\tilde{x}$ for at least one $i \in \underline{p}$ with the corresponding component $\tilde{u}_i$ of $\tilde{u}$ positive (and $N(\tilde{x}, \tilde{u}) > 0$), or $\xi \rightarrow G_{j_m}(\xi, \tilde{t}^m)$ is strictly $(\hat{\phi}_m, \eta, \hat{\rho}_m, \theta, \tilde{m})$-pseudosonvex at $\tilde{x}$ for at least one $m \in \underline{\tilde{\nu}_0}$, or $\xi \rightarrow \tilde{w}_m H_{k_m}(\xi, \tilde{s}^m)$ is strictly $(\breve{\phi}_m, \eta, \breve{\rho}_m, \theta, \tilde{m})$-pseudosonvex at $\tilde{x}$ for at least one $m \in \underline{\tilde{\nu}} \backslash \underline{\tilde{\nu}_0}$, or

$$\rho^*(x^*, \tilde{x}) + \sum_{m=1}^{\tilde{\nu}_0} \tilde{v}_m \hat{\rho}_m(x^*, \tilde{x}) + \sum_{m=\tilde{\nu}_0+1}^{\tilde{\nu}} \tilde{w}_m \breve{\rho}_m(x^*, \tilde{x}) > 0, \text{ where}$$

$$\rho^*(x^*, \tilde{x}) = \sum_{i=1}^{p} \tilde{u}_i [D(\tilde{x}, \tilde{u}) \bar{\rho}_i(x^*, \tilde{x}) + N(\tilde{x}, \tilde{u}) \tilde{\rho}_i(x^*, \tilde{x})].$$

(b) The assumptions specified in part (b) of Theorem 1.17 are satisfied for the feasible solution $\tilde{S}$ of (DII). Moreover, $\phi(a) > 0 \Rightarrow a > 0$, $f_i$ is strictly $(\phi, \eta, \bar{\rho}_i, \theta), \tilde{m})$-sonvex at $\tilde{x}$ for at least one $i \in \underline{p}$ with the corresponding component $\tilde{u}_i$ of $\tilde{u}$ positive, or $-g_i$ is strictly $(\phi, \eta, \tilde{\rho}_i, \theta, \tilde{m})$-sonvex at $\tilde{x}$ for at least one $i \in \underline{p}$ with the corresponding component $\tilde{u}_i$ of $\tilde{u}$ positive (and $N(\tilde{x}, \tilde{u}) > 0$), or $\xi \rightarrow G_{j_m}(\xi, \tilde{t}^m)$ is strictly $(\hat{\phi}, \eta, \hat{\rho}, \theta, \tilde{m})$-pseudosonvex at $\tilde{x}$ for at least one $m \in \underline{\tilde{\nu}_0}$, or $\xi \rightarrow \tilde{w}_m H_{k_m}(\xi, \tilde{s}^m)$ is strictly $(\breve{\phi}_m, \eta, \breve{\rho}_m, \theta, \tilde{m})$-pseudosonvex at $\tilde{x}$ for at least one $m \in \underline{\tilde{\nu}} \backslash \underline{\tilde{\nu}_0}$, or $\rho^*(x^*, \tilde{x}) + \hat{\rho}_m(x^*, \tilde{x}) + \sum_{m=\tilde{\nu}_0+1}^{\tilde{\nu}} \tilde{w}_m \breve{\rho}_m(x^*, \tilde{x}) > 0$.

(c) The assumptions specified in part (c) of Theorem 1.17 are satisfied for the feasible solution $\tilde{S}$ of (DII). Moreover, $\phi(a) > 0 \Rightarrow a > 0$, $f_i$ is strictly $(\phi, \eta, \bar{\rho}_i, \theta), \tilde{m})$-sonvex at $\tilde{x}$ for at least one $i \in \underline{p}$ with the corresponding component $\tilde{u}_i$ of $\tilde{u}$ positive, or $-g_i$ is strictly $(\phi, \eta, \tilde{\rho}_i, \theta, \tilde{m})$-sonvex at $\tilde{x}$ for at least one $i \in \underline{p}$ with the corresponding component $\tilde{u}_i$ of $\tilde{u}$ positive (and $N(\tilde{x}, \tilde{u}) > 0$), or $\xi \rightarrow G_{j_m}(\xi, \tilde{t}^m)$ is strictly $(\hat{\phi}_m, \eta, \hat{\rho}_m, \theta, \tilde{m})$-pseudosonvex at $\tilde{x}$ for at least one $m \in \underline{\tilde{\nu}_0}$, or $\xi \rightarrow \tilde{w}_m H_{k_m}(\xi, \tilde{s}^m)$ is strictly $(\breve{\phi}, \eta, \breve{\rho}, \theta, \tilde{m})$-pseudosonvex at $\tilde{x}$ for at least one $m \in \underline{\tilde{\nu}} \backslash \underline{\tilde{\nu}_0}$, or $\rho^*(x^*, \tilde{x}) + \sum_{m=1}^{\tilde{\nu}_0} \tilde{v}_m \hat{\rho}_m(x^*, \tilde{x}) + \tilde{w}_m \breve{\rho}_m(x^*, \tilde{x}) > 0$.

(d) The assumptions specified in part (d) of Theorem 1.17 are satisfied for the feasible solution $\tilde{S}$ of (DII). Moreover, $\phi(a) > 0 \Rightarrow a > 0$, $f_i$ is strictly $(\phi, \eta, \bar{\rho}_i, \theta), \tilde{m})$-sonvex at $\tilde{x}$ for at least one $i \in \underline{p}$ with the corresponding component $\tilde{u}_i$ of $\tilde{u}$ positive, or $-g_i$ is strictly $(\phi,\eta,\tilde{\rho}_i,\theta,\tilde{m})$-sonvex at $\tilde{x}$ for at least one $i \in \underline{p}$ with the corresponding component $\tilde{u}_i$ of $\tilde{u}$ positive (and $N(\tilde{x},\tilde{u}) > 0$), or $\xi \to G_{j_m}(\xi,\tilde{t}^m)$ is strictly $(\hat{\phi},\eta,\hat{\rho},\theta,\tilde{m})$-pseudosonvex at $\tilde{x}$ for at least one $m \in \underline{\tilde{\nu}_0}$, or $\xi \to \tilde{w}_m H_{k_m}(\xi,\tilde{s}^m)$ is strictly $(\breve{\phi},\eta,\breve{\rho},\theta,\tilde{m})$-pseudosonvex at $\tilde{x}$ for at least one $m \in \underline{\tilde{\nu}}\backslash\underline{\tilde{\nu}_0}$, or $\rho^*(x^*,\tilde{x}) + \hat{\rho}(x^*,\tilde{x}) + \breve{\rho}_m(x^*,\tilde{x}) > 0$.

(e) The assumptions specified in part (e) of Theorem 1.17 are satisfied for the feasible solution $\tilde{S}$ of (DII). Moreover, $\phi(a) > 0 \Rightarrow a > 0$, $f_i$ is strictly $(\phi,\eta,\bar{\rho}_i,\theta),\tilde{m})$-sonvex at $\tilde{x}$ for at least one $i \in \underline{p}$ with the corresponding component $\tilde{u}_i$ of $\tilde{u}$ positive, or $-g_i$ is strictly $(\phi,\eta,\tilde{\rho}_i,\theta,\tilde{m})$-sonvex at $\tilde{x}$ for at least one $i \in \underline{p}$ with the corresponding component $\tilde{u}_i$ of $\tilde{u}$ positive (and $N(\tilde{x},\tilde{u}) > 0$), or $\xi \to \mathbb{G}(\xi,\tilde{v},\tilde{w})$ is strictly $(\hat{\phi},\eta,\hat{\rho},\theta,\tilde{m})$-pseudosonvex at $\tilde{x}$ or $\rho^*(x^*,\tilde{x}) + \hat{\rho}(x^*,\tilde{x}) + > 0$.

Then $\tilde{x} = x^*$ and $\varphi(x^*) = \psi_{II}(\tilde{x},\tilde{z},\tilde{u},\tilde{v},\tilde{w})$.

*Proof.* The proof is similar to that of Theorem 1.15. □

**Theorem 1.20.** (Weak Duality) Let $x$ and $\mathscr{S} \equiv (y,z,u,v,w)$ be arbitrary feasible solutions of (P) and (DII), respectively, and assume that any one of the following five sets of hypotheses is satisfied:

(a) (i) $\xi \to \mathscr{E}(\xi,y,u)$ is $(\bar{\phi},\eta,\bar{\rho},\theta,\tilde{m})$-pseudosonvex at $y$, and $\bar{\phi}(a) \geq 0 \Rightarrow a \geq 0$;

(ii) for each $t \in T_j$, $j \in \bar{q}$, $G_j(x,t)$ is $(\hat{\phi}_j,\eta,\hat{\rho}_j,\theta,\tilde{m})$-quasisonvex at $y$, $\hat{\phi}_j$ is increasing, and $\hat{\phi}_j(0) = 0$;

(iii) for each $s \in S_k$, $k \in \bar{r}$, $\xi \to \mathscr{D}_k(\xi,v)$ is $(\breve{\phi}_k,\eta,\breve{\rho}_k,\theta,\tilde{m})$-quasisonvex at $y$, $\breve{\phi}_k$ is increasing, and $\breve{\phi}_k(0) = 0$;

(iv) $$\bar{\rho}(x,y) + \sum_{m=1}^{\nu_0} v_m\hat{\rho}_m(x,y) + \sum_{m=\nu_0+1}^{\nu} w_m\breve{\rho}_m(x,y) \geq 0;$$

(b) (i) $\xi \to \mathscr{E}(\xi,y,u)$ is $(\bar{\phi},\eta,\bar{\rho},\theta,\tilde{m})$-pseudosonvex at $y$, and $\bar{\phi}(a) \geq 0 \Rightarrow a \geq 0$;

(ii) $\xi \to \mathscr{C}(\xi,v)$ is $(\hat{\phi},\eta,\hat{\rho},\theta,\tilde{m})$-quasisonvex at $y$, $\hat{\phi}$ is increasing, and $\hat{\phi}(0) = 0$;

(iii) for each $k \in K_*$, $\xi \to \mathscr{D}_k(\xi,w)$ is $(\breve{\phi}_k,\eta,\breve{\rho}_k,\theta,\tilde{m})$-quasisonvex at $y$, $\breve{\phi}_k$ is increasing, and $\breve{\phi}_k(0) = 0$;

(iv) $$\bar{\rho}(x,y) + \hat{\rho}(x,y) + \sum_{m=\nu_0+1}^{\nu} w_m\breve{\rho}_m(x,y) \geq 0;$$

(c) (i) $\xi \to \mathscr{E}(\xi,y,u)$ is $(\bar{\phi},\eta,\bar{\rho},\theta,\tilde{m})$-pseudosonvex at $y$, and $\bar{\phi}(a) \geq 0 \Rightarrow a \geq 0$;

(ii) for each $m \in J_+$, $G_{j_m}$ is $(\hat{\phi}_m,\eta,\hat{\rho}_m,\theta,\tilde{m})$-quasisonvex at $y$, $\hat{\phi}_j$ is increasing, and $\hat{\phi}_j(0) = 0$;

(iii) $\xi \to \mathscr{D}(\xi,w)$ is $(\breve{\phi},\eta,\breve{\rho},\theta,\tilde{m})$-quasisonvex at $y$, $\breve{\phi}$ is increasing, and $\breve{\phi}(0) = 0$;

(iv) $\bar{\rho}(x,y)+\sum_{m=1}^{v_0} v_m\hat{\rho}_m(x,y)+\breve{\rho}(x,y)\geq 0$;

(d) (i) $\xi \rightarrow \mathscr{E}(\xi,y,u)$ is $(\bar{\phi},\eta,\bar{\rho},\theta,\tilde{m})$-pseudosonvex at $y$, and $\bar{\phi}(a)\geq 0 \Rightarrow a\geq 0$;

(ii) $\xi \rightarrow \mathscr{C}(\xi,v)$ is $(\hat{\phi},\eta,\hat{\rho},\theta,\tilde{m})$-quasisonvex at $y$, $\hat{\phi}$ is increasing, and $\hat{\phi}(0)=0$;

(iii) $\xi \rightarrow \mathscr{D}(\xi,w)$ is $(\breve{\phi},\eta,\breve{\rho},\theta,\tilde{m})$-quasisonvex at $y$, $\breve{\phi}$ is increasing, and $\breve{\phi}(0)=0$;

(iv) $\bar{\rho}(x,y)+\hat{\rho}(x,x^*)+\breve{\rho}(x,y)\geq 0$;

(e) (i) $\xi \rightarrow \mathscr{E}(\xi,y,u)$ is $(\bar{\phi},\eta,\bar{\rho},\theta,\tilde{m})$-pseudosonvex at $y$, and $\bar{\phi}(a)\geq 0 \Rightarrow a\geq 0$;

(ii) $\xi \rightarrow \mathscr{G}(\xi,v,w)$ is $(\hat{\phi},\eta,\hat{\rho},\theta,\tilde{m})$-quasisonvex at $y$, $\hat{\phi}$ is increasing, and $\hat{\phi}(0)=0$;

(iii) $\bar{\rho}(x,y)+\hat{\rho}(x,y)\geq 0$.

Then $\varphi(x)\geq \psi_{II}(y,z,u,v,w)$.

*Proof.* (a) : In view of our assumptions specified in (ii) and (iii), (24) and (25) remain valid for the present case. From (19), (20), (24), (25), and (iv) we deduce that

$$\Big\langle \sum_{i=1}^{p} u_i[D(y,u)\nabla f_i(y)-N(y,u)\nabla g_i(y)],\eta(x,y)\Big\rangle+\frac{1}{2}\Big\langle z,\sum_{i=1}^{p} u_i[D(y,u)\nabla^2 f_i(y) -N(y,u)\nabla^2 g_i(y)]z\Big\rangle \geq \Big[\sum_{m=1}^{v_0} v_m\hat{\rho}_m(x,y)+\sum_{m=v_0+1}^{v} w_m\breve{\rho}_k(x,y)\Big]\|\theta(x,y)\|^m \geq -\bar{\rho}(x,y)\|\theta(x,y)\|^m,$$

which in view of (i) implies that

$$\bar{\phi}\big(\mathscr{E}(x,y,u)-\mathscr{E}(y,y,u)\big)\geq 0.$$

Based on the properties of the function $\bar{\phi}$, the last inequality yields

$$\mathscr{E}(x,y,u)\geq \mathscr{E}(y,y,u)=0,$$

where the equality follows from the definitions of $N(y,u)$ and $D(y,u)$. As shown in the proof of Theorem 1.17, this inequality leads to the conclusion that $\varphi(x)\geq \psi_{II}(y,z,u,v,w)$.

(b) - (e) : The proofs are similar to that of part (a).

□

**Theorem 1.21.** (Weak Duality) Let $x$ and $\mathscr{S}\equiv(y,z,u,v,w)$ be arbitrary feasible solutions of (P) and (DII), respectively, and assume that any one of the following seven sets of hypotheses is satisfied:

(a) (i) $\xi \rightarrow \mathscr{E}(\xi,y,u)$ is prestrictly $(\bar{\phi},\eta,\bar{\rho},\theta,\tilde{m})$-quasisonvex at $y$ and $\bar{\phi}(a)\geq 0 \Rightarrow a\geq 0$;

(ii) for each $m \in J_+ \equiv J_+(v)$, $G_{j_m}$ is strictly $(\hat{\phi}_m, \eta, \hat{\rho}_m, \theta, \tilde{m})$-pseudosonvex at $y$, $\hat{\phi}_j$ is increasing, and $\hat{\phi}_j(0) = 0$;

(iii) for each $m \in K_* \equiv K_*(w)$, $\xi \to \mathscr{D}_{k_m}(\xi, w)$ is $(\breve{\phi}_m, \eta, \breve{\rho}_k, \theta, \tilde{m})$-quasisonvex at $y$, $\breve{\phi}_k$ is increasing, and $\breve{\phi}_k(0) = 0$;

(iv) $\bar{\rho}(x,y) + \sum_{m=1}^{\nu_0} v_m \hat{\rho}_m(x,y) + \sum_{m=\nu_0+1}^{\nu} w_m \breve{\rho}_m(x,y) \geq 0$;

(b) (i) $\xi \to \mathscr{E}(\xi, y, u)$ is prestrictly $(\bar{\phi}, \eta, \bar{\rho}, \theta, \tilde{m})$-quasisonvex at $y$, and $\bar{\phi}(a) \geq 0 \Rightarrow a \geq 0$;

(ii) $\xi \to \mathscr{C}(\xi, v)$ is strictly $(\hat{\phi}, \eta, \hat{\rho}, \theta, \tilde{m})$-pseudosonvex at $y$, $\hat{\phi}$ is increasing, and $\hat{\phi}(0) = 0$;

(iii) for each $m \in K_*$, $\xi \to \mathscr{D}_{k_m}(\xi, w)$ is $(\breve{\phi}_m, \eta, \breve{\rho}_m, \theta, \tilde{m})$-quasisonvex at $y$, $\breve{\phi}_k$ is increasing, and $\breve{\phi}_k(0) = 0$;

(iv) $\bar{\rho}(x,y) + \hat{\rho}(x,y) + \sum_{m=\nu_0+1}^{\nu} w_m \breve{\rho}_m(x,y) \geq 0$;

(c) (i) $\xi \to \mathscr{E}(\xi, y, u)$ is prestrictly $(\bar{\phi}, \eta, \bar{\rho}, \theta, \tilde{m})$-quasisonvex at $y$ and $\bar{\phi}(a) \geq 0 \Rightarrow a \geq 0$;

(ii) for each $m \in J_+$, $G_{j_m}$ is $(\hat{\phi}_j, \eta, \hat{\rho}_j, \theta, m)$-quasisonvex at $y$, $\hat{\phi}_j$ is increasing, and $\hat{\phi}_j(0) = 0$;

(iii) for each $m \in K_*$, $\xi \to \mathscr{D}_{k_m}(\xi, w)$ is strictly $(\breve{\phi}_k, \eta, \breve{\rho}_k, \theta, \tilde{m})$-pseudosonvex at $y$, $\breve{\phi}_m$ is increasing, and $\breve{\phi}_m(0) = 0$;

(iv) $\bar{\rho}(x,y) + \sum_{m=1}^{\nu_0} v_m \hat{\rho}_m(x,y) + \sum_{m=\nu_0+1}^{\nu} w_m \breve{\rho}_m(x,y) \geq 0$;

(d) (i) $\xi \to \mathscr{E}(\xi, y, u)$ is prestrictly $(\bar{\phi}, \eta, \bar{\rho}, \theta, \tilde{m})$-quasisonvex at $y$ and $\bar{\phi}(a) \geq 0 \Rightarrow a \geq 0$;

(ii) for each $m \in J_+$, $\xi \to G_{j_m}$ is $(\hat{\phi}_m, \eta, \hat{\rho}_m, \theta, \tilde{m})$-quasisonvex at $y$, $\hat{\phi}_j$ is increasing, and $\hat{\phi}_j(0) = 0$;

(iii) $\xi \to \mathscr{D}(\xi, w)$ is strictly $(\breve{\phi}, \eta, \breve{\rho}, \theta, \tilde{m})$-pseudosonvex at $y$, $\breve{\phi}$ is increasing, and $\breve{\phi}(0) = 0$;

(iv) $\bar{\rho}(x,y) + \sum_{m=1}^{\nu_0} v_m \hat{\rho}_m(x,y) + \breve{\rho}(x,y) \geq 0$;

(e) (i) $\xi \to \mathscr{E}(\xi, y, u)$ is prestrictly $(\bar{\phi}, \eta, \bar{\rho}, \theta, \tilde{m})$-quasisonvex at $y$ and $\bar{\phi}(a) \geq 0 \Rightarrow a \geq 0$;

(ii) $\xi \to \mathscr{C}(\xi, v)$ is strictly $(\hat{\phi}, \eta, \hat{\rho}, \theta, \tilde{m})$-pseudosonvex at $y$, $\hat{\phi}$ is increasing, and $\hat{\phi}(0) = 0$;

(iii) $\xi \to \mathscr{D}(\xi, w)$ is $(\breve{\phi}, \eta, \breve{\rho}, \theta, \tilde{m})$-quasisonvex at $y$, $\breve{\phi}$ is increasing, and $\breve{\phi}(0) = 0$;

(iv) $\bar{\rho}(x,y) + \hat{\rho}(x,y) + \breve{\rho}(x,y) \geq 0$;

(f) (i) $\xi \to \mathscr{E}(\xi, y, u)$ is prestrictly $(\bar{\phi}, \eta, \bar{\rho}, \theta, \tilde{m})$-quasisonvex at $y$ and $\bar{\phi}(a) \geq 0 \Rightarrow a \geq 0$;

(ii) $\xi \to \mathscr{C}(\xi, v)$ is $(\hat{\phi}, \eta, \hat{\rho}, \theta, \tilde{m})$-quasisonvex at $y$, $\hat{\phi}$ is increasing, and $\hat{\phi}(0) = 0$;

(iii) $\xi \to \mathscr{D}(\xi, w)$ is strictly $(\breve{\phi}, \eta, \breve{\rho}, \theta, \tilde{m})$-pseudosonvex at $y$, $\breve{\phi}$ is increasing, and $\breve{\phi}(0) = 0$;

(iv) $\bar{\rho}(x,y) + \hat{\rho}(x,y) + \breve{\rho}(x,y) \geq 0$;

(g) (i) $\xi \to \mathscr{E}(\xi, y, u)$ is prestrictly $(\bar{\phi}, \eta, \bar{\rho}, \theta, \tilde{m})$-quasisonvex at $y$ and $\bar{\phi}(a) \geq 0 \Rightarrow a \geq 0$;

(ii) $\xi \to \mathscr{G}(\xi, v, w)$ is strictly $(\hat{\phi}, \eta, \hat{\rho}, \theta, \tilde{m})$-pseudosonvex at $y$, $\hat{\phi}$ is increasing, and $\hat{\phi}(0) = 0$;

(iii) $\bar{\rho}(x,y) + \hat{\rho}(x,y) \geq 0$.

Then $\varphi(x) \geq \psi_{II}(y, z, u, v, w)$.

*Proof.* (a) : Since for each $j \in J_+$, $G_j(x) - G_j(y) \leq 0$ and hence $\hat{\phi}_j\big(G_j(x) - G_j(y)\big) \leq 0$, (ii) implies that

$$\langle \nabla G_{j_m}(y), \eta(x,y)\rangle + \frac{1}{2}\langle z, \nabla^2 G_{j_m}(y) z\rangle < -\hat{\rho}_j(x,y)\|\theta(x,y)\|^{\tilde{m}}, \, m \in \underline{\nu_0}).$$

As $v_m \geq 0$ for each $m \in \underline{\nu_0}$ and $v_m = 0$ for each $m \in \underline{\nu} \setminus \underline{\nu_0}$, the above inequalities yield

$$\begin{aligned}&\Big\langle \sum_{m=1}^{\nu_0} v_m \nabla G_{j_m}(y, t^m), \eta(x,y)\Big\rangle \\ &+\frac{1}{2}\Big\langle z, \sum_{m=1}^{\nu_0} v_m \nabla^2 G_{j_m}(y, t^m) z\Big\rangle \\ &< -\sum_{m=1}^{\nu_0} v_m \hat{\rho}_m(x,y)\|\theta(x,y)\|^{\tilde{m}}.\end{aligned}$$

Now combining this inequality with (19), (20), and (25) (which is valid for the present case because of (iii)), and using (iv), we obtain

$$\begin{aligned}&\Big\langle \sum_{i=1}^{p} u_i[D(y,u)\nabla f_i(y) - N(y,u)\nabla g_i(y)], \eta(x,y)\Big\rangle \\ &\qquad + \frac{1}{2}\Big\langle z, \sum_{i=1}^{p} u_i[D(y,u)\nabla^2 f_i(y) \\ &- N(y,u)\nabla^2 g_i(y)]z\Big\rangle \geq \Big[\sum_{m=1}^{\nu_0} v_m \hat{\rho}_m(x,y) + \sum_{m=\nu_0+1}^{\nu} w_m \breve{\rho}_m(x,y)\Big]\|\theta(x,y)\|^{\tilde{m}} \\ &\qquad > -\bar{\rho}(x,y)\|\theta(x,y)\|^{\tilde{m}},\end{aligned}$$

which in view of (i) implies that

$$\bar{\phi}\big(\mathscr{E}(x,y,u)-\mathscr{E}(y,y,u)\big)\geq 0.$$

The rest of the proof is identical to that of Theorem 1.17.

(b) - (g) : The proofs are similar to that of part (a). □

## 5. Duality Model III

Based on the duality model II, we construct the following parameter-free duality models for establishing some duality theorems applying the generalized $(\phi,\eta,\rho,\theta,\tilde{m})$-sonvexity hypotheses, which are upgrades to the previous section.

$$(DIII)\quad \text{Maximize}\quad \psi_{III}(y,z,u,v,w)=\frac{\sum_{i=1}^{p}u_i f_i(y)}{\sum_{i=1}^{p}u_i g_i(y)}$$

subject to

$$\sum_{i=1}^{p}u_i[D(y,u)\nabla f_i(y)-N(y,u)\nabla g_i(y)]+\sum_{m=1}^{\nu_0}v_m\nabla G_{j_m}(y,t^m)$$
$$+\sum_{m=\nu_0+1}^{\nu}w_m\nabla H_{k_m}(y,s^m)=0, \tag{1.26}$$

$$\Big\langle z,\Big\{\sum_{i=1}^{p}u_i[D(y,u)\nabla^2 f_i(y)-N(y,u)\nabla^2 g_i(y)]+\sum_{m=1}^{\nu_0}v_m\nabla^2 G_{j_m}(y,t^m)$$
$$+\sum_{m=\nu_0+1}^{\nu}w_m\nabla^2 H_{k_m}(y,s^m)\Big\}z\Big\rangle\geq 0, \tag{1.27}$$

$$u_i[D(y,u)f_i(y)-N(y,u)g_i(y)]\geq 0,\ i\in\underline{p}, \tag{1.28}$$

$$v_m G_{j_m}(y,t^m)\geq 0,\ m\in\underline{\nu_0}), \tag{1.29}$$

$$w_m H_{k_m}(y,s^m)\geq 0,\ m\in\underline{\nu}\setminus\underline{\nu_0}, \tag{1.30}$$

$$y\in X,\ z\in C(y),\ u\in U,\ v\in\mathbb{R}^q_+,\ w\in\mathbb{R}^r; \tag{1.31}$$

$$(\tilde{D}III)\quad \text{Maximize}\quad \psi_{III}(y,z,u,v,w)=\frac{\sum_{i=1}^{p}u_i f_i(y)}{\sum_{i=1}^{p}u_i g_i(y)}$$

subject to (8) and (27) - (31).

The remarks made earlier about the relationships between $(DI)$ and $(\tilde{D}I)$ are, of course, also valid for $(DIII)$ and $(\tilde{D}III)$. The next theorem shows that $(DIII)$ is a dual problem for $(P)$.

**Theorem 1.22.** (Weak Duality) Let $x$ and $\mathscr{S} = (y, z, u, v, w)$ be arbitrary feasible solutions of (P) and (DIII), respectively, and assume that any one of the following five sets of hypotheses is satisfied:

(a) (i) for each $i \in I_+ \equiv I_+(u)$, $\xi \to \mathscr{E}_i(\xi, y, u)$ is prestrictly $(\bar{\phi}_i, \eta, \bar{\rho}_i, \theta, \tilde{m})$-quasisonvex at $y$, $\bar{\phi}_i$ is strictly increasing, and $\bar{\phi}_i(0) = 0$;

(ii) for each $m \in J_+ \equiv J_+(v)$, $G_{j_m}$ is $(\hat{\phi}_m, \eta, \hat{\rho}_j, \theta, \tilde{m})$-quasisonvex at $y$, $\hat{\phi}_j$ is increasing, and $\hat{\phi}_j(0) = 0$;

(iii) for each $m \in K_* \equiv K_*(w)$, $\xi \to \mathscr{D}_{k_m}(\xi, w)$ is $(\breve{\phi}_k, \eta, \breve{\rho}_k, \theta, \tilde{m})$-quasisonvex at $y$, $\breve{\phi}_k$ is increasing, and $\breve{\phi}_k(0) = 0$;

(iv) $\rho^\circ(x,y) + \sum_{m=1}^{\nu_0} v_m \hat{\rho}_m(x,y) + \sum_{m=\nu_0+1}^{\nu} w_m \breve{\rho}_k(x,y) > 0$, where $\rho^\circ(x,y) = \sum_{i \in I_+} u_i \bar{\rho}_i(x,y)$;

(b) (i) for each $i \in I_+$, $\xi \to \mathscr{E}_i(\xi, y, u)$ is prestrictly $(\bar{\phi}_i, \eta, \bar{\rho}_i, \theta, \tilde{m})$-quasisonvex at $y$, $\bar{\phi}_i$ is strictly increasing, and $\bar{\phi}_i(0) = 0$;

(ii) $\xi \to \mathscr{C}(\xi, v)$ is $(\hat{\phi}, \eta, \hat{\rho}, \theta, \tilde{m})$-quasisonvex at $y$, $\hat{\phi}$ is increasing, and $\hat{\phi}(0) = 0$;

(iii) for each $m \in K_*$, $\xi \to \mathscr{D}_m(\xi, w)$ is $(\breve{\phi}_k, \eta, \breve{\rho}_k, \theta, \tilde{m})$-quasisonvex at $y$, $\breve{\phi}_k$ is increasing, and $\breve{\phi}_k(0) = 0$;

(iv) $\rho^\circ(x,y) + \hat{\rho}(x,y) + \sum_{m=\nu_0+1}^{\nu} w_m \breve{\rho}_m(x,y) > 0$;

(c) (i) for each $i \in I_+$, $\xi \to \mathscr{E}_i(\xi, y, u)$ is prestrictly $(\bar{\phi}_i, \eta, \bar{\rho}_i, \theta, \tilde{m})$-quasisonvex at $y$, $\bar{\phi}_i$ is strictly increasing, and $\bar{\phi}_i(0) = 0$;

(ii) for each $m \in J_+$, $G_{j_m}$ is $(\hat{\phi}_m, \eta, \hat{\rho}_m, \theta, \tilde{m})$-quasisonvex at $y$, $\hat{\phi}_j$ is increasing, and $\hat{\phi}_j(0) = 0$;

(iii) $\xi \to \mathscr{D}(\xi, w)$ is $(\breve{\phi}, \eta, \breve{\rho}, \theta, \tilde{m})$-quasisonvex at $y$, $\breve{\phi}$ is increasing, and $\breve{\phi}(0) = 0$;

(iv) $\rho^\circ(x,y) + \sum_{m=1}^{\nu_0)} v_m \hat{\rho}_m(x,y) + \breve{\rho}(x,y) > 0$;

(d) (i) for each $i \in I_+$, $\xi \to \mathscr{E}_i(\xi, y, u)$ is prestrictly $(\bar{\phi}_i, \eta, \bar{\rho}_i, \theta, \tilde{m})$-quasisonvex at $y$, $\bar{\phi}_i$ is strictly increasing, and $\bar{\phi}_i(0) = 0$;

(ii) $\xi \to \mathscr{C}(\xi, v)$ is $(\hat{\phi}, \eta, \hat{\rho}, \theta, \tilde{m})$-quasisonvex at $y$, $\hat{\phi}$ is increasing, and $\hat{\phi}(0) = 0$;

(iii) $\xi \to \mathscr{D}(\xi, w)$ is $(\breve{\phi}, \eta, \breve{\rho}, \theta, \tilde{m})$-quasisonvex at $y$, $\breve{\phi}$ is increasing, and $\breve{\phi}(0) = 0$;

(iv) $\rho^\circ(x,y)+\hat{\rho}(x,y)+\breve{\rho}(x,y)>0$;

(e) (i) for each $i\in I_+$, $\xi\to\mathscr{E}_i(\xi,y,u)$ is prestrictly $(\bar{\phi}_i,\eta,\bar{\rho}_i,\theta,\tilde{m})$-quasisonvex at $y$, $\bar{\phi}_i$ is strictly increasing, and $\bar{\phi}_i(0)=0$;

(ii) $\xi\to\mathscr{G}(\xi,v,w)$ is $(\hat{\phi},\eta,\hat{\rho},\theta,\tilde{m})$-quasisonvex at $y$, $\hat{\phi}$ is increasing, and $\hat{\phi}(0)=0$;

(iii) $\rho^\circ(x,y)+\hat{\rho}(x,y)>0$.

Then $\varphi(x)\geq\psi_{III}(y,z,u,v,w)$.

*Proof.* (a) : Suppose that to the contrary that $\varphi(x)<\psi_{III}(y,z,u,v,w)$. This implies that for each $i\in\underline{p}$, $\mathscr{E}_i(x,y,u)<0$. Since $\mathscr{E}_i(y,y,u)\geq 0$ by (28), it follows that $\mathscr{E}_i(x,y,u)<\mathscr{E}_i(y,y,u)$, and hence for each $i\in I_+$, $\bar{\phi}_i\big(\mathscr{E}_i(x,y,u)-\mathscr{E}_i(y,y,u)\big)<0$, which by virtue of (i) implies that

$$\langle D(y,u)\nabla f_i(y)-N(y,u)\nabla g_i(y),\eta(x,y)\rangle+\frac{1}{2}\langle z,[D(y,u)\nabla^2 f_i(y)-N(y,u)\nabla^2 g_i(y)]z\rangle]\leq-\bar{\rho}_i(x,y)\|\theta(x,y)\|^m.$$

Since $u\geq 0$ and $\sum_{i=1}^{p}u_i=1$, the above inequalities yield

$$\Big\langle\sum_{i=1}^{p}u_i[D(y,u)\nabla f_i(y)-N(y,u)\nabla g_i(y)],\eta(x,y)\Big\rangle+\frac{1}{2}\Big\langle z,\sum_{i=1}^{p}u_i[D(y,u)\nabla^2 f_i(y)-N(y,u)\nabla^2 g_i(y)]z\Big\rangle\leq-\rho^\circ(x,y)\|\theta(x,y)\|^m.\quad(1.32)$$

Since for each $j\in J_+$, $G_j(x,t)-G_j(y,t)\leq 0$ and hence $\hat{\phi}_j\big(G_j(x,t)-G_j(y,t)\big)\leq 0$, (ii) implies that

$$\langle\nabla G_j(y,t),\eta(x,y)\rangle+\frac{1}{2}\langle z,\nabla^2 G_j(y,t)z\rangle\leq-\hat{\rho}_j(x,y)\|\theta(x,y)\|^{\tilde{m}}.$$

As $v_m\geq 0$ for each $m\in\underline{v_0}$, the above inequalities yield

$$\Big\langle\sum_{m=1}^{v_0}v_m\nabla G_{j_m}(y,t^m),\eta(x,y)\Big\rangle+\frac{1}{2}\Big\langle z,\sum_{m=1}^{v_0}v_m\nabla^2 G_{j_m}(y)z\Big\rangle\leq-\sum_{m=1}^{v_0}v_m\hat{\rho}_m(x,y)\|\theta(x,y)\|^{\tilde{m}}.$$

Now combining this inequality with (19), (20), and (25) (which is valid for the present case because of (iii)), and using (iv), we obtain

$$\Big\langle\sum_{i=1}^{p}u_i[D(y,u)\nabla f_i(y)-N(y,u)\nabla g_i(y)],\eta(x,y)\Big\rangle+\frac{1}{2}\Big\langle z,\sum_{i=1}^{p}u_i[D(y,u)\nabla^2 f_i(y)-N(y,u)\nabla^2 g_i(y)]z\Big\rangle\geq\Big[\sum_{m=1}^{v_0}v_m\hat{\rho}_m(x,y)+\sum_{m=v_0+1}^{v}w_m\breve{\rho}_m(x,y)\Big]\|\theta(x,y)\|^{\tilde{m}}>-\rho^\circ(x,y)\|\theta(x,y)\|^{\tilde{m}},$$

which contradicts (32). Therefore, we conclude that $\varphi(x) \geq \psi_{III}(y,z,u,v,w)$.

(b) - (e) : The proofs are similar to that of part (a).

□

**Theorem 1.23.** (Strong Duality) Let $x^*$ be a normal optimal solution of (P) and assume that any one of the five sets of conditions specified in Theorem 1.22 is satisfied for all feasible solutions of (DIII). Then for each $z^* \in C(x^*)$, there exist $u^* \in U$, $v^* \in \mathbb{R}^q_+$, and $w^* \in \mathbb{R}^r$ such that $(x^*,z^*,u^*,v^*,w^*)$ is an optimal solution of (DIII) and $\varphi(x^*) = \psi_{III}(x^*,z^*,u^*,v^*,w^*)$.

*Proof.* The proof is similar to that of Theorem 1.15. □

## 6. Duality Model IV

Finally, we examine duality models , which are certain variants of $(DI)$ and $(\tilde{D}I)$, as follows:

$$(DIV) \qquad \text{Maximize} \quad \psi_{IV}(y,z,u,v,w) =$$

$$\frac{\sum_{i=1}^p u_i f_i(y) + \sum_{m=1}^{\nu_0} v_m G_{j_m}(y,t^m) + \sum_{m=\nu_0+1}^{\nu} w_m H_{k_m}(y,s^m)}{\sum_{i=1}^p u_i g_i(y)}$$

subject to

$$D(y,u)\Big[\sum_{i=1}^p u_i \nabla f_i(y) + \sum_{m=1}^{\nu_0} v_m \nabla G_{j_m}(y,t^m) + \sum_{m=\nu_0+1}^{\nu} w_m \nabla H_{k_m}(y,s^m)\Big] - [N(y,u) + \mathscr{G}(y,v,w)] \sum_{i=1}^p u_i \nabla g_i(y) = 0, \quad (1.33)$$

$$\Big\langle z, \Big\{D(y,u)\Big[\sum_{i=1}^p u_i \nabla^2 f_i(y) + \sum_{m=1}^{\nu_0} v_m \nabla^2 G_{j_m}(y,t^m) + \sum_{m=\nu_0+1}^{\nu} w_m \nabla^2 H_{k_m}(y,s^m)\Big] - [N(y,u) + \mathscr{G}(y,v,w)] \sum_{i=1}^p u_i \nabla^2 g_i(y)\Big\} z\Big\rangle \geq 0, \quad (1.34)$$

$$\sum_{m=1}^{\nu_0} v_m G_{j_m}(y,t^m) + \sum_{m=\nu_0+1}^{\nu} w_m H_{k_m}(y,s^m) \geq 0, \quad (1.35)$$

$$y \in X,\ z \in C(y),\ u \in U,\ v \in \underline{\nu_0},\ w \in \underline{\nu} \setminus \underline{\nu_0}; \quad (1.36)$$

$$(\tilde{D}IV) \qquad \text{Maximize} \quad \psi_{IV}(y,z,u,v,w)$$

$$= \frac{\sum_{i=1}^p u_i f_i(y) + \sum_{m=1}^{\nu_0} v_m G_{j_m}(y,t^m) + \sum_{m=\nu_0+1}^{\nu} w_m H_{k_m}(y,s^m)}{\sum_{i=1}^p u_i g_i(y)}$$

subject to (34) - (36) and

$$\Big\langle D(y,u)\Big[\sum_{i=1}^{p} u_i[\nabla f_i(y) + \sum_{m=1}^{\nu_0} v_m \nabla G_{j_m}(y,t^m)$$

$$+ \sum_{m=\nu_0+1}^{\nu} w_m \nabla H_{k_m}(y,s^m)\Big]$$

$$- [N(y,u) + \mathscr{G}(y,v,w)] \sum_{i=1}^{p} u_i \nabla g_i(y), \eta(x,y)\Big\rangle \geq 0 \text{ for all } x \in \mathbb{F},$$

where $\eta$ is a function from $X \times X$ to $\mathbb{R}^n$.
The following two theorems show that $(DIV)$ is a dual problem for $(P)$.

**Theorem 1.24.** (Weak Duality) Let $x$ and $\mathscr{S} \equiv (y,z,u,v,w)$ be arbitrary feasible solutions of (P) and (DIV), respectively, and assume that $N(y,u) \geq 0$, $D(y,u) > 0$, and that either one of the following two sets of hypotheses is satisfied:

(a) (i) for each $i \in I_+ \equiv I_+(u)$, $f_i$ is $(\phi,\eta,\bar{\rho}_i,\theta,\tilde{m})$-sonvex and $-g_i$ is $(\phi,\eta,\tilde{\rho}_i,\theta,\tilde{m})$-sonvex at $y$,

(ii) for each $m \in \underline{\nu_0}$, $G_{j_m}$ is $(\phi,\eta,\hat{\rho}_m,\theta,\tilde{m})$-sonvex at $y$;

(iii) for each $m \in \underline{\nu} \setminus \underline{\nu_0}$, $\xi \to \mathscr{D}_{k_m}(\xi,w)$ is $(\phi,\eta,\breve{\rho}_k,\theta,\tilde{m})$-sonvex at $y$;

(iv) $\phi$ is superlinear and $\phi(a) \geq 0 \Rightarrow a \geq 0$;

(v) $D(y,u)\Big[\sum_{i\in I_+} u_i\bar{\rho}_i(x,y) + \sum_{m=1}^{\nu_0} v_m\hat{\rho}_m(x,y) + \sum_{m=\nu_0+1}^{\nu} w_m\breve{\rho}_m(x,y)\Big]$
$+ [N(y,u) + \mathscr{G}(y,u,w)] \sum_{i\in I_+} u_i\tilde{\rho}_i(x,y) \geq 0$;

(b) the Lagrangian-type function

$$\xi \to \Lambda(\xi,y,u,v,w) = D(y,u)\Big[\sum_{i=1}^{p} u_i f_i(\xi) + \sum_{m=1}^{\nu_0} v_m G_{j_m}(\xi,t^m)$$

$$+ \sum_{m=\nu_0+1}^{\nu} w_m H_{k_m}(\xi,s^m)\Big] - [N(y,u) + \mathscr{G}(y,v,w)] \sum_{i=1}^{p} u_i g_i(\xi),$$

is $(\phi,\eta,\rho,\theta,\tilde{m})$-pseudosonvex at $y$, $\rho(x,y) \geq 0$, and $\phi(a) \geq 0 \Rightarrow a \geq 0$.

Then $\varphi(x) \geq \psi_{IV}(y,z,u,v,w)$.

*Proof.* (a): In light of assumptions in (i) - (iii), (33) - (36) hold. Multiplying (33) by $u_i D(y,u)$ and then summing over $i \in I_+$, (34) by $u_i[N(y,u)+\mathscr{G}(y,v,w)]$ and then summing $i \in I_+$, (35) by $v_m D(y,u)$ and then summing over $m \in \underline{v_0}$, and by $w_m D(y,u)$ and then summing over $m \in \underline{v} \setminus \underline{v_0}$, and adding the resulting inequalities, using the superlinearity of $\phi$, we obtain

$$\phi\Big(D(y,u)\Big[\sum_{i=1}^{p} u_i f_i(x) + \sum_{m=1}^{v_0} v_m G_{j_m}(x,t^m) + \sum_{m=v_0+1}^{v} w_m H_{k_m}(x,s^m)\Big]$$
$$- [N(y,u)+\mathscr{G}(y,v,w)]\sum_{i=1}^{p} u_i g_i(x) - \Big\{D(y,u)\Big[\sum_{i=1}^{p} u_i f_i(y) + \sum_{m=1}^{v_0} v_m G_{j_m}(y,t^m)$$
$$+ \sum_{m=v_0+1}^{v} w_m H_{k_m}(y,s^m)\Big] - [N(y,u)+\mathscr{G}(y,v,w)]\sum_{i=1}^{p} u_i g_i(y)\Big\}\Big)$$
$$\geq \Big\langle D(y,u)\Big[\sum_{i=1}^{p} u_i \nabla f_i(y) + \sum_{m=1}^{v_0} v_m \nabla G_{j_m}(y,t^m)$$
$$+ \sum_{m=v_0+1}^{v} w_m \nabla H_{k_m}(y,s^m)\Big] - [N(y,u)+\mathscr{G}(y,v,w)]\sum_{i=1}^{p} u_i \nabla g_i(y), \eta(x,y)\Big\rangle$$
$$+ \frac{1}{2}\Big\langle z, \Big\{D(y,u)\Big[\sum_{i=1}^{p} u_i \nabla^2 f_i(y) + \sum_{m=1}^{v_0} v_m \nabla^2 G_{j_m}(y,s^m)$$
$$+ \sum_{m=v_0+1}^{v} w_m \nabla^2 H_{k_m}(y,s^m)\Big] - [N(y,u)+\mathscr{G}(y,v,w)]\sum_{i=1}^{p} u_i \nabla^2 g_i(y)\Big\}z\Big\rangle$$
$$+ \Big\{D(y,u)\Big[\sum_{i\in I_+} u_i \bar{\rho}_i(x,y) + \sum_{m=1}^{v_0} v_m \hat{\rho}_m(x,y)$$
$$+ \sum_{m=v_0+1}^{v} w_m \breve{\rho}_m(x,y)\Big] + [N(y,u)+\mathscr{G}(y,v,w)]\sum_{i\in I_+} u_i \tilde{\rho}_i(x,y)\Big\}\|\theta(x,y)\|^{\tilde{m}}.$$

Because of the dual feasibility of $(y,z,u,v,w)$, (33), (34), and (v), the above inequality becomes

$$\phi\Big(D(y,u)\Big[\sum_{i=1}^{p} u_i f_i(x) + \sum_{m=1}^{v_0} v_m G_{j_m}(x,t^m) + \sum_{m=v_0+1}^{v} w_m H_{k_m}(x,s^m)\Big]$$
$$- [N(y,u)+\mathscr{G}(y,v,w)]\sum_{i=1}^{p} u_i g_i(x) - \Big\{D(y,u)\Big[\sum_{i=1}^{p} u_i f_i(y) + \sum_{m=1}^{v_0} v_m G_{j_m}(y,t^m) +$$
$$\sum_{m=v_0+1}^{v} w_m H_{k_m}(y,s^m)\Big] - [N(y,u)+\mathscr{G}(y,v,w)]\sum_{i=1}^{p} u_i g_i(y)\Big\}\Big) \geq 0.$$

In view of the definitions of $D(y,u)$, $N(y,u)$, and $\mathscr{G}(y,v,w)$, the right-hand side of this inequality is equal to zero. Since $x \in \mathbb{F}$, the above inequality reduces to

$$\sum_{i=1}^{p} u_i\{D(y,u) f_i(x) - [N(y,u)+\mathscr{G}(y,v,w)] g_i(x)\} \geq 0. \tag{1.37}$$

Now using (37) and Lemma 1.12, we obtain the weak duality inequality as follows:

$$\begin{aligned}\varphi(x) &= \max_{a\in U} \frac{\sum_{i=1}^{p} a_i f_i(x)}{\sum_{i=1}^{p} a_i g_i(x)}\\ &\geq \frac{\sum_{i=1}^{p} u_i f_i(x)}{\sum_{i=1}^{p} u_i g_i(x)} \geq \frac{N(y,u)+\mathcal{G}(y,v,w)}{D(y,u)} = \psi_{IV}(y,z,u,v,w).\end{aligned}$$

(b) : Using the dual feasibility of $\mathcal{S}$, nonnegativity of $\rho(x,y)$, (33), and (34), we obtain the following inequality:

$$\begin{aligned}&\langle \nabla\Lambda(y,y,u,v,w), \eta(x,y)\rangle + \frac{1}{2}\langle z, \nabla^2\Lambda(y,y,u,v,w)z\rangle \geq 0\\ &\geq -\rho(x,y)\|\theta(x,y)\|^{\tilde{m}},\end{aligned}$$

which in view of our $(\phi,\eta,\rho,\theta,\tilde{m})$-pseudosonvexity assumption implies that

$$\phi\big(\Lambda(x,y,u,v,w)-\Lambda(y,y,u,v,w)\big) \geq 0.$$

But $\phi(a) \geq 0 \Rightarrow a \geq 0$, and hence we have

$$\Lambda(x,y,u,v,w) \geq \Lambda(y,y,u,v,w) = 0,$$

where the equality follows from the definitions of $D(y,u)$, $N(y,u)$, and $\mathcal{G}(y,v,w)$. Because $x \in \mathbb{F}$ and $v \geq 0$, we get

$$\sum_{i=1}^{p} u_i\{D(y,u)f_i(x) - [N(y,u)+\mathcal{G}(y,v,w)]g_i(x)\} \geq 0,$$

which is precisely (37). This inequality leads to the desired conclusion that $\varphi(x) \geq \psi_{IV}(y,z,u,v,w)$. □

## 7. General Remarks

Based on the generalized $(\phi,\eta,\zeta,\tau,\rho,\theta,\tilde{m})$-sonvexity, we have established several second-order parameter - free duality models for a class of semiinfinite discrete minmax fractional programming problems, and related results that encompass most of the investigations available in the current literature. Furthermore, the presentation approach adopted in this chapter as well as the main results explored here will prove useful investigating other related categories of nonlinear programming problems and utilizing similar generalized convexity concepts.

# Bibliography


[1] C. R. Bector, S. Chandra, S. Gupta and S. K. Suneja, Univex Sets, Functions, and Univex Nonlinear Programming, in *Generalized Convexity*, S. Komlòsi, T. Rapcsák, and S. Schaible (eds.), Springer-Verlag, New York, 1994, pp. 3 - 19.

[2] H. Gfrerer, Second-order optimality conditions for scalar and vector optimization problems in Banach spaces, *SIAM J. Optim.* **45** (2006), 972 - 997.

[3] M. A. Hanson, On sufficiency of the Kuhn-Tucker conditions, *J. Math. Anal. Appl.* **80** (1981), 545 - 550.

[4] M. A. Hanson, Second order invexity and duality in mathematical programming, *Opsearch* **30** (1993), 313 - 320.

[5] L. R. Huang and K. F. Ng, Second-order necessary and sufficient conditions in nonsmooth optimization, *Math. Prog.* **66** (1994), 379 - 402.

[6] V. I. Ivanov, Second-order optimality conditions in nonsmooth optimization, *C. R. Acad. Bulgare Sci.* **60** (2007), 1053 - 1058.

[7] V. I. Ivanov, Second-order optimality conditions for inequality constrained problems with locally Lipschitz data, *Optim. Lett.* **4** (2010), 597 - 608.

[8] V. Jeyakumar and X. Wang, Approximate Hessian matrices and second-order optimality conditions for nonlinear programming problems with $C^1$ data, *J. Austral. Math. Soc.* **40** (1999), 403 - 420.

[9] B. Jiménez and V. Novo, First and second order sufficient conditions for strict minimality in multiobjective programming, *Numer. Funct. Anal. Optim.* **23** (2002), 303 - 322.

[10] B. Jiménez and V. Novo, First and second order sufficient conditions for strict minimality in nonsmooth vector optimization, *J. Math. Anal. Appl.* **284** (2003), 496 - 510.

[11] B. Jiménez and V. Novo, Second-order necessary conditions in set constrained differentiable vector optimization, *Math. Methods Oper. Res.* **58** (2003), 299 - 317.

[12] B. Jiménez and V. Novo, Optimality conditions in differentiable optimization via second-order tangent sets, *Appl. Math. Optim.* **49** (2004), 123 - 144.

[13] H. Kawasaki, Second-order necessary conditions of the Kuhn-Tucker type under new constraint qualifications, *J. Optim. Theory Appl.* **57** (1988), 253 - 264.

[14] P. Q. Khanh and N. D. Tuan, First and second-order optimality conditions using approximations for nonsmooth vector optimization in Banach spaces, *J. Optim. Theory Appl.* **130** (2006), 289 - 308.

[15] P. Q. Khanh and N. D. Tuan, Optimality conditions for nonsmooth multiobjective optimization using Hadamard directional derivatives, *J. Optim. Theory Appl.* **133** (2007), 341 - 357.

[16] P. Q. Khanh and N. D. Tuan, First and second-order approximations as derivatives of mappings in optimality conditions for nonsmooth vector optimization, *Appl. Math. Optim.* **58** (2008), 147 - 166.

[17] K. Kisiel, Characterization theorems associated with $(F,\rho)$-convex and $(F,\rho)$-pseudoconvex functions, *Numer. Funct. Anal. Optim.* **30** (2009), 773 - 783.

[18] W. Krabs, Dualität bei diskreter rationaler Approximationen, *Internat. Ser. Numer. Anal.* **7** (1967), 33 - 41.

[19] G. M. Lee and M. H. Kim, On second order necessary optimality conditions for vector optimization problems, *J. Korean Math. Soc.* **40** (2003), 287 - 305.

[20] L. Liu and M. Křížek, The second order optimality conditions for nonlinear mathematical programming with $C^{1,1}$ data, *Appl. Math.* **42** (1997), 311 - 320.

[21] J. Liu and B. G. Liu, On second-order sufficient conditions in smooth nonlinear programs, *WSSIAA* **5** (1995), 239 - 254.

[22] L. Liu, P. Neittaanmäki, and M. Křížek, Second-order optimality conditions for nondominated solutions of multiobjective programming with $C^{1,1}$ data, *Appl. Math.* **45** (2000), 381 - 397.

[23] H. Maurer, First- and second-order sufficient optimality conditions in mathematical programming and optimal control, *Math. Prog. Study* **14** (1981), 163 - 177.

[24] H. Maurer and J. Zowe, First and second order necessary and sufficient optimality conditions for infinite-dimensional programming problems, *Math. Prog.* **16** (1979), 98 - 110.

[25] G. P. McCormick, Second order conditions for constrained minima, *SIAM J. Appl. Math.* **15** (1967), 641 - 652.

[26] E. J. Messerli and E. Polak, On Second-order necessary conditions of optimality, *SIAM J. Control* **7** (1969), 272 - 291.

[27] B. Mond, Second order duality for nonlinear programs, *Opsearch* **11** (1974), 90 - 99.

[28] B. Mond and J. Zhang *Duality for multiobjective programming involving second-order V-invex functions,* in *Proceedings of the Optimization Miniconference II* (B. M. Glover and V. Jeyakumar, eds.), University of New South Wales, Sydney, Australia, 1995, pp. 89 - 100.

[29] J. P. Penot, Second order conditions for optimization problems with constraints, *SIAM J. Control Optim.* **37** (1998), 303 - 318.

[30] J. J. Rückman and A. Shapiro, Second-order optimality conditions in generalized semi-infinite programming, *Set-Valued Anal.* **9** (2001), 169 - 186.

[31] P. H. Sach, Second-order necessary optimality conditions for optimization problems involving set-valued maps, *Appl. Math. Optim.* **22** (1990), 189 - 209.

[32] S. Schaible, Fractional programming: A recent survey, *J. Stat. Manag. Syst.* **5** (2002), 63 - 86.

[33] S. Schaible and J. Shi, Recent developments in fractional programming: single ratio and max-min case, *Nonlinear Analysis and Convex Analysis*, Yokohama Publ., Yokohama, 2004, pp. 493 - 506.

[34] I. M. Stancu-Minasian, *Fractional Programming: Theory, Models and Applications*, Kluwer, Dordrecht, 1997.

[35] I. M. Stancu-Minasian, A sixth bibliography of fractional programming, *Optimization* **55**(2006), 405 - 428.

[36] M. Studniarski, Second-order necessary conditions for optimality in nonsmooth nonlinear programming, *J. Math. Anal. Appl.* **154** (1991), 303 - 317.

[37] R. U. Verma, Hybrid $(\mathscr{G},\beta,\phi,\rho,\theta,\tilde{p},\tilde{r})$-univexities of higher-orders and applications to minimax fractional programming, *Transactions on Mathematical Programming and Applications* **1** (5) (2013), 63 - 86.

[38] R. U. Verma and G. J. Zalmai, Parameter - free duality models and applications to semiinfinite minmax fractional programming based on higher-order $(\phi,\eta,\rho,\theta,m)$-sonvexities, *OPSEARCH* **55**(1) (2018), 381 - 410.

[39] R. U. Verma and G. J. Zalmai, Higher order parametric optimality conditions in semi-infinite discrete minmax fractional programming based on generalized sonvexities, *Communications on Applied Nonlinear Analysis* **23** (1) (2016), 11 - 33.

[40] R. U. Verma and G. J. Zalmai, Second-order parametric optimality conditions in discrete minmax fractional programming, *Communications on Applied Nonlinear Analysis* **23** (3) (2016), 1 - 32.

[41] R. U. Verma and G. J. Zalmai, Hybrid second-order parametric optimality conditions in discrete minmax fractional programming, *Communications on Applied Nonlinear Analysis* **23** (2) (2016), 63 - 94.

[42] G. J. Zalmai, Generalized second-Order $(\mathscr{F}, \beta, \phi, \rho, \theta)$-univex functions and parametric duality models in semiinfinite discrete minmax fractional programming, *Advances in Nonlinear Variational Inequalities* **15** (2)(2012), 63 - 91.

# Chapter 2

# Hybrid Duality Models

In this chapter, hybrid duality models are presented based on generalized second-order Hanson-Antczak-type $(\alpha, \beta, \gamma, \xi, \eta, \rho, \theta)$-V-sonvexities [14], leading to several second-order parametric duality results using new duality models under various Hanson-Antczak-type generalized $(\alpha, \beta, \gamma, \xi, \eta, \rho, \theta)$-V-sonvexity constraints for a semiinfinite multiobjective fractional programming problem. The results thus obtained, applying Hanson-Antczak-type generalized $(\alpha, \beta, \gamma, \xi, \eta, \rho, \theta)$-V-sonvexities, are new and encompass most of the duality models and duality results including, but not limited to, first-order duality models [13] in the literature.

## 1. Introduction

In this section, we first present the second-order new classes of generalized Hanson-Antczak-type $(\alpha,\beta,\gamma,\xi,\eta,\rho,\theta)$-V-invex functions, based on recently defined in [12]. The second-order invexties are also referred to as 'sonvexities' in the literature. Furthermore, applying the new class of Hanson-Antczak-type V-sonvexities, we construct a number of second-order parametric duality models as well as establish numerous duality results for the following semiinfinite multiobjective fractional programming problem:

$$(P) \qquad \text{Minimize } \varphi(x) = \big(\varphi_1(x), \ldots, \varphi_p(x)\big) = \left(\frac{f_1(x)}{g_1(x)}, \ldots, \frac{f_p(x)}{g_p(x)}\right)$$

subject to

$$G_j(x,t) \leqq 0 \text{ for all } t \in T_j, \ j \in \underline{q},$$

$$H_k(x,s) = 0 \text{ for all } s \in S_k, \ k \in \underline{r},$$

$$x \in X,$$

where $p$, $q$, and $r$ are positive integers, $X$ is a nonempty open convex subset of $\mathbb{R}^n$ ($n$-dimensional Euclidean space), for each $j \in \underline{q} \equiv \{1,2,\ldots,q\}$ and $k \in \underline{r}$, $T_j$ and $S_k$ are compact subsets of complete metric spaces, for each $i \in \underline{p}$, $f_i$ and $g_i$ are real-valued functions defined on $X$, for each $j \in \underline{q}$, $G_j(\cdot,t)$ is a real-valued function defined on $X$ for all $t \in T_j$, for each $k \in \underline{r}$, $H_k(\cdot,s)$ is a real-valued function defined on $X$ for all $s \in S_k$, for each $j \in \underline{q}$ and $k \in \underline{r}$, $G_j(x,\cdot)$ and $H_k(x,\cdot)$ are continuous real-valued functions defined, respectively, on $T_j$ and $S_k$ for all $x \in X$, and for each $i \in \underline{p}$, $g_i(x) > 0$ for all $x$ satisfying the constraints of $(P)$.

The present chapter is based on the famous recent publication of Zalmai [14] on multiobjective fractional programming problems. In [12], a detailed information about multiobjective fractional programming is presented, the current status of semiinfinite programming is briefly explored and numerous key references are cited, further generalizations of the concept of V-invexity defined recently by Antczak [2] are proposed, and a fairly large number of sets of global parametric sufficient efficiency results under various generalized $HA(\alpha,\beta,\gamma,\xi,\eta,\rho,\theta)$-V-invexity assumptions are established. In light of the efficiency results established in [12], several first-order parametric duality models are constructed in [13] and appropriate duality theorems are proved under numerous generalized $HA(\alpha,\beta,\gamma,\xi,\eta,\rho,\theta)$-V-invexity conditions. For more details on the necessary background material and preliminaries, the reader is referred to [12, 13]. Now we shall first present the second-order versions of the generalized V-invex functions defined in [12], and then, inspired by these new classes of functions, we shall construct several second-order parametric duality models for $(P)$ and establish appropriate duality theorems.

Second-order duality for a conventional nonlinear programming problem of the form

$$(P_0) \qquad \text{Minimize } f(x) \text{ subject to } g_i(x) \leq 0, \ i \in \underline{m}, \ x \in \mathbb{R}^n,$$

where $f$ and $g_i$, $i \in \underline{m}$, are real-valued functions defined on $\mathbb{R}^n$, was initially considered by Mangasarian [5]. The idea underlying his approach to constructing a second-order dual problem was based on taking linear and quadratic approximations of the objective and constraint functions about an arbitrary but fixed point, forming the Wolfe dual of the approximated problem, and then letting the fixed point to vary. More specifically, he formulated the following second-order dual problem for $(P_0)$:

$$(D_0) \qquad \text{Maximize } f(y) + \sum_{i=1}^{m} u_i g_i(y) - \frac{1}{2}\Big\langle z, \Big[\nabla^2 f(y) + \sum_{i=1}^{m} u_i \nabla^2 g_i(y)\Big] z\Big\rangle$$

subject to

$$\nabla f(y) + \sum_{i=1}^{m} u_i \nabla g_i(y) + \Big[\nabla^2 f(y) + \sum_{i=1}^{m} u_i \nabla^2 g_i(y)\Big] z = 0,$$

$$y \in \mathbb{R}^n, \ u \in \mathbb{R}^m, \ u \geq 0, \ z \in \mathbb{R}^n,$$

where $\nabla f(y)$ and $\nabla^2 f(y)$ are, respectively, the gradient and Hessian of the function $f : \mathbb{R}^n \to \mathbb{R}$ evaluated at $y$, and $\langle a,b \rangle$ denotes the inner product of the vectors $a$ and $b$. Imposing somewhat complicated conditions on $f$, $g_i$, $i \in \underline{m}$, and $z$, he proved weak, strong, and converse duality theorems for $(P_0)$ and $(D_0)$. Reconsidering Mangasarian's second-order dual problem, Mond [6] established some duality results under relatively simpler conditions involving a certain second-order generalization of the concept of convexity, pointed out some possible computational advantages of second-order duality results, and also studied a pair of second-order symmetric dual problems. Subsequently, Mond's original notion of second-order convexity was generalized by other authors in different ways and utilized for establishing various second-order duality results for several classes of nonlinear programming problems. For brief accounts of the evolution of these generalized second-order convexity concepts, the reader is referred to [3, 4, 8, 9], and for a fairly extensive list of references pertaining to second- and higher-order duality results, the reader may consult [11].

Our observation indicates that semiinfinite multiobjective programming is a rapidly fast-cxpanding field of research in mathematical fractional programming, while several second-order duality results exist in the related literature for several classes of mathematical programming problems with a finite number of constraints. Motivated by these concerns, we apply generalized second-order $HA(\alpha,\beta,\gamma,\xi,\eta,\rho,\theta)$-V-sonvex functions to semiinfinite multiobjective programming problems.

Clearly, all the second-order duality results established in this chapter can be modified and presented for each one of the seven classes of nonlinear programming problems designated as $(P1)-(P7)$ in [12], which are special cases of $(P)$. In general this chapters offers a challenging opportunity for interdisciplinary research for all research areas even beyond the mathematical programming.

## 2. Preliminaries

In this section, we emphasize the role of second-order certain classes of generalized convex functions as well as first-order generalized convex functions introduced by Zalmai [14] and [13], which will be valuabe for the next generation research for all researchers. For more details, we refer [12].

We first recall the definitions of the second-order versions of $(\alpha,\eta)$-V-invex, $(\alpha,\eta)$-V-pseudoinvex, and $(\alpha,\eta)$-V-quasiinvex functions which were first defined in [8].

Let the function $F=(F_1,\ldots,F_p):X\to\mathbb{R}^p$ be twice differentiable at $x^*$.

**Definition 2.1.** The function $F$ is said to be second-order $(\alpha,\eta)$-V-invex at $x^*$ if there exist functions $\alpha_i : X\times X\to\mathbb{R}_+\backslash\{0\}\equiv(0,\infty)$, $i\in\underline{p}$, and $\eta : X\times X\to\mathbb{R}^n$ such that for each $x\in X$, $z\in\mathbb{R}^n$, and $i\in\underline{p}$,

$$F_i(x)-F_i(x^*)\geqq\langle\alpha_i(x,x^*)[\nabla F_i(x^*)+\nabla^2F_i(x^*)],\eta(x,x^*)\rangle-\frac{1}{2}\langle z,\nabla^2F_i(x^*)z\rangle.$$

**Definition 2.2.** The function $F$ is said to be second-order $(\beta,\eta)$-V-pseudoinvex at $x^*$ if there exist functions $\beta_i : X\times X\to\mathbb{R}_+\backslash\{0\}$, $i\in\underline{p}$, and $\eta : X\times X\to\mathbb{R}^n$ such that for each $x\in X$ and $z\in\mathbb{R}^n$,

$$\Big\langle\sum_{i=1}^{p}[\nabla F_i(x^*)+\nabla^2F_i(x^*)]z,\eta(x,x^*)\Big\rangle\geqq 0\;\Rightarrow$$

$$\sum_{i=1}^{p}\beta_i(x,x^*)F_i(x)\geq\sum_{i=1}^{p}\beta_i(x,x^*)\big[F_i(x^*)-\frac{1}{2}\langle z,\nabla^2F_i(x^*)z\rangle\big].$$

**Definition 2.3.** The function $F$ is said to be second-order $(\gamma,\eta)$-V-quasiinvex at $x^*$ if there exist functions $\gamma_i : X\times X\to\mathbb{R}_+\backslash\{0\}, i\in\underline{p}$, and $\eta : X\times X\to\mathbb{R}^n$ such that for each $x\in X$

and $z \in \mathbb{R}^n$,

$$\sum_{i=1}^{p} \gamma_i(x,x^*)F_i(x) \leqq \sum_{i=1}^{p} \gamma_i(x,x^*)[F_i(x^*) - \frac{1}{2}\langle z, \nabla^2 F_i(x^*)z\rangle] \Rightarrow$$
$$\Big\langle \sum_{i=1}^{p} [\nabla F_i(x^*) + \nabla^2 F_i(x^*)], \eta(x,x^*) \Big\rangle \leqq 0.$$

The following generalizations of the notions of second-order $(\alpha, \eta)$-V-invexity, $(\alpha, \eta)$-V-pseudoinvexity, and $(\alpha, \eta)$-V-quasiinvexity for vector-valued functions will be used in this chapter. We shall use the word sonvex for second-order invex and call the related generalized invex functions HA($\alpha$, $\beta$, $\gamma$, $\xi$, $\eta$, $\rho$, $\theta$)-V-sonvex, $HA(\alpha,\beta,\gamma,\xi,\eta,\rho,\theta)$-V-pseudosonvex, and HA($\alpha$, $\beta$, $\gamma$, $\xi$, $\eta$, $\rho$, $\theta$)-V-quasisonvex functions. They are the second-order counterparts of the classes of generalized HA($\alpha$, $\beta$, $\gamma$, $\xi$, $\eta$, $\rho$, $\theta$)-V-invex functions defined in [12].

**Definition 2.4** The function $F$ is sa.id to be (strictly) $HA(\alpha,\beta,\gamma,\xi,\eta,\rho,\theta)$-V-sonvex at $x^* \in X$ if there exist functions $\alpha : X \times X \to \mathbb{R}$, $\beta : X \times X \to \mathbb{R}$, $\gamma_i : X \times X \to \mathbb{R}_+$, $\xi_i : X \times X \to \mathbb{R}_+ \backslash \{0\}$, $i \in \underline{p}$, $\eta : X \times X \to \mathbb{R}^n$, $\rho_i : X \times X \to \mathbb{R}$, $i \in \underline{p}$, and $\theta : X \times X \to \mathbb{R}^n$ such that for all $x \in X (x \neq x^*)$, $z \in \mathbb{R}^n$, and $i \in \underline{p}$,

$$\frac{1}{\alpha(x,x^*)}\gamma_i(x,x^*)\Big(e^{\alpha(x,x^*)[F_i(x)-F_i(x^*)+\frac{1}{2}\langle z,\nabla^2 F_i(x^*)z\rangle]} - 1\Big)(>) \geqq$$
$$\frac{1}{\beta(x,x^*)}\langle \xi_i(x,x^*)[\nabla F_i(x^*) + \nabla^2 F_i(x^*)z], e^{\beta(x,x^*)\eta(x,x^*)} - \mathbf{1}\rangle + \rho_i(x,x^*)\|\theta(x,x^*)\|^2$$
$$\text{if } \alpha(x,x^*) \neq 0 \text{ and } \beta(x,x^*) \neq 0 \text{ for all } x \in X,$$

$$\frac{1}{\alpha(x,x^*)}\gamma_i(x,x^*)\Big(e^{\alpha(x,x^*)[F_i(x)-F_i(x^*)+\frac{1}{2}\langle z,\nabla^2 F_i(x^*)z\rangle]} - 1\Big)(>) \geqq$$
$$\langle \xi_i(x,x^*)[\nabla F_i(x^*) + \nabla^2 F_i(x^*)z], \eta(x,x^*)\rangle + \rho_i(x,x^*)\|\theta(x,x^*)\|^2$$
$$\text{if } \alpha(x,x^*) \neq 0 \text{ and } \beta(x,x^*) = 0 \text{ for all } x \in X,$$

$$\gamma_i(x,x^*)[F_i(x) - F_i(x^*) + \frac{1}{2}\langle z, \nabla^2 F_i(x^*)z\rangle](>) \geqq$$
$$\frac{1}{\beta(x,x^*)}\langle \xi_i(x,x^*)[\nabla F_i(x^*) + \nabla^2 F_i(x^*)z], e^{\beta(x,x^*)\eta(x,x^*)} - \mathbf{1}\rangle + \rho_i(x,x^*)\|\theta(x,x^*)\|^2$$
$$\text{if } \alpha(x,x^*) = 0 \text{ and } \beta(x,x^*) \neq 0 \text{ for all } x \in X,$$

$$\gamma_i(x,x^*)[F_i(x) - F_i(x^*) + \frac{1}{2}\langle z, \nabla^2 F_i(x^*)z\rangle](>) \geqq$$
$$\langle \xi_i(x,x^*)[\nabla F_i(x^*) + \nabla^2 F_i(x^*)z], \eta(x,x^*)\rangle + \rho_i(x,x^*)\|\theta(x,x^*)\|^2$$
$$\text{if } \alpha(x,x^*) = 0 \text{ and } \beta(x,x^*) = 0 \text{ for all } x \in X,$$

where

$$\left(e^{\beta(x,x^*)\eta(x,x^*)}-\mathbf{1}\right)\equiv\left(e^{\beta(x,x^*)\eta_1(x,x^*)}-1,\ldots,e^{\beta(x,x^*)\eta_n(x,x^*)}-1\right)$$

and $\|\cdot\|$ is a norm on $\mathbb{R}^n$.

The function $F$ is said to be (strictly) $HA(\alpha,\beta,\gamma,\xi,\eta,\rho,\theta)$-V-sonvex on $X$ if it is (strictly) $HA(\alpha,\beta,\gamma,\xi,\eta,\rho,\theta)$-V-sonvex at each point $x^*\in X$.

**Definition 2.5.** The function $F$ is said to be (strictly) $HA(\alpha,\beta,\gamma,\xi,\eta,\rho,\theta)$-V-pseudosonvex at $x^*\in X$ if there exist functions $\alpha: X\times X\to\mathbb{R}$, $\beta: X\times X\to\mathbb{R}$, $\gamma: X\times X\to\mathbb{R}_+$, $\xi_i: X\times X\to\mathbb{R}_+\backslash\{0\}$, $i\in \underline{p}$, $\eta: X\times X\to\mathbb{R}^n$, $\rho: X\times X\to\mathbb{R}$, and $\theta: X\times X\to\mathbb{R}^n$ such that for all $x\in X(x\neq x^*)$ and $z\in\mathbb{R}^n$,

$$\frac{1}{\beta(x,x^*)}\left\langle\sum_{i=1}^{p}[\nabla F_i(x^*)+\nabla^2F_i(x^*)z],e^{\beta(x,x^*)\eta(x,x^*)}-\mathbf{1}\right\rangle\geqq-\rho(x,x^*)\|\theta(x,x^*)\|^2$$
$$\Rightarrow\ \frac{1}{\alpha(x,x^*)}\gamma(x,x^*)\left(e^{\alpha(x,x^*)\sum_{i=1}^{p}\xi_i(x,x^*)[F_i(x)-F_i(x^*)+\frac{1}{2}\langle z,\nabla^2F_i(x^*)z\rangle]}-1\right)(>)\geqq 0$$
$$\text{if } \alpha(x,x^*)\neq 0 \text{ and } \beta(x,x^*)\neq 0 \text{ for all } x\in X,$$

$$\left\langle\sum_{i=1}^{p}[\nabla F_i(x^*)+\nabla^2F_i(x^*)z],\eta(x,x^*)\right\rangle\geqq-\rho(x,x^*)\|\theta(x,x^*)\|^2$$
$$\Rightarrow\ \frac{1}{\alpha(x,x^*)}\gamma(x,x^*)\left(e^{\alpha(x,x^*)\sum_{i=1}^{p}\xi_i(x,x^*)[F_i(x)-F_i(x^*)+\frac{1}{2}\langle z,\nabla^2F_i(x^*)z\rangle]}-1\right)(>)\geqq 0$$
$$\text{if } \alpha(x,x^*)\neq 0 \text{ and } \beta(x,x^*)= 0 \text{ for all } x\in X,$$

$$\frac{1}{\beta(x,x^*)}\left\langle\sum_{i=1}^{p}[\nabla F_i(x^*)+\nabla^2F_i(x^*)z],e^{\beta(x,x^*)\eta(x,x^*)}-\mathbf{1}\right\rangle\geqq-\rho(x,x^*)\|\theta(x,x^*)\|^2$$
$$\Rightarrow\ \gamma(x,x^*)\sum_{i=1}^{p}\xi_i(x,x^*)[F_i(x)-F_i(x^*)+\frac{1}{2}\langle z,\nabla^2F_i(x^*)z\rangle](>)\geqq 0$$
$$\text{if } \alpha(x,x^*)= 0 \text{ and } \beta(x,x^*)\neq 0 \text{ for all } x\in X,$$

$$\left\langle\sum_{i=1}^{p}[\nabla F_i(x^*)+\nabla^2F_i(x^*)z],\eta(x,x^*)\right\rangle\geqq-\rho(x,x^*)\|\theta(x,x^*)\|^2$$
$$\Rightarrow\ \gamma(x,x^*)\sum_{i=1}^{p}\xi_i(x,x^*)[F_i(x)-F_i(x^*)+\frac{1}{2}\langle z,\nabla^2F_i(x^*)z\rangle](>)\geqq 0$$
$$\text{if } \alpha(x,x^*)= 0 \text{ and } \beta(x,x^*)= 0 \text{ for all } x\in X.$$

The function $F$ is said to be (strictly) $HA(\alpha,\beta,\gamma,\xi,\eta,\rho,\theta)$-V-pseudosonvex on $X$ if it is (strictly) $HA(\alpha,\beta,\gamma,\xi,\eta,\rho,\theta)$-V-pseudosonvex at each point $x^*\in X$.

**Definition 2.6.** The function $F$ is said to be (prestrictly) $HA(\alpha,\beta,\gamma,\xi,\eta,\rho,\theta)$-V-quasisonvex at $x^*\in X$ if there exist functions $\alpha: X\times X\to\mathbb{R}$, $\beta: X\times X\to\mathbb{R}$, $\gamma: X\times X\to$

$\mathbb{R}_+$, $\xi_i : X \times X \to \mathbb{R}_+ \backslash \{0\}$, $i \in \underline{p}$, $\eta : X \times X \to \mathbb{R}^n$, $\rho : X \times X \to \mathbb{R}$, and $\theta : X \times X \to \mathbb{R}^n$ such that for all $x \in X$ and $z \in \mathbb{R}^n$,

$$\frac{1}{\alpha(x,x^*)}\gamma(x,x^*)\Big(e^{\alpha(x,x^*)\sum_{i=1}^p \xi_i(x,x^*)[F_i(x)-F_i(x^*)+\frac{1}{2}\langle z,\nabla^2 F_i(x^*)z\rangle]}-1\Big)(<) \leqq 0$$
$$\Rightarrow \frac{1}{\beta(x,x^*)}\Big\langle \sum_{i=1}^p [\nabla F_i(x^*)+\nabla^2 F_i(x^*)z], e^{\beta(x,x^*)\eta(x,x^*)}-\mathbf{1}\Big\rangle \leqq -\rho(x,x^*)\|\theta(x,x^*)\|^2$$
$$\text{if } \alpha(x,x^*) \neq 0 \text{ and } \beta(x,x^*) \neq 0 \text{ for all } x \in X,$$

$$\frac{1}{\alpha(x,x^*)}\gamma(x,x^*)\Big(e^{\alpha(x,x^*)\sum_{i=1}^p \xi_i(x,x^*)[F_i(x)-F_i(x^*)+\frac{1}{2}\langle z,\nabla^2 F_i(x^*)z\rangle]}-1\Big)(<) \leqq 0$$
$$\Rightarrow \Big\langle \sum_{i=1}^p [\nabla F_i(x^*)+\nabla^2 F_i(x^*)z], \eta(x,x^*)\Big\rangle \leqq -\rho(x,x^*)\|\theta(x,x^*)\|^2$$
$$\text{if } \alpha(x,x^*) \neq 0 \text{ and } \beta(x,x^*) = 0 \text{ for all } x \in X,$$

$$\gamma(x,x^*)\sum_{i=1}^p \xi_i(x,x^*)[F_i(x)-F_i(x^*)+\frac{1}{2}\langle z,\nabla^2 F_i(x^*)z\rangle](<) \leqq 0$$
$$\Rightarrow \frac{1}{\beta(x,x^*)}\Big\langle \sum_{i=1}^p [\nabla F_i(x^*)+\nabla^2 F_i(x^*)z], e^{\beta(x,x^*)\eta(x,x^*)}-\mathbf{1}\Big\rangle \leqq -\rho(x,x^*)\|\theta(x,x^*)\|^2$$
$$\text{if } \alpha(x,x^*) = 0 \text{ and } \beta(x,x^*) \neq 0 \text{ for all } x \in X,$$

$$\gamma(x,x^*)\sum_{i=1}^p \xi_i(x,x^*)[F_i(x)-F_i(x^*)+\frac{1}{2}\langle z,\nabla^2 F_i(x^*)z\rangle](<) \leqq 0$$
$$\Rightarrow \Big\langle \sum_{i=1}^p [\nabla F_i(x^*)+\nabla^2 F_i(x^*)z], \eta(x,x^*)\Big\rangle \leqq -\rho(x,x^*)\|\theta(x,x^*)\|^2$$
$$\text{if } \alpha(x,x^*) = 0 \text{ and } \beta(x,x^*) = 0 \text{ for all } x \in X.$$

The function $F$ is said to be (prestrictly) $HA(\alpha,\beta,\gamma,\xi,\eta,\rho,\theta)$-V-quasisonvex on $X$ if it is (prestrictly) $HA(\alpha,\beta,\gamma,\xi,\eta,\rho,\theta)$-V-quasisonvex at each point $x^* \in X$.

We observe that during the proofs of the duality theorems, sometimes it may be more appropriate and convenient to use certain alternative but equivalent forms of the above definitions. These are obtained by considering the contrapositive statements. For example, $HA(\alpha,\beta,\gamma,\eta,\rho,\theta)$-V-pseudosonvexity (when $\alpha(x,x^*) \neq 0$ and $\beta(x,x^*) \neq 0$ for all $x \in X$) can be defined in the following equivalent manner:
The function $f$ is said to be $HA(\alpha,\beta,\gamma,\xi,\eta,\rho,\theta)$-V-pseudosonvex at $x^* \in X$ if there exist functions $\alpha : X \times X \to \mathbb{R}$, $\beta : X \times X \to \mathbb{R}$, $\gamma : X \times X \to \mathbb{R}_+$, $\xi_i : X \times X \to \mathbb{R}_+ \backslash \{0\}$, $i \in \underline{p}$, $\eta : X \times X \to \mathbb{R}^n$, $\rho : X \times X \to \mathbb{R}$, and $\theta : X \times X \to \mathbb{R}^n$ such that for all $x \in X$ and $z \in \mathbb{R}^n$,

$$\frac{1}{\alpha(x,x^*)}\gamma(x,x^*)\Big(e^{\alpha(x,x^*)\sum_{i=1}^p \xi_i(x,x^*)[F_i(x)-F_i(x^*)+\frac{1}{2}\langle z,\nabla^2 F_i(x^*)z\rangle]}-1\Big) < 0$$
$$\Rightarrow \frac{1}{\beta(x,x^*)}\Big\langle \sum_{i=1}^p [\nabla F_i(x^*)+\nabla^2 F_i(x^*)z], e^{\beta(x,x^*)\eta(x,x^*)}-\mathbf{1}\Big\rangle < -\rho(x,x^*)\|\theta(x,x^*)\|^2.$$

In the sequel, we need a consistent notation for vector inequalities. For $a,b \in \mathbb{R}^m$, the following order notation will be used: $a \geqq b$ if and only if $a_i \geqq b_i$ for all $i \in \underline{m}$; $a \geqslant b$ if and only if $a_i \geqq b_i$ for all $i \in \underline{m}$, but $a \neq b$; $a > b$ if and only if $a_i > b_i$ for all $i \in \underline{m}$; and $a \ngeqslant b$ is the negation of $a \geqslant b$.

Consider the multiobjective problem

$$(P^*) \qquad \underset{x \in \mathbb{F}}{\text{Minimize}}\, F(x) = (F_1(x), \ldots, F_p(x)),$$

where $F_i$, $i \in \underline{p}$, are real-valued functions defined on $\mathbb{R}^n$.

An element $x^\circ \in \mathbb{F}$ is said to be an *efficient (Pareto optimal, nondominated, noninferior)* solution of $(P^*)$ if there exists no $x \in \mathbb{F}$ such that $F(x) \leqslant F(x^\circ)$. There are several versions of the concept of efficiency that are currently used in the area of multiobjective programming. However, throughout this chapter we shall deal exclusively with the efficient solutions of $(P)$ in the sense defined above.

We conclude this section by recalling a set of parametric necessary efficiency conditions for $(P)$ which will be needed for proving strong duality theorems.

**Theorem 2.7.** [14] Let $x^* \in \mathbb{F}$, $\lambda^* = \varphi(x^*)$, for each $i \in \underline{p}$, let $f_i$ and $g_i$ be continuously differentiable at $x^*$, for each $j \in \underline{q}$, let the function $\zeta \to G_j(\zeta,t)$ be continuously differentiable at $x^*$ for all $t \in T_j$, and for each $k \in \underline{r}$, let the function $\zeta \to H_k(\zeta,s)$ be continuously differentiable at $x^*$ for all $s \in S_k$. If $x^*$ is an efficient solution of (P), if the generalized Guignard constraint qualification holds at $x^*$, and if for each $i_0 \in \underline{p}$, the set $cone(\{\nabla G_j(x^*,t) : t \in \hat{T}_j(x^*), j \in \underline{q}\} \cup \{\nabla f_i(x^*) - \lambda_i^* \nabla g_i(x^*) : i \in \underline{p}, i \neq i_0\}) + span(\{\nabla H_k(x^*,s) : s \in S_k, k \in \underline{r}\})$ is closed, then there exist $u^* \in U$ and integers $\nu_0$ and $\nu$, with $0 \leqq \nu_0 \leqq \nu \leqq n+1$, such that there exist $\nu_0$ indices $j_m$, with $1 \leqq j_m \leqq q$, together with $\nu_0$ points $t^m \in \hat{T}_{j_m}(x^*)$, $m \in \underline{\nu_0}$, $\nu - \nu_0$ indices $k_m$, with $1 \leqq k_m \leqq r$, together with $\nu - \nu_0$ points $s^m \in S_{k_m}$ for $m \in \underline{\nu} \backslash \underline{\nu_0}$, and $\nu$ real numbers $v_m^*$, with $v_m^* > 0$ for $m \in \underline{\nu_0}$, with the property that

$$\sum_{i=1}^{p} u_i^* [\nabla f_i(x^*) - \lambda_i^* \nabla g_i(x^*)] + \sum_{m=1}^{\nu_0} v_m^* \nabla G_{j_m}(x^*,t^m) + \sum_{m=\nu_0+1}^{\nu} v_m^* \nabla H_{k_m}(x^*,s^m) = 0,$$

where $cone(V)$ is the conic hull of the set $V \subset \mathbb{R}^n$ (i.e., the smallest convex cone containing $V$), $span(V)$ is the linear hull of $V$ (i.e., the smallest subspace containing $V$), $\hat{T}_j(x^*) = \{t \in T_j : G_j(x^*,t) = 0\}$, $U = \{u \in \mathbb{R}^p : u > 0, \sum_{i=1}^{p} u_i = 1\}$, and $\underline{\nu^*} \backslash \underline{\nu_0^*}$ is the complement of the set $\underline{\nu_0^*}$ relative to the set $\underline{\nu^*}$.

For brevity, we shall henceforth refer to an efficient solution $x^* \in \mathbb{F}$ as a *normal* efficient solution of $(P)$ if the generalized Guignard constraint qualification is satisfied at $x^*$ and for each $i_0 \in \underline{p}$, the set $cone(\{\nabla G_j(x^*,t) : t \in \hat{T}_j(x^*), j \in \underline{q}\} \cup \{\nabla f_i(x^*) - \lambda_i^* \nabla g_i(x^*) : i \in \underline{p}, i \neq i_0\}) + span(\{\nabla H_k(x^*,s) : s \in S_k, k \in \underline{r}\})$ is closed, where $\lambda_i^* = f_i(x^*)/g_i(x^*)$, $i \in \underline{p}$.

In the rest of this chapter, we assume that the functions $f_i$, $g_i$, $i \in \underline{p}$, $\zeta \to G_j(\zeta,t)$, and $\zeta \to H_k(\zeta,s)$, are twice continuously differentiable on $X$ for all $t \in T_j$, $j \in \underline{q}$, and $s \in S_k$, $k \in \underline{r}$.

Our observation indicates with regard to the choice of the type of generalized $HA(\alpha,\beta,\gamma,\xi,\eta,\rho,\theta)$-V-sonvex functions, specified in Definitions 2.4 - 2.6, to be used in the statements and proofs of our second-order duality theorems, we shall consistently use the cases in which the functions $\alpha$ and $\beta$ are nonzero for all $(x,y) \in X \times X$. All the duality results established in this chapter would be extremely useful for all research enthusiasts as well as the general research professionals from all fields of research.

## 3. Duality Model I

In this section, we consider two dual problems with relatively simple constraint structures and prove weak, strong, and strict converse duality theorems under an appropriate $HA(\alpha,\beta,\gamma,\xi,\eta,\rho,\theta)$-V-pseudosonvexity condition. More general duality models and results for $(P)$ will be discussed in the subsequent sections.

Let

$$\begin{aligned}\mathbb{H} = \Big\{(y,z,u,v,\lambda,\nu,\nu_0,J_{\nu_0},K_{\nu\backslash\nu_0},\bar{t},\bar{s}) : y \in X;\ z \in \mathbb{R}^n;\ u \in U;\ \lambda \in \mathbb{R}^p;\\ 0 \leqq \nu_0 \leqq \nu \leqq n+1;\ v \in \mathbb{R}^\nu, v_i > 0, 1 \leqq i \leqq \nu_0;\\ J_{\nu_0} = (j_1, j_2, \ldots, j_{\nu_0}), 1 \leqq j_i \leqq q;\ K_{\nu\backslash\nu_0} = (k_{\nu_0+1}, \ldots, k_\nu),\\ 1 \leqq k_i \leqq r;\ \bar{t} = (t^1, t^2, \ldots, t^{\nu_0}), t^i \in T_{j_i};\ \bar{s} = (s^{\nu_0+1}, \ldots, s^\nu), s^i \in S_{k_i}\Big\}.\end{aligned}$$

Consider the following two problems:

$$(DI) \qquad \sup_{(y,z,u,v,\lambda,\nu,\nu_0,J_{\nu_0},K_{\nu\backslash\nu_0},\bar{t},\bar{s}) \in \mathbb{H}} \lambda = (\lambda_1, \ldots, \lambda_p)$$

subject to

$$\begin{aligned}&\sum_{i=1}^p u_i[\nabla f_i(y) - \lambda_i \nabla g_i(y)] + \sum_{m=1}^{\nu_0} v_m \nabla G_{j_m}(y,t^m) + \sum_{m=\nu_0+1}^{\nu} v_m \nabla H_{k_m}(y,s^m) +\\ &\Big\{\sum_{i=1}^p u_i[\nabla^2 f_i(y) - \lambda_i \nabla^2 g_i(y)] + \sum_{m=1}^{\nu_0} v_m \nabla^2 G_{j_m}(y,t^m) + \sum_{m=\nu_0+1}^{\nu} v_m \nabla^2 H_{k_m}(y,s^m)\Big\} z = 0, \quad (2.1)\end{aligned}$$

$$\begin{aligned}&\sum_{i=1}^p u_i[f_i(y) - \lambda_i g_i(y)] + \sum_{m=1}^{\nu_0} v_m G_{j_m}(y,t^m) + \sum_{m=\nu_0+1}^{\nu} v_m H_{k_m}(y,s^m)\\ &\qquad - \frac{1}{2}\Big\langle z, \Big\{\sum_{i=1}^p u_i[\nabla^2 f_i(y)\\ &\qquad - \lambda_i \nabla^2 g_i(y)] + \sum_{m=1}^{\nu_0} v_m \nabla^2 G_{j_m}(y,t^m)\\ &\qquad + \sum_{m=\nu_0+1}^{\nu} v_m \nabla^2 H_{k_m}(y,s^m)\Big\} z\Big\rangle \geqq 0 \quad (2.2)\end{aligned}$$

for all $x \in \mathbb{F}$;

$$(\tilde{D}I) \qquad \sup_{(y,z,u,v,\lambda,\nu,\nu_0,J_{\nu_0},K_{\nu\backslash\nu_0},\bar{t},\bar{s}) \in \mathbb{H}} \lambda = (\lambda_1, \ldots, \lambda_p)$$

subject to (2) and

$$\frac{1}{\beta(x,y)}\Big\langle \sum_{i=1}^{p} u_i[\nabla f_i(y) - \lambda_i \nabla g_i(y)] + \sum_{m=1}^{\nu_0} v_m \nabla G_{j_m}(y,t^m) + \sum_{m=\nu_0+1}^{\nu} v_m \nabla H_{k_m}(y,s^m) + \Big\{\sum_{i=1}^{p} u_i[\nabla^2 f_i(y) - \lambda_i \nabla^2 g_i(y)] + \sum_{m=1}^{\nu_0} v_m \nabla^2 G_{j_m}(y,t^m) + \sum_{m=\nu_0+1}^{\nu} v_m \nabla^2 H_{k_m}(y,s^m)\Big\} z, e^{\beta(x,y)\eta(x,y)} - \mathbf{1}\Big\rangle \geqq 0 \text{ for all } x \in \mathbb{F}, \quad (2.3)$$

where $\beta$ and $\eta$ are functions from $X \times X$ to $\mathbb{R}$ and $\mathbb{R}^n$, respectively.

Comparing $(DI)$ and $(\tilde{D}I)$, we see that $(\tilde{D}I)$ is relatively more general than $(DI)$ in the sense that any feasible solution of $(DI)$ is also feasible for $(\tilde{D}I)$, but the converse is not necessarily true. Furthermore, we observe that (1) is a system of $n$ equations, whereas (3) is a single inequality. Clearly, from a computational point of view, $(DI)$ is preferable to $(\tilde{D}I)$ because of the dependence of (3) on the feasible set of $(P)$.

Despite these apparent differences, it turns out that the statements and proofs of all the duality theorems for $(P)-(DI)$ and $(P)-(\tilde{D}I)$ are almost identical and, therefore, we shall consider only the pair $(P)-(DI)$.

The next two theorems show that $(DI)$ is a dual problem for $(P)$.

**Theorem 2.8.** Let $x$ and $w \equiv (y,z,u,v,\lambda,\nu,\nu_0,J_{\nu_0},K_{\nu\setminus\nu_0},\bar{t},\bar{s})$ be arbitrary feasible solutions of (P) and (DI), respectively, and assume that the function
$L(\cdot,u,v,\lambda,\bar{t},\bar{s}) : X \to \mathbb{R}^p$ defined by

$$L(\zeta,u,v,\lambda,\bar{t},\bar{s}) = \big(L_1(\zeta,u,v,\lambda,\bar{t},\bar{s}),\ldots,L_p(\zeta,u,v,\lambda,\bar{t},\bar{s})\big)$$

is $HA(\alpha,\beta,\gamma,\xi,\eta,0,\theta)$-V-pseudosonvex at $y$ and $\gamma(x,y) > 0$, where

$$L_i(\zeta,u,v,\lambda,\bar{t},\bar{s}) = u_i\Big[f_i(\zeta) - \lambda_i g_i(\zeta) + \sum_{m=1}^{\nu_0} v_m G_{j_m}(\zeta,t^m) + \sum_{m=\nu_0+1}^{\nu} v_m H_{k_m}(\zeta,s^m)\Big], \quad i \in \underline{p}.$$

Then $\varphi(x) \nleq \lambda$.

*Proof.* Using the $HA(\alpha,\beta,\gamma,\xi,\eta,0,\theta)$-V-pseudosonvexity assumption and (1) we conclude that

$$\frac{1}{\alpha(x,y)}\gamma(x,y)\Big(e^{\alpha(x,y)\sum_{i=1}^{p}\xi_i(x,y)[L_i(x,u,v,\lambda,\bar{t},\bar{s}) - L_i(y,u,v,\lambda,\bar{t},\bar{s}) + \frac{1}{2}\langle z,\nabla^2 L_i(y,u,v,\lambda,\bar{t},\bar{s})z\rangle]} - 1\Big) \geqq 0.$$

We need to consider two cases: $\alpha(x,y) > 0$ and $\alpha(x,y) < 0$. If we assume that $\alpha(x,y) > 0$ and recall that $\gamma(x,y) > 0$, then we see that the above inequality becomes

$$e^{\alpha(x,y)\sum_{i=1}^{p}\xi_i(x,y)\left[L_i(x,u,v,\lambda,\bar{t},\bar{s}) - L_i(y,u,v,\lambda,\bar{t},\bar{s}) + \frac{1}{2}\left\langle z,\nabla^2 L_i(y,u,v,\lambda,\bar{t},\bar{s})z\right\rangle\right]} \geqq 1.$$

This implies that

$$\sum_{i=1}^{p}\xi_i(x,y)L_i(x,u,v,\lambda,\bar{t},\bar{s}) \geqq \sum_{i=1}^{p}\xi_i(x,y)\big[L_i(y,u,v,\lambda,\bar{t},\bar{s})$$

$$-\frac{1}{2}\langle z,\nabla^2 L_i(y,u,v,\lambda,\bar{t},\bar{s})z\rangle\big] \geqq 0,$$

where the second inequality follows from (2). Inasmuch as $x \in \mathbb{F}$ and $v_m > 0$, $m \in \underline{\nu_0}$, the above inequality reduces to

$$\sum_{i=1}^{q} u_i\xi_i(x,y)[f_i(x) - \lambda_i g_i(x)] \geqq 0. \tag{2.4}$$

Since $u > 0$ and $\xi_i(x,y) > 0$, $i \in \underline{p}$, (4) implies that

$$\big(f_1(x) - \lambda_1 g_1(x),\ldots,f_p(x) - \lambda_p g_p(x)\big) \nleq (0,\ldots,0),$$

which in turn implies that

$$\varphi(x) = \Big(\frac{f_1(x)}{g_1(x)},\ldots,\frac{f_p(x)}{g_p(x)}\Big) \nleq (\lambda_1,\ldots,\lambda_p) = \lambda.$$

If we assume that $\alpha(x,y) < 0$, we arrive at the same conclusion. □

**Theorem 2.9.** (Strong Duality) Let $x^*$ be a normal efficient solution of (P) and assume that the conditions set forth in Theorem 2.8 are satisfied for all feasible solutions of (DI). Then there exist

$$u^*, v^*, \lambda^*, \nu^*, \nu_0^*, J_{\nu_0^*}, K_{\nu^*\setminus\nu_0^*}, \bar{t}^*, and\, \bar{s}^*$$

such that

$$(x^*, z^* = 0, u^*, v^*, \lambda^*, \nu^*, \nu_0^*, J_{\nu_0^*}, K_{\nu^*\setminus\nu_0^*}, \bar{t}^*, \bar{s}^*)$$

is an efficient solution of (DI) and $\varphi(x^*) = \lambda^*$.

*Proof.* Since $x^*$ is a normal efficient solution of $(P)$, by Theorem 2.8, there exist

$$u^*, v^*, \lambda^*, \nu^*, \nu_0^*, J_{\nu_0^*}, K_{\nu^*\setminus\nu_0^*}, \bar{t}^*, and\, \bar{s}^*$$

such that

$$(x^*, z^* = 0, u^*, v^*, \lambda^*, \nu^*, \nu_0^*, J_{\nu_0^*}, K_{\nu^*\setminus\nu_0^*}, \bar{t}^*, \bar{s}^*)$$

is a feasible solution of $(DI)$ and $\varphi(x^*) = \lambda^*$. If it were not efficient, then there would exist a feasible solution

$$(\tilde{x}, \tilde{z}, \tilde{u}, \tilde{v}, \tilde{\lambda}, \tilde{\nu}, \tilde{\nu}_0, J_{\tilde{\nu}_0}, K_{\tilde{\nu}\setminus\tilde{\nu}_0}, \tilde{\bar{t}}, \tilde{\bar{s}})$$

of $(DI)$ such that $\tilde{\lambda} \geqslant \lambda^* = \varphi(x^*)$, contradicting Theorem 2.8. Therefore,

$$(x^*, z^* = 0, u^*, v^*, \lambda^*, \nu^*, \nu_0^*, J_{\nu_0^*}, K_{\nu^*\setminus\nu_0^*}, \bar{t}^*, \bar{s}^*)$$

is an efficient solution of $(DI)$. □

We also have the following converse duality result for $(P)$ and $(DI)$.

**Theorem 2.10.** (Strict Converse Duality) Let $x^*$ and $\tilde{w} \equiv (\tilde{x}, \tilde{z}, \tilde{u}, \tilde{v}, \tilde{\lambda}, \tilde{\nu}, \tilde{\nu}_0, J_{\tilde{\nu}_0}, K_{\tilde{\nu}\backslash\tilde{\nu}_0}, \bar{\tilde{t}}, \bar{\tilde{s}})$ be arbitrary feasible solutions of (P) and (DI), respectively, such that

$$\sum_{i=1}^{p} \tilde{u}_i \xi_i(x^*, \tilde{x})[f_i(x^*) - \tilde{\lambda}_i g_i(x^*)] = 0, \tag{2.5}$$

where for each $i \in \underline{p}$, $\xi_i$ is a function from $X \times X$ to $\mathbb{R}_+ \backslash \{0\}$. Furthermore, assume that the function $\zeta \to L(\zeta, \tilde{u}, \tilde{v}, \tilde{\lambda}, \bar{\tilde{t}}, \bar{\tilde{s}})$ is strictly $HA(\alpha, \beta, \gamma, \xi, \eta, 0, \theta)$-V-pseudosonvex at $\tilde{x}$. Then $\tilde{x} = x^*$ and $\varphi(x^*) = \tilde{\lambda}$.

*Proof.* (a): Suppose to the contrary that $\tilde{x} \neq x^*$. Now proceeding as in the proof of Theorem 2.8 (with $x$ replaced by $x^*$ and $w$ by $\tilde{w}$), we arrive at the strict inequality

$$\sum_{i=1}^{p} \tilde{u}_i \xi_i(x^*, \tilde{x})[f_i(x^*) - \tilde{\lambda}_i g_i(x^*)] > 0,$$

which contradicts (5). Therefore, we conclude that $\tilde{x} = x^*$ and $\varphi(x^*) = \tilde{\lambda}$.

□

# 4. Duality Model II

In this section, we consider certain variants of $(DI)$ and $(\tilde{D}I)$ that allow for more general $HA(\alpha, \beta, \gamma, \xi, \eta, \rho, \theta)$-V-sonvexity conditions under which duality can be established. These duality models can be stated as follows:

$$(DII) \qquad \sup_{(y,z,u,v,\lambda,\nu,\nu_0,J_{\nu_0},K_{\nu\backslash\nu_0},\bar{t},\bar{s}) \in \mathbb{H}} \quad \lambda = (\lambda_1, \dots, \lambda_p)$$

subject to

$$\sum_{i=1}^{p} u_i[\nabla f_i(y) - \lambda_i \nabla g_i(y)] + \sum_{m=1}^{\nu_0} v_m \nabla G_{j_m}(y, t^m) + \sum_{m=\nu_0+1}^{\nu} v_m \nabla H_{k_m}(y, s^m) + \Big\{ \sum_{i=1}^{p} u_i[\nabla^2 f_i(y) - \lambda_i \nabla^2 g_i(y)] + \sum_{m=1}^{\nu_0} v_m \nabla^2 G_{j_m}(y, t^m) + \sum_{m=\nu_0+1}^{\nu} v_m \nabla^2 H_{k_m}(y, s^m) \Big\} z = 0, \tag{2.6}$$

$$f_i(y) - \lambda_i g_i(y) - \frac{1}{2}\langle z, [\nabla^2 f_i(y) - \lambda_i \nabla^2 g_i(y)]z \rangle \geqq 0, \quad i \in \underline{p}, \tag{2.7}$$

$$G_{j_m}(y, t^m) - \frac{1}{2}\langle z, \nabla^2 G_{j_m}(y, t^m) z \rangle \geqq 0, \quad m \in \underline{\nu_0}, \tag{2.8}$$

$$v_m H_{k_m}(y, s^m) - \frac{1}{2}\langle z, v_m \nabla^2 H_{k_m}(y, s^m) z \rangle \geqq 0, \quad m \in \underline{\nu} \backslash \underline{\nu_0}; \tag{2.9}$$

$$(\tilde{D}II) \qquad \sup_{(y,z,u,v,\lambda,\nu,\nu_0,J_{\nu_0},K_{\nu\backslash\nu_0},\bar{t},\bar{s})\in\mathbb{H}} \lambda = (\lambda_1,\ldots,\lambda_p)$$

subject to (3) and (7) - (9).

The remarks and observations made earlier about the relationships between $(DI)$ and $(\tilde{D}I)$ are, of course, also valid for $(DII)$ and $(\tilde{D}II)$. Since the constraint inequalities of $(DII)$ are formed by splitting the inequality (2) into three inequalities (7) - (9), it is clear that Theorems 2.8 - 2.10 are valid for the pair $(P)-(DII)$. Below, we shall establish some duality results in which appropriate generalized $HA(\alpha,\beta,\gamma,\xi,\eta,\rho,\theta)$-V-sonvexity requirements will be placed on certain vector functions including $(\mathscr{E}_1(\cdot,u,\lambda),\ldots,\mathscr{E}_p(\cdot,u,\lambda))$, where for each $i\in \underline{p}$, the component function $\mathscr{E}_i(\cdot,u,\lambda)$ is defined, for fixed $\lambda$ and $u$, on $X$ by

$$\mathscr{E}_i(\zeta,u,\lambda) = u_i[f_i(\zeta)-\lambda_i g_i(\zeta)].$$

The following theorems show that $(DII)$ is a dual problem for $(P)$.

**Theorem 2.11.** Let $x$ and $w\equiv(y,z,u,v,\lambda,\nu,\nu_0,J_{\nu_0},K_{\nu\backslash\nu_0},\bar{t},\bar{s})$ be arbitrary feasible solutions of (P) and (DII), respectively, and assume that any one of the following four sets of hypotheses is satisfied:

(a) (i) $(\mathscr{E}_1(\cdot,u,\lambda),\ldots,\mathscr{E}_p(\cdot,u,\lambda))$ is $HA(\alpha,\beta,\bar{\gamma},\xi,\eta,\bar{\rho},\theta)$ -V-pseudosonvex at $y$ and $\bar{\gamma}(x,y)>0$;

(ii) $(v_1G_{j_1}(\cdot,t^1),\ldots,v_{\nu_0}G_{j_{\nu_0}}(\cdot,t^{\nu_0}))$ is $HA(\alpha,\beta,\tilde{\gamma},\pi,\eta,\tilde{\rho},\theta)$-V-quasisonvex at $y$;

(iii) $(v_{\nu_0+1}H_{k_{\nu_0+1}}(\cdot,s^{\nu_0+1}),\ldots,v_\nu H_{k_\nu}(\cdot,s^\nu))$ is $HA(\alpha,\beta,\check{\gamma},\delta,\eta,\check{\rho},\theta)$ -V-quasisonvex at $y$;

(iv) $\bar{\rho}(x,y)+\tilde{\rho}(x,y)+\check{\rho}(x,y)\geqq 0$;

(b) (i) $(\mathscr{E}_1(\cdot,u,\lambda),\ldots,\mathscr{E}_p(\cdot,u,\lambda))$ is prestrictly $HA(\alpha,\beta,\bar{\gamma},\xi,\eta,\bar{\rho},\theta)$ -V-quasisonvex at $y$ and $\bar{\gamma}(x,y)>0$;

(ii) $(v_1G_{j_1}(\cdot,t^1),\ldots,v_{\nu_0}G_{j_{\nu_0}}(\cdot,t^{\nu_0}))$ is $HA(\alpha,\beta,\tilde{\gamma},\pi,\eta,\tilde{\rho},\theta)$-V-quasisonvex at $y$;

(iii) $(v_{\nu_0+1}H_{k_{\nu_0+1}}(\cdot,s^{\nu_0+1}),\ldots,v_\nu H_{k_\nu}(\cdot,s^\nu))$ is $HA(\alpha,\beta,\check{\gamma},\delta,\eta,\check{\rho},\theta)$ -V-quasisonvex at $y$;

(iv) $\bar{\rho}(x,y)+\tilde{\rho}(x,y)+\check{\rho}(x,y)>0$;

(c) (i) $(\mathscr{E}_1(\cdot,u,\lambda),\ldots,\mathscr{E}_p(\cdot,u,\lambda))$ is prestrictly $HA(\alpha,\beta,\bar{\gamma},\xi,\eta,\bar{\rho},\theta)$ -V-quasisonvex at $y$ and $\bar{\gamma}(x,y)>0$;

(ii) $(v_1G_{j_1}(\cdot,t^1),\ldots,v_{\nu_0}G_{j_{\nu_0}}(\cdot,t^{\nu_0}))$ is strictly HA($\alpha$, $\beta$, $\tilde{\gamma}$, $\pi$, $\eta$, $\tilde{\rho}$, $\theta$) -V-pseudosonvex at $y$;

(iii) $(v_{\nu_0+1}H_{k_{\nu_0+1}}(\cdot,s^{\nu_0+1}),\ldots,v_\nu H_{k_\nu}(\cdot,s^\nu))$ is $HA(\alpha,\beta,\check{\gamma},\delta,\eta,\check{\rho},\theta)$ -V-quasisonvex at $y$;

(iv) $\bar{\rho}(x,y)+\tilde{\rho}(x,y)+\check{\rho}(x,y)\geqq 0$;

(d) (i) $\big(\mathscr{E}_1(\cdot,u,\lambda),\ldots,\mathscr{E}_p(\cdot,u,\lambda)\big)$ is prestrictly $HA(\alpha,\beta,\bar{\gamma},\xi,\eta,\bar{\rho},\theta)$ -V-quasisonvex at $y$ and $\bar{\gamma}(x,y)>0$;

(ii) $\big(v_1 G_{j_1}(\cdot,t^1),\ldots,v_{v_0}G_{j_{v_0}}(\cdot,t^{v_0})\big)$ is HA($\alpha$, $\beta$, $\tilde{\gamma}$, $\pi$, $\eta$, $\tilde{\rho}$, $\theta$)-V-quasisonvex at $y$;

(iii) $\big(v_{v_0+1}H_{k_{v_0+1}}(\cdot,s^{v_0+1}),\ldots,v_v H_{k_v}(\cdot,s^v)\big)$ is strictly HA($\alpha$, $\beta$, $\breve{\gamma}$, $\delta$, $\eta$, $\breve{\rho}$, $\theta$) -V-pseudosonvex at $y$;

(iv) $\bar{\rho}(x,y)+\tilde{\rho}(x,y)+\breve{\rho}(x,y)\geqq 0$.

Then $\varphi(x)\nleq\lambda$.

*Proof.* (a): Applying (8) and the primal feasibility of $x$, we have

$$G_{j_m}(x,t^m)\leqq 0\leqq G_{j_m}(y,t^m)-\frac{1}{2}\langle z,\nabla^2 G_{j_m}(y,t^m)z\rangle \text{ for each } m\in\underline{v_0}.$$

Since $\pi_m(x,y)>0$ for each $m\in\underline{v_0}$, it follows that

$$\sum_{m=1}^{v_0} v_m\pi_m(x,y)G_{j_m}(x,t^m)\leqq\sum_{m=1}^{v_0} v_m\pi_m(x,x)\big[G_{j_m}(y,t^m)-\frac{1}{2}\langle z,\nabla^2 G_{j_m}(y,t^m)z\rangle\big],$$

and hence

$$\frac{1}{\alpha(x,y)}\tilde{\gamma}(x,y)\Big(e^{\alpha(x,y)\sum_{m=1}^{v_0}v_m\pi_m(x,y)\{G_{j_m}(x,t^m)-[G_{j_m}(y,t^m)-\frac{1}{2}\langle z,\nabla^2 G_{j_m}(y,t^m)z\rangle]\}}-1\Big)\leqq 0,$$

which in view of (ii) implies that

$$\Big\langle\sum_{m=1}^{v_0}\big[v_m\nabla G_{j_m}(y,t^m)+v_m\nabla^2 G_{j_m}(y,t^m)z\big],e^{\beta(x,y)\eta(x,y)}-\mathbf{1}\Big\rangle\leqq-\tilde{\rho}(x,y)\|\theta(x,y)\|^2. \quad (2.10)$$

Similarly, we can show that our assumptions in (iii) combined with the primal feasibility of $x$ and (9) lead to the following inequality:

$$\Big\langle\sum_{m=v_0+1}^{v}\big[v_m\nabla H_{k_m}(y,s^m)+v_m\nabla^2 H_{k_m}(y,s^m)z\big],e^{\beta(x,y)\eta(x,y)}-\mathbf{1}\Big\rangle\leqq-\breve{\rho}(x,y)\|\theta(x,y)\|^2. \quad (2.11)$$

Now because of (10), (11), and (iv), (6) yields

$$\frac{1}{\beta(x,y)}\Big\langle\sum_{i=1}^{p}\big[\nabla\mathscr{E}_i(y,u,\lambda)+\nabla^2\mathscr{E}_i(y,u,\lambda)z\big],e^{\beta(x,y)\eta(x,y)}-\mathbf{1}\Big\rangle\geqq-\bar{\rho}(x,y)\|\theta(x,y)\|^2,$$

which in view of (i) implies that

$$\frac{1}{\alpha(x,y)}\bar{\gamma}(x,y)\Big(e^{\alpha(x,y)\sum_{i=1}^{p}\xi_i(x,y)\{\mathscr{E}_i(x,u,\lambda)-[\mathscr{E}_i(y,u,\lambda)-\frac{1}{2}\langle z,\nabla^2\mathscr{E}_i(y,u,\lambda)z\rangle]\}}-1\Big)\geqq 0.$$

Since $\bar{\gamma}(x,y)>0$, this inequality implies that

$$\sum_{i=1}^{p}u_i\xi_i(x,y)\big[f_i(x)-\lambda_i g_i(x)\big]$$
$$\geqq\sum_{i=1}^{p}u_i\xi_i(x,y)\big\{f_i(y)-\lambda_i g_i(y)-\frac{1}{2}\langle z,[\nabla^2 f_i(y)-\lambda_i\nabla^2 g_i(y)]z\rangle\big\}\geqq 0,$$

where the last inequality follows from (7). Since for each $i \in \underline{p}$, $u_i\xi_i(x,y) > 0$, this inequality implies that

$$(f_1(x) - \lambda_1 g_1(x), \ldots, f_p(x) - \lambda_p g_p(x)) \nleq (0, \ldots, 0),$$

which in turn implies that

$$\varphi(x) = \left(\frac{f_1(x)}{g_1(x)}, \ldots, \frac{f_p(x)}{g_p(x)}\right) \nleq (\lambda_1, \ldots, \lambda_p) = \lambda.$$

(b) : The proof is similar to that of part (a).

(c) : Suppose to the contrary that $\varphi(x) \leqslant \lambda$. Then $f_i(x) - \lambda_i g_i(x) \leqq 0$, $i \in p$, with strict inequality holding for at least one index $\ell \in \underline{p}$. Since $u > 0$, $\xi_i(x,y) > 0$, $i \in \underline{p}$, and (7) holds, we see that

$$\sum_{i=1}^{p} \xi_i(x,y) u_i [f_i(x) - \lambda_i g_i(x)]$$
$$< 0 \leqq \sum_{i=1}^{p} \xi_i(x,y) u_i \{f_i(y) - \lambda_i g_i(y) - \frac{1}{2}\langle z, [\nabla^2 f_i(y) - \lambda_i \nabla^2 g_i(y)] z\rangle\},$$

and hence

$$\frac{1}{\alpha(x,y)}\bar{\gamma}(x,y)\left(e^{\alpha(x,y)\sum_{i=1}^{p}\xi_i(x,y)\{\mathscr{E}_i(x,u,\lambda) - [\mathscr{E}_i(y,u,\lambda) - \frac{1}{2}\langle z, \nabla^2\mathscr{E}_i(y,u,\lambda)z\rangle]\}} - 1\right) < 0,$$

which in view of (i) implies that

$$\frac{1}{\beta(x,y)}\left\langle \sum_{i=1}^{p} [\nabla\mathscr{E}_i(y,u,\lambda) + \nabla^2\mathscr{E}_i(y,u,\lambda)z], e^{\beta(x,y)\eta(x,y)} - \mathbf{1}\right\rangle \leqq -\bar{\rho}(x,y)\|\theta(x,y)\|^2. \quad (2.12)$$

Because of (8) and the primal feasibility of $x$, we have

$$G_{j_m}(x,t^m) \leqq 0 \leqq G_{j_m}(y,t^m) - \frac{1}{2}\langle z, \nabla^2 G_{j_m}(y,t^m)z\rangle \text{ for each } m \in \underline{v_0},$$

and hence

$$\sum_{m=1}^{v_0} v_m \pi_m(x,y) G_{j_m}(x,t^m) \leqq \sum_{m=1}^{v_0} v_m \pi_m(x,y)\left[G_{j_m}(y,t^m) - \frac{1}{2}\langle z, \nabla^2 G_{j_m}(y,t^m)z\rangle\right].$$

From this we see that

$$\frac{1}{\alpha(x,y)}\tilde{\gamma}(x,y)\left(e^{\alpha(x,y)\sum_{m=1}^{v_0} v_m\pi_m(x,y)\{G_{j_m}(x,t^m) - [G_{j_m}(y,t^m) - \frac{1}{2}\langle z, \nabla^2 G_{j_m}(y,t^m)z\rangle]\}} - 1\right) \leqq 0,$$

which in view of (ii) implies that

$$\frac{1}{\beta(x,y)}\left\langle \sum_{m=1}^{v_0} [v_m \nabla G_{j_m}(y,t^m) + v_m \nabla^2 G_{j_m}(y,t^m)z], e^{\beta(x,y)\eta(x,y)} - \mathbf{1}\right\rangle$$
$$< -\tilde{\rho}(x,y)\|\theta(x,y)\|^2. \quad (2.13)$$

Similarly, we can show that our assumptions in (iii) combined with the primal feasibility of $x$ and (9) lead to the following inequality:

$$\frac{1}{\beta(x,y)}\Big\langle \sum_{m=\nu_0+1}^{\nu} \big[v_m \nabla H_{k_m}(y,s^m) + v_m \nabla^2 H_{k_m}(y,s^m) z\big], e^{\beta(x,y)\eta(x,y)} - \mathbf{1}\Big\rangle$$
$$\leqq -\breve{\rho}(x,y)\|\theta(x,y)\|^2. \tag{2.14}$$

Now combining (12) - (14) and using (iv), we get

$$\frac{1}{\beta(x,y)}\Big\langle \sum_{i=1}^{p} u_i[\nabla f_i(y) - \lambda_i \nabla g_i(y)] + \sum_{m=1}^{\nu_0} v_m \nabla G_{j_m}(y,t^m) + \sum_{m=\nu_0+1}^{\nu} v_m \nabla H_{k_m}(y,s^m) +$$
$$\Big\{\sum_{i=1}^{p} u_i[\nabla^2 f_i(y) - \lambda_i \nabla^2 g_i(y)] + \sum_{m=1}^{\nu_0} v_m \nabla^2 G_{j_m}(y,t^m) +$$
$$\sum_{m=\nu_0+1}^{\nu} v_m \nabla^2 H_{k_m}(y,s^m)\Big\} z, e^{\beta(x,y)\eta(x,y)} - \mathbf{1}\Big\rangle < -[\bar{\rho}(x,y) + \tilde{\rho}(x,y) + \breve{\rho}(x,y)]\|\theta(x,y)\|^2 \leqq 0,$$

which contradicts (6). Therefore, we conclude that $\varphi(x) \nleq \lambda$.

(d) : The proof is similar to that of part (c).

□

**Theorem 2.12.** (Strong Duality) Let $x^*$ be a normal efficient solution of (P) and assume that for each feasible solution $(y,z,u,v,\lambda,\nu,\nu_0,\ J_{\nu_0},K_{\nu\setminus\nu_0},\ \bar{t},\bar{s})$ of (DII), any one of the four sets of conditions specified in Theorem 2.11 is satisfied. Then there exist $u^*, v^*, \lambda^*$, $\nu^*, \nu_0^*, J_{\nu_0^*},\ K_{\nu^*\setminus\nu_0^*}, \bar{t}^*$, and $\bar{s}^*$ such that $(x^*, z^* = 0, u^*, v^*, \lambda^*, \nu^*, \nu_0^*,\ J_{\nu_0^*},\ K_{\nu^*\setminus\nu_0^*}, \bar{t}^*, \bar{s}^*)$ is an efficient solution of (DII) and $\varphi(x^*) = \lambda^*$.

*Proof.* The proof is similar to that of Theorem 2.11. □

**Theorem 2.13.** (Strict Converse Duality) Let $x^*$ and $\tilde{w} \equiv (\tilde{x},\tilde{z},\tilde{u},\tilde{v},\tilde{\lambda},\tilde{\nu},\tilde{\nu}_0,J_{\tilde{\nu}_0},K_{\tilde{\nu}\setminus\tilde{\nu}_0},\bar{\tilde{t}},\tilde{\bar{s}})$ be arbitrary feasible solutions of (P) and (DII), respectively, such that

$$\sum_{i=1}^{p} \tilde{u}_i \xi_i(x^*,\tilde{x})[f_i(x^*) - \tilde{\lambda}_i g_i(x^*)] = 0, \tag{2.15}$$

where $\xi_i$, $i \in \underline{p}$, are given real-valued functions from $X \times X$ to $\mathbb{R}_+ \setminus \{0\}$. Furthermore, assume that any one of the following four sets of hypotheses is satisfied:

(a) The assumptions specified in part (a) of Theorem 2.11 are satisfied for the feasible solution $\tilde{w}$ of (DII), and the function $\big(\mathscr{E}_1(\cdot,\tilde{u},\tilde{\lambda}),\ldots,\mathscr{E}_p(\cdot,\tilde{u},\tilde{\lambda})\big)$ is strictly HA($\alpha$, $\beta$, $\bar{\gamma}$, $\xi$, $\eta$, $\bar{\rho}$, $\theta$) -V-pseudosonvex at $\tilde{x}$.

(b) The assumptions specified in part (b) of Theorem 2.11 are satisfied for the feasible solution $\tilde{w}$ of (DII), and the function $\big(\mathscr{E}_1(\cdot,\tilde{u},\tilde{\lambda}),\ldots,\mathscr{E}_p(\cdot,\tilde{u},\tilde{\lambda})\big)$ is HA($\alpha$, $\beta$, $\bar{\gamma}$, $\xi$, $\eta$, $\bar{\rho}$, $\theta$) -V-quasisonvex at $\tilde{x}$.

(c) The assumptions specified in part (c) of Theorem 2.11 are satisfied for the feasible solution $\tilde{w}$ of (DII), and the function $\left(\mathscr{E}_1(\cdot,\tilde{u},\tilde{\lambda}),\ldots,\mathscr{E}_p(\cdot,\tilde{u},\tilde{\lambda})\right)$ is HA($\alpha$, $\beta$, $\bar{\gamma}$, $xi$, $\eta$, $\bar{\rho}$, $\theta$) -V-quasisonvex at $\tilde{x}$.

(d) The assumptions specified in part (d) of Theorem 2.11 are satisfied for the feasible solution $\tilde{w}$ of (DII), and the function $\left(\mathscr{E}_1(\cdot,\tilde{u},\tilde{\lambda}),\ldots,\mathscr{E}_p(\cdot,\tilde{u},\tilde{\lambda})\right)$ is HA($\alpha$, $\beta$, $\bar{\gamma}$, $\xi$, $\eta$, $\bar{\rho}$, $\theta$) -V-quasisonvex at $\tilde{x}$.

Then $\tilde{x}=x^*$ and $\varphi(x^*)=\tilde{\lambda}$.

*Proof.* (a) and (b) : The proofs are similar to that of Theorem 2.12.

(c) : Suppose to the contrary that $\tilde{x}\neq x^*$. Now proceeding as in the proof of part (c) of Theorem 2.11 (with $x$ replaced by $x^*$ and $w$ by $\tilde{w}$), we arrive at (13) and (14) which can be combined as follows:

$$-\frac{1}{\beta(x^*,\tilde{x})}\Big\langle\sum_{m=1}^{\tilde{\nu}_0}\tilde{v}_m\left[\nabla G_{j_m}(\tilde{x},\tilde{t}^m)+\nabla^2 G_{j_m}(\tilde{x},\tilde{t}^m)\tilde{z}\right]+$$
$$\sum_{m=\tilde{\nu}_0+1}^{\tilde{\nu}}\tilde{v}_m\left[\nabla H_{k_m}(\tilde{x},\tilde{s}^m)+\nabla^2 H_{k_m}(\tilde{x},\tilde{s}^m)\tilde{z}\right],e^{\beta(x^*,\tilde{x})\eta(x^*,\tilde{x})}-\mathbf{1}\Big\rangle$$
$$>[\tilde{\rho}(x^*,\tilde{x})+\breve{\rho}(x^*,\tilde{x})]\|\theta(x^*,\tilde{x})\|^2.$$

Combining this inequality with (6) and using (iv), we obtain

$$\frac{1}{\beta(x^*,\tilde{x})}\Big\langle\sum_{i=1}^{p}[\nabla\mathscr{E}_i(\tilde{x},\tilde{u},\tilde{\lambda})+\nabla^2\mathscr{E}_i(\tilde{x},\tilde{u},\tilde{\lambda})\tilde{z}],e^{\beta(x^*,\tilde{x})\eta(x^*,\tilde{x})}-\mathbf{1}\Big\rangle$$
$$>-\bar{\rho}(x^*,\tilde{x})\|\theta(x^*,\tilde{x})\|^2,$$

which by virtue of our $HA(\alpha,\beta,\bar{\gamma},\xi,\eta,\bar{\rho},\theta)$-V-quasisonvexity hypothesis implies that

$$\frac{1}{\alpha(x^*,\tilde{x})}\bar{\gamma}(x^*,\tilde{x})\left(e^{\alpha(x^*,\tilde{x})\sum_{i=1}^{p}\xi_i(x^*,\tilde{x})\{\mathscr{E}_i(x^*,\tilde{u},\tilde{\lambda})-[\mathscr{E}_i(\tilde{x},\tilde{u},\tilde{\lambda})-\frac{1}{2}\langle z,\nabla^2\mathscr{E}_i(\tilde{x},\tilde{u},\tilde{\lambda})\tilde{z}\rangle]\}}-1\right)>0.$$

Since $\bar{\gamma}(x^*,\tilde{x})>0$, this inequality implies that

$$\sum_{i=1}^{p}\xi_i(x^*,\tilde{x})\mathscr{E}_i(x^*,\tilde{u},\tilde{\lambda})>\sum_{i=1}^{p}\xi_i(x^*,\tilde{x})\left[\mathscr{E}_i(\tilde{x},\tilde{u},\tilde{\lambda})-\frac{1}{2}\langle\tilde{z},\nabla^2\mathscr{E}_i(\tilde{x},\tilde{u},\tilde{\lambda})\tilde{z}\rangle\right]\geqq 0,$$

where the second inequality follows from (7). This inequality contradicts (15) and, therefore, we must have $\tilde{x}=x^*$ and $\varphi(x^*)=\tilde{\lambda}$.

(d) : The proof is similar to that of part (c).

□

## 5. Duality Model III

In this section, we discuss several families of duality results under various generalized $HA(\alpha,\beta,\gamma,\xi,\eta,\rho,\theta)$-V-sonvexity hypotheses imposed on certain vector functions whose components are formed by considering different combinations of the problem functions. This is accomplished by employing a partitioning scheme which was originally proposed in [7] for the purpose of constructing generalized dual problems for nonlinear programming problems. For this we need some additional notation.

Let $\nu_0$ and $\nu$ be integers, with $1 \leqq \nu_0 \leqq \nu \leqq n+1$, and let $\{J_0, J_1, \ldots, J_M\}$ and $\{K_0, K_1, \ldots, K_M\}$ be partitions of the sets $\underline{\nu_0}$ and $\underline{\nu}\backslash\underline{\nu_0}$, respectively; thus, $J_i \subseteq \underline{\nu_0}$ for each $i \in \underline{M} \cup \{0\}$, $J_i \cap J_j = \emptyset$ for each $i, j \in \underline{M} \cup \{0\}$ with $i \neq j$, and $\cup_{i=0}^{M} J_i = \underline{\nu_0}$. Obviously, similar properties hold for $\{K_0, K_1, \ldots, K_M\}$. Moreover, if $m_1$ and $m_2$ are the numbers of the partitioning sets of $\underline{\nu_0}$ and $\underline{\nu}\backslash\underline{\nu_0}$, respectively, then $M = \max\{m_1, m_2\}$ and $J_i = \emptyset$ or $K_i = \emptyset$ for $i > \min\{m_1, m_2\}$.

In addition, we use the real-valued functions $\Phi_i(\cdot, u, v, \lambda, \bar{t}, \bar{s})$ and $\Lambda_\tau(\cdot, v, \bar{t}, \bar{s})$ defined, for fixed $u, v, \lambda, \bar{t} \equiv (t^1, t^2, \ldots, t^{\nu_0})$, and $\bar{s} \equiv (s^{\nu_0+1}, s^{\nu_0+2}, \ldots, s^{\nu})$, on $X$ as follows:

$$\Phi_i(\zeta, u, v, \lambda, \bar{t}, \bar{s}) = u_i\Big[f_i(\zeta) - \lambda_i g_i(\zeta) + \sum_{m\in J_0} v_m G_{j_m}(\zeta, t^m) + \sum_{m\in K_0} v_m H_{k_m}(\zeta, s^m)\Big], \quad i \in \underline{p},$$

$$\Lambda_\tau(\zeta, v, \bar{t}, \bar{s}) = \sum_{m\in J_\tau} v_m G_{j_m}(\zeta, t^m) + \sum_{m\in K_\tau} v_m H_{k_m}(\zeta, s^m), \qquad \tau \in \underline{M}.$$

Making use of the sets and functions defined above, we can state our general duality models as follows:

$$(DIII) \qquad \sup_{(y,z,u,v,\lambda,\nu,\nu_0,J_{\nu_0},K_{\nu\backslash\nu_0},\bar{t},\bar{s})\in\mathbb{H}} \lambda = (\lambda_1, \cdots, \lambda_p)$$

subject to

$$\sum_{i=1}^{p} u_i[\nabla f_i(y) - \lambda_i \nabla g_i(y)] + \sum_{m=1}^{\nu_0} v_m \nabla G_{j_m}(y, t^m) + \sum_{m=\nu_0+1}^{\nu} v_m \nabla H_{k_m}(y, s^m) + \Big\{\sum_{i=1}^{p} u_i[\nabla^2 f_i(y) - \lambda_i \nabla^2 g_i(y)] + \sum_{m=1}^{\nu_0} v_m \nabla^2 G_{j_m}(y, t^m) + \sum_{m=\nu_0+1}^{\nu} v_m \nabla^2 H_{k_m}(y, s^m)\Big\} z = 0, \quad (2.16)$$

$$f_i(y) - \lambda_i g_i(y) + \sum_{m\in J_0} v_m G_{j_m}(y, t^m) + \sum_{m\in K_0} v_m H_{k_m}(y, s^m) - \frac{1}{2}\Big\langle z, \Big[\nabla^2 f_i(y) - \lambda_i \nabla^2 g_i(y) + \sum_{m\in J_0} v_m \nabla^2 G_{j_m}(y, t^m) + \sum_{m\in K_0} v_m \nabla^2 H_{k_m}(y, s^m)\Big] z\Big\rangle \geqq 0, \ i \in \underline{p}, \quad (2.17)$$

$$\sum_{m\in J_\tau} v_m G_{j_m}(y,t^m) + \sum_{m\in K_\tau} v_m H_{k_m}(y,s^m) - \frac{1}{2}\Big\langle z, \Big[\sum_{m\in J_\tau} v_m \nabla^2 G_{j_m}(y,t^m) + \sum_{m\in K_\tau} v_m \nabla^2 H_{k_m}(y,s^m)\Big] z\Big\rangle \geqq 0,\ \ \tau \in \underline{M}, \quad (2.18)$$

for all $x \in \mathbb{F}$;

$(\tilde{D}III)$ $\sup_{(y,z,u,v,\lambda,\nu,\nu_0,J_{\nu_0},K_{\nu\backslash\nu_0},\bar{t},\bar{s})\in\mathbb{H}} \lambda = (\lambda_1,\ldots,\lambda_p)$

subject to (3), (17), and (18).

The remarks and observations made earlier about the relationships between $(DI)$ and $(\tilde{D}I)$ are, of course, also valid for $(DIII)$ and $(\tilde{D}III)$.

The next two theorems show that $(DIII)$ is a dual problem for $(P)$.

**Theorem 2.14.** Let $x$ and $(y,z,u,v,\lambda,\nu,\nu_0,J_{\nu_0},K_{\nu\backslash\nu_0},\bar{t},\bar{s})$ be arbitrary feasible solutions of (P) and (DIII), respectively, and assume that any one of the following three sets of hypotheses is satisfied:

(a) (i) $\big(\Phi_1(\cdot,u,v,\lambda,\bar{t},\bar{s}),\ldots,\Phi_p(\cdot,u,v,\lambda,\bar{t},\bar{s})\big)$ is HA($\alpha$, $\beta$, $\bar{\gamma}$, $\xi$, $\eta$, $\bar{\rho}$, $\theta$) -V - pseudosonvex at $y$ and $\bar{\gamma}(x,y) > 0$;

(ii) $\big(\Lambda_1(\cdot,v,\bar{t},\bar{s}),\ldots,\Lambda_M(\cdot,v,\bar{t},\bar{s})\big)$ is $HA(\alpha,\beta,\tilde{\gamma},\pi,\eta,\tilde{\rho},\theta)$-V-quasisonvex at $y$;

(iii) $\bar{\rho}(x,y)+\tilde{\rho}(x,y) \geqq 0$;

(b) (i) $\big(\Phi_1(\cdot,u,v,\lambda,\bar{t},\bar{s}),\ldots,\Phi_p(\cdot,u,v,\lambda,\bar{t},\bar{s})\big)$ is prestrictly $HA(\alpha,\beta,\bar{\gamma},\xi,\eta,\bar{\rho},\theta)$-V-quasisonvex at $y$ and $\bar{\gamma}(x,y) > 0$;

(ii) $\big(\Lambda_1(\cdot,v,\bar{t},\bar{s}),\ldots,\Lambda_M(\cdot,v,\bar{t},\bar{s})\big)$ is $HA(\alpha,\beta,\tilde{\gamma},\pi,\eta,\tilde{\rho},\theta)$-V-quasisonvex at $y$;

(iii) $\bar{\rho}(x,y)+\tilde{\rho}(x,y) > 0$;

(c) (i) $\big(\Phi_1(\cdot,u,v,\lambda,\bar{t},\bar{s}),\ldots,\Phi_p(\cdot,u,v,\lambda,\bar{t},\bar{s})\big)$ is prestrictly $HA(\alpha,\beta,\bar{\gamma},\xi,\eta,\bar{\rho},\theta)$-V-quasisonvex at $y$ and $\bar{\gamma}(x,y) > 0$;

(ii) $\big(\Lambda_1(\cdot,v,\bar{t},\bar{s}),\ldots,\Lambda_M(\cdot,v,\bar{t},\bar{s})\big)$ is strictly HA($\alpha$, $\beta$, $\tilde{\gamma}$, $\pi$, $\eta$, $\tilde{\rho}$, $\theta$) - V-pseudosonvex at $y$;

(iii) $\bar{\rho}(x,y)+\tilde{\rho}(x,y) \geqq 0$.

Then $\varphi(x) \nleq \lambda$.

*Proof.* (a): It is clear that (16) can be expressed as follows:

$$\frac{1}{\beta(x,y)}\Big\langle \sum_{i=1}^{p} u_i[\nabla f_i(y) - \lambda_i \nabla g_i(y)] + \sum_{m\in J_0} v_m \nabla G_{j_m}(y,t^m) + \sum_{m\in K_0} v_m \nabla H_{k_m}(y,s^m) +$$
$$\Big\{\sum_{i=1}^{p} u_i[\nabla^2 f_i(y) - \lambda_i \nabla^2 g_i(y)] + \sum_{m\in J_0} v_m \nabla^2 G_{j_m}(y,t^m) + \sum_{m\in K_0} v_m \nabla^2 H_{k_m}(y,s^m)\Big\} z$$
$$+ \sum_{\tau=1}^{M}\Big\{\sum_{m\in J_\tau} v_m \nabla G_{j_m}(y,t^m) + \sum_{m\in K_\tau} v_m \nabla H_{k_m}(y,s^m) + \Big[\sum_{m\in J_\tau} v_m \nabla^2 G_{j_m}(y,t^m) +$$
$$\sum_{m\in K_\tau} v_m \nabla^2 H_{k_m}(y,s^m)\Big] z\Big\}, e^{\beta(x,y)\eta(x,y)} - \mathbf{1}\Big\rangle = 0. \quad (2.19)$$

Since $x \in \mathbb{F}$, $v_m > 0$, $m \in \underline{v_0}$, and (18) holds, it follows that for each $\tau \in \underline{M}$,

$$\begin{aligned}\Lambda_\tau(x,v,\bar{t},\bar{s}) &= \sum_{m\in J_\tau} v_m G_{j_m}(x,t^m) + \sum_{m\in K_\tau} v_m H_{k_m}(x,s^m)\\ &\leqq 0\\ &\leqq \Lambda_\tau(y,v,\bar{t},\bar{s}) - \frac{1}{2}\Big\langle z, \nabla^2\Lambda_\tau(y,v,\bar{t},\bar{s})z\Big\rangle.\end{aligned}$$

As $\pi_\tau(x,y) > 0$ for each $\tau \in M$, the above inequalities yield

$$\sum_{\tau=1}^{M} \pi_\tau(x,y)\Lambda_\tau(x,v,\bar{t},\bar{s}) \leqq \sum_{\tau=1}^{M} \pi_\tau(x,y)\big[\Lambda_\tau(y,v,\bar{t},\bar{s}) - \frac{1}{2}\langle z, \nabla^2\Lambda_\tau(y,v,\bar{t},\bar{s})z\rangle\big],$$

and hence

$$\frac{1}{\alpha(x,y)}\tilde{\gamma}(x,y)\Big(e^{\alpha(x,y)\sum_{\tau=1}^{M}\pi_\tau(x,y)[\Lambda_\tau(x,v,\bar{t},\bar{s}) - \Lambda_\tau(y,v,\bar{t},\bar{s}) + \frac{1}{2}\langle z,\nabla^2\Lambda_\tau(y,v,\bar{t},\bar{s})z\rangle]} - 1\Big) \leqq 0,$$

which in view of (ii) implies that

$$\frac{1}{\beta(x,y)}\Big\langle \sum_{\tau=1}^{M}\Big\{\sum_{m\in J_\tau} v_m \nabla G_{j_m}(y,t^m) + \sum_{m\in K_\tau} v_m \nabla H_{k_m}(y,s^m) + \Big[\sum_{m\in J_\tau} v_m \nabla^2 G_{j_m}(y,t^m)$$
$$+ \sum_{m\in K_\tau} v_m \nabla^2 H_{k_m}(y,s^m)\Big] z\Big\}, e^{\beta(x,y)\eta(x,y)} - \mathbf{1}\Big\rangle \leqq -\tilde{\rho}(x,y)\|\theta(x,y)\|^2. \quad (2.20)$$

Combining (19) and (20), and using (iii) we get

$$\frac{1}{\beta(x,y)}\Big\langle \sum_{i=1}^{p} u_i[\nabla f_i(y) - \lambda_i \nabla g_i(y)] + \sum_{m\in J_0} v_m \nabla G_{j_m}(y,t^m)$$
$$+ \sum_{m\in K_0} v_m \nabla H_{k_m}(y,s^m) + \Big\{\sum_{i=1}^{p} u_i[\nabla^2 f_i(y) - \lambda_i \nabla^2 g_i(y)] + \sum_{m\in J_0} v_m \nabla^2 G_{j_m}(y,t^m)$$
$$+ \sum_{m\in K_0} v_m \nabla^2 H_{k_m}(y,s^m\Big\} z, e^{\beta(x,y)\eta(x,y)} - \mathbf{1}\Big\rangle \geqq \tilde{\rho}(x,y)\|\theta(x,y)\|^2 \geqq -\bar{\rho}(x,y)\|\theta(x,y)\|^2,$$

which by virtue of (i) implies that

$$\frac{1}{\alpha(x,y)}\bar{\gamma}(x,y)$$
$$\left(e^{\alpha(x,y)\sum_{i=1}^{p}\xi_i(x,y)[\Phi_i(x,u,v,\lambda,\bar{t},\bar{s})-\Phi_i(y,u,v,\lambda,\bar{t},\bar{s})+\frac{1}{2}\langle z,\nabla^2\Phi(y,u,v,\lambda,\bar{t},\bar{s})z\rangle]}-1\right)\geqq 0.$$

Since $\bar{\gamma}(x,y)>0$, we can deduce from this inequality that

$$\sum_{i-1}^{p}\xi_i(x,y)\Phi_i(x,u,v,\lambda,\bar{t},\bar{s})\geqq\sum_{i=1}^{p}\xi_i(x,y)\big[\Phi_i(y,u,v,\lambda,\bar{t},\bar{s})$$
$$-\ \frac{1}{2}\langle z,\nabla^2\Phi(y,u,v,\lambda,\bar{t},\bar{s})z\big]\geqq 0,$$

where the second inequality follows from (17). Therefore, we have

$$\begin{aligned}0&\leqq\sum_{i=1}^{p}\xi_i(x,y)\Phi_i(x,u,v,\lambda,\bar{t},\bar{s})\\&=\sum_{i=1}^{p}u_i\xi_i(x,y)\Big[f_i(x)-\lambda_i g_i(x)]+\sum_{m\in J_0}v_mG_{j_m}(x,t^m)+\sum_{m\in K_0}v_mH_{k_m}(x,s^m)\Big]\\&\leqq\sum_{i=1}^{p}u_i\xi_i(x,y)[f_i(x)-\lambda_i g_i(x)]\ \text{(by the primal feasibility of } x).\end{aligned}$$

As shown in the proof of Theorem 2.11, this inequality leads to the conclusion that $\varphi(x)\nleq\lambda$.

(b): The proof is similar to that of part (a).

(c) : Suppose to the contrary that $\varphi(x)\leqslant\lambda$. This implies that for each $i\in\underline{p}$,

$$f_i(x)-\lambda_i g_i(x)\leqq 0, \tag{2.21}$$

with strict inequality holding for at least one index $\ell\in\underline{p}$. Keeping in mind that $v_m>0$ for each $m\in\underline{v_0}$, we have

$$\begin{aligned}\Phi_i(x,u,v,\lambda,\bar{t},\bar{s})&\leqq\sum_{i=1}^{p}u_i[f_i(x)-\lambda_i g_i(x)]\ \text{(by the primal feasibility of } x)\\&<0\ \text{(by (21) and the fact that } u>0)\\&\leqq\Phi_i(y,u,v,\lambda,\bar{t},\bar{s})-\frac{1}{2}\langle z,\nabla^2\Phi_i(y,u,v,\lambda,\bar{t},\bar{s})z\rangle\ \text{(by (17))}\end{aligned}$$

and hence

$$\sum_{i=1}^{p}\xi_i(x,y)\Phi_i(x,u,v,\lambda,\bar{t},\bar{s})<\sum_{i=1}^{p}\xi_i(x,y)\Big[\Phi_i(y,u,v,\lambda,\bar{t},\bar{s})$$
$$-\ \frac{1}{2}\Big\langle z,\nabla^2\Phi_i(y,u,v,\lambda,\bar{t},\bar{s})z\Big\rangle\Big],$$

and

$$\frac{1}{\alpha(x,y)}\bar{\gamma}(x,y)$$
$$\left(e^{\alpha(x,y)\sum_{i=1}^{p}\xi_i(x,y)[\Phi_i(x,u,v,\lambda,\bar{t},\bar{s})-\Phi_i(y,u,v,\lambda,\bar{t},\bar{s})+\frac{1}{2}\langle z,\nabla^2\Phi(y,u,v,\lambda,\bar{t},\bar{s})z\rangle]}-1\right)<0,$$

which in view of (i) implies that

$$\frac{1}{\beta(x,y)}\Big\langle\sum_{i=1}^{p}u_i[\nabla f_i(y)-\lambda_i\nabla g_i(y)]+\sum_{m\in J_0}v_m\nabla G_{j_m}(y,t^m)$$
$$+\sum_{m\in K_0}v_m\nabla H_{k_m}(y,s^m)+\Big\{\sum_{i=1}^{p}u_i[\nabla^2 f_i(y)-\lambda_i\nabla^2 g_i(y)]+\sum_{m\in J_0}v_m\nabla^2 G_{j_m}(y,t^m)$$
$$+\sum_{m\in K_0}v_m\nabla^2 H_{k_m}(y,s^m\Big\}z,e^{\beta(x,y)\eta(x,y)}-\mathbf{1}\Big\rangle\leqq-\bar{\rho}(x,y)\|\theta(x,y)\|^2. \quad (2.22)$$

As shown in the proof of part (a), for each $\tau\in\underline{M}$, we have

$$\Lambda_\tau(x,v,\bar{t},\bar{s})\leqq\Lambda_\tau(y,v,\bar{t},\bar{s})-\frac{1}{2}\langle z,\nabla^2\Lambda_\tau(y,v,\bar{t},\bar{s})z\rangle,$$

which in view of the properties of the functions $\pi_\tau,\ \tau\in M$, can be expressed as

$$\sum_{\tau=1}^{M}\pi_\tau(x,y)\Lambda_\tau(x,v,\bar{t},\bar{s})\leqq\sum_{\tau=1}^{M}\pi_\tau(x,y)[\Lambda_\tau(y,v,\bar{t},\bar{s})-\frac{1}{2}\langle z,\nabla^2\Lambda_\tau(y,v,\bar{t},\bar{s})z\rangle]$$

and hence

$$\frac{1}{\alpha(x,y)}\tilde{\gamma}(x,y)\left(e^{\alpha(x,y)\sum_{\tau=1}^{M}\pi_\tau(x,y)[\Lambda_\tau(x,v,\bar{t},\bar{s})-\Lambda_\tau(y,v,\bar{t},\bar{s})+\frac{1}{2}\langle z,\nabla^2\Lambda_\tau(y,v,\bar{t},\bar{s})z\rangle]}-1\right)\leqq0.$$

In view of (ii) this implies that

$$\frac{1}{\beta(x,y)}\Big\langle\sum_{\tau=1}^{M}\Big\{\sum_{m\in J_\tau}v_m\nabla G_{j_m}(y,t^m)+\sum_{m\in K_\tau}v_m\nabla H_{k_m}(y,s^m)+$$
$$\Big[\sum_{m\in J_\tau}v_m\nabla^2 G_{j_m}(y,t^m)+\sum_{m\in K_\tau}v_m\nabla^2 H_{k_m}(y,s^m)\Big]z\Big\},e^{\beta(x,y)\eta(x,y)}-\mathbf{1}\Big\rangle$$
$$<-\tilde{\rho}(x,y)\|\theta(x,y)\|^2.$$

Combining this inequality with (19) and using (iii), we get

$$\frac{1}{\beta(x,y)}\Big\langle\sum_{i=1}^{p}u_i[\nabla f_i(y)-\lambda_i\nabla g_i(y)]+\sum_{m\in J_0}v_m\nabla G_{j_m}(y,t^m)$$
$$+\sum_{m\in K_0}v_m\nabla H_{k_m}(y,s^m)+\Big\{\sum_{i=1}^{p}u_i[\nabla^2 f_i(y)-\lambda_i\nabla^2 g_i(y)]+\sum_{m\in J_0}v_m\nabla^2 G_{j_m}(y,t^m)$$
$$+\sum_{m\in K_0}v_m\nabla^2 H_{k_m}(y,s^m\Big\}z,e^{\beta(x,y)\eta(x,y)}-\mathbf{1}\Big\rangle>\tilde{\rho}(x,y)\|\theta(x,y)\|^2\geqq-\bar{\rho}(x,y)\|\theta(x,y)\|^2,$$

which contradicts (22). Therefore, we conclude that $\varphi(x)\nleq\lambda$. □

**Theorem 2.15.** (Strong Duality) Let $x^*$ be a normal efficient solution of (P) and assume that for each feasible solution $(y,z,u,v,\lambda,\nu,\nu_0, J_{\nu_0},K_{\nu\backslash\nu_0}, \bar{t},\bar{s})$ of (DIII), any one of the three sets of conditions specified in Theorem 2.14 is satisfied. Then there exist $u^*,v^*,\lambda^*$, $\nu^*,\nu_0^*,J_{\nu_0^*}$, $K_{\nu^*\backslash\nu_0^*},\bar{t}^*$, and $\bar{s}^*$ such that $(x^*,z^*=0,u^*,v^*,\lambda^*,\nu^*,\nu_0^*, J_{\nu_0^*}, K_{\nu^*\backslash\nu_0^*},\bar{t}^*,\bar{s}^*)$ is an efficient solution of (DIII) and $\varphi(x^*)=\lambda^*$.

*Proof.* The proof is similar to that of Theorem 2.12. □

**Theorem 2.16.** (Strict Converse Duality) Let $x^*$ and $\tilde{w}\equiv(\tilde{x},\tilde{z},\tilde{u},\tilde{v},\tilde{\lambda},\tilde{\nu},\tilde{\nu}_0, J_{\tilde{\nu}_0}, K_{\tilde{\nu}\backslash\tilde{\nu}_0},\bar{\tilde{t}},\bar{\tilde{s}})$ be arbitrary feasible solutions of (P) and (DIII), respectively, such that

$$\sum_{i=1}^{p}\tilde{u}_i\xi_i(x^*,\tilde{x})[f_i(x^*)-\tilde{\lambda}_i g_i(x^*)]=0, \tag{2.23}$$

where $\xi_i$, $i\in \underline{p}$, are given real-valued functions from $X\times X$ to $\mathbb{R}_+\backslash\{0\}$. Furthermore, assume that any one of the following three sets of hypotheses is satisfied:

(a) The assumptions specified in part (a) of Theorem 2.14 are satisfied for the feasible solution $\tilde{w}$ of (DIII), and the function $\big(\Phi_1(\cdot,\tilde{u},\tilde{v},\tilde{\lambda},\bar{\tilde{t}},\bar{\tilde{s}}),\ldots,\Phi_p(\cdot,\tilde{u},\tilde{v},\tilde{\lambda},\bar{\tilde{t}},\bar{\tilde{s}})\big)$ is strictly $HA(\alpha,\beta,\bar{\gamma},\xi,\eta,\bar{\rho},\theta)$-V-pseudosonvex at $\tilde{x}$.

(b) The assumptions specified in part (b) of Theorem 2.14 are satisfied for the feasible solution $\tilde{w}$ of (DIII), and the function $\big(\Phi_1(\cdot,\tilde{u},\tilde{v},\tilde{\lambda},\bar{\tilde{t}},\bar{\tilde{s}}),\ldots,\Phi_p(\cdot,\tilde{u},\tilde{v},\tilde{\lambda},\bar{\tilde{t}},\bar{\tilde{s}})\big)$ is $HA(\alpha,\beta,\bar{\gamma},\xi,\eta,\bar{\rho},\theta)$-V-quasisonvex $\tilde{x}$.

(c) The assumptions specified in part (c) of Theorem 2.14 are satisfied for the feasible solution $\tilde{w}$ of (DIII), and the function $\big(\Phi_1(\cdot,\tilde{u},\tilde{v},\tilde{\lambda},\bar{\tilde{t}},\bar{\tilde{s}}),\ldots,\Phi_p(\cdot,\tilde{u},\tilde{v},\tilde{\lambda},\bar{\tilde{t}},\bar{\tilde{s}})\big)$ is $HA(\alpha,\beta,\bar{\gamma},\xi,\eta,\bar{\rho},\theta)$-V-quasisonvex $\tilde{x}$.

Then $\tilde{x}=x^*$ and $\varphi(x^*)=\tilde{\lambda}$.

*Proof.* The proof is similar to that of Theorem 2.13. □

Each one of the three sets of results given in Theorems 2.14 - 2.16 can be viewed as a family of duality results whose members can easily be identified by appropriate choices of the partitioning sets $J_\mu$ and $K_\mu$, $\mu\in \underline{M}\cup\{0\}$.

## 6. Duality Model IV

In this section we discuss another collection of duality results for $(P)$ which are different from those stated in Theorems 2.14 - 2.16. In formulating these duality results, we utilize a partition of $\underline{p}$ in addition to those of $\underline{\nu_0}$ and $\underline{\nu}\backslash\underline{\nu_0}$. It appears that this partitioning scheme was first proposed in [10] for a multiobjective fractional programming problem with a finite number of constraints. In our theorems, we impose appropriate generalized $HA(\alpha,\beta,\gamma,\xi,\eta,\rho,\theta)$-V-sonvexity requirements on certain vector functions whose components comprise some combinations of the functions $f_i$, $g_i$, $i\in\underline{p}$, $G_j(\cdot,t)$, $j\in\underline{q}$, and $H_k(\cdot,s)$, $k\in\underline{r}$.

Let $\{I_0, I_1, \ldots, I_d\}$, $\{J_0, J_1, \ldots, J_e\}$, and $\{K_0, K_1, \ldots, K_e\}$ be partitions of $\underline{p}$, $\underline{\nu_0}$, and $\underline{\nu}\backslash\underline{\nu_0}$, respectively, such that $D = \{0, 1, 2, \ldots, d\} \subset E = \{0, 1, \ldots, e\}$, and let the function $\Pi_\tau(\cdot, u, v, \lambda, \bar{t}, \bar{s}) : X \to \mathbb{R}$ be defined, for fixed $u, v, \lambda, \bar{t}$, and $\bar{s}$ by

$$\Pi_\tau(\zeta, u, v, \lambda, \bar{t}, \bar{s}) = \sum_{i \in I_\tau} u_i [f_i(\zeta) - \lambda_i g_i(\zeta)] + \sum_{m \in J_\tau} v_m G_{j_m}(\zeta, t^m) + \sum_{m \in K_\tau} v_m H_{k_m}(\zeta, s^m), \quad \tau \in D.$$

Making use of the sets and functions defined above, we consider the following second-order duality models:

$$(DIV) \qquad \sup_{(y,z,u,v,\lambda,\nu,\nu_0,J_{\nu_0},K_{\nu\backslash\nu_0},\bar{t},\bar{s}) \in \mathbb{H}} \lambda = (\lambda_1, \cdots, \lambda_p)$$

subject to

$$\sum_{i=1}^{p} u_i [\nabla f_i(y) - \lambda_i \nabla g_i(y)] + \sum_{m=1}^{\nu_0} v_m \nabla G_{j_m}(y, t^m) + \sum_{m=\nu_0+1}^{\nu} v_m \nabla H_{k_m}(y, s^m) + \Big\{ \sum_{i=1}^{p} u_i [\nabla^2 f_i(y) - \lambda_i \nabla^2 g_i(y)] + \sum_{m=1}^{\nu_0} v_m \nabla^2 G_{j_m}(y, t^m) + \sum_{m=\nu_0+1}^{\nu} v_m \nabla^2 H_{k_m}(y, s^m) \Big\} z = 0, \quad (2.24)$$

$$\sum_{i \in I_\tau} u_i [f_i(y) - \lambda_i g_i(y)] + \sum_{m \in J_\tau} v_m G_{j_m}(y, t^m) + \sum_{m \in K_\tau} v_m H_{k_m}(y, s^m) - \frac{1}{2} \Big\langle \omega(x, y), \Big\{ \sum_{i \in I_\tau} u_i [\nabla^2 f_i(y) - \lambda_i \nabla^2 g_i(y)] + \sum_{m \in J_\tau} v_m \nabla^2 G_{j_m}(y, t^m) + \sum_{m \in K_\tau} v_m \nabla^2 H_{k_m}(y, s^m) \Big\} z \Big\rangle \geqq 0, \quad \tau \in D, and\ x \in \mathbb{F}, \quad (2.25)$$

$$\sum_{m \in J_\tau} v_m G_{j_m}(y, t^m) + \sum_{m \in K_\tau} v_m H_{k_m}(y, s^m) - \frac{1}{2} \Big\langle z, \Big[ \sum_{m \in J_\tau} v_m \nabla^2 G_{j_m}(y, t^m) + \sum_{m \in K_\tau} v_m \nabla^2 H_{k_m}(y, s^m) \Big] z \Big\rangle \geqq 0, \quad \tau \in E \backslash D, x \in \mathbb{F}; \quad (2.26)$$

$$(\tilde{D}IV) \qquad \sup_{(y,z,u,v,\lambda,\nu,\nu_0,J_{\nu_0},K_{\nu\backslash\nu_0},\bar{t},\bar{s}) \in \mathbb{H}} \lambda = (\lambda_1, \ldots, \lambda_p)$$

subject to (3), (25), and (26).

The remarks and observations made earlier about the relationships between $(DI)$ and $(\tilde{D}I)$ are, of course, also valid for $(DIV)$ and $(\tilde{D}IV)$.

The next two theorems show that $(DIV)$ is a dual problem for $(P)$.

**Theorem 2.17.** (Weak Duality) Let $x$ and $(y, z, u, v, \lambda, \nu, \nu_0, J_{\nu_0}, K_{\nu\backslash\nu_0}, \bar{t}, \bar{s})$ be arbitrary feasible solutions of (P) and (DIV), respectively, and assume that any one of the following three sets of hypotheses is satisfied:

(a) (i) $\left(\Pi_0(\cdot,u,v,\lambda,\bar{t},\bar{s}),\ldots,\Pi_d(\cdot,u,v,\lambda,\bar{t},\bar{s})\right)$ is $HA(\alpha,\beta,\bar{\gamma},\xi,\eta,\bar{\rho},\theta)$-V-pseudosonvex at $y$ and $\bar{\gamma}(x,y)>0$;

(ii) $\left(\Lambda_{d+1}(\cdot,v,\bar{t},\bar{s}),\ldots,\Lambda_e(\cdot,v,\bar{t},\bar{s})\right)$ is $HA(\alpha,\beta,\tilde{\gamma},\pi,\eta,\omega,\tilde{\rho},\theta)$-V-quasisonvex at $y$;

(iii) $\bar{\rho}(x,y)+\tilde{\rho}(x,y)\geqq 0$;

(b) (i) $\left(\Pi_0(\cdot,u,v,\lambda,\bar{t},\bar{s}),\ldots,\Pi_d(\cdot,u,v,\lambda,\bar{t},\bar{s})\right)$ is prestrictly $HA(\alpha,\beta,\bar{\gamma},\xi,\eta,\bar{\rho},\theta)$ -V-quasisonvex at $y$ and $\bar{\gamma}(x,y)>0$;

(ii) $\left(\Lambda_{d+1}(\cdot,v,\bar{t},\bar{s}),\ldots,\Lambda_e(\cdot,v,\bar{t},\bar{s})\right)$ is strictly $HA(\alpha,\beta,\tilde{\gamma},\pi,\eta,\tilde{\rho},\theta)$ -V-pseudosonvex at $y$;

(iii) $\bar{\rho}(x,y)+\tilde{\rho}(x,y)\geqq 0$;

(c) (i) $\left(\Pi_0(\cdot,u,v,\lambda,\bar{t},\bar{s}),\ldots,\Pi_d(\cdot,u,v,\lambda,\bar{t},\bar{s})\right)$ is prestrictly $HA(\alpha,\beta,\bar{\gamma},\xi,\eta,\bar{\rho},\theta)$ -V-
quasisonvex at $y$ and $\bar{\gamma}(x,y)>0$;

(ii) $\left(\Lambda_{d+1}(\cdot,v,\bar{t},\bar{s}),\ldots,\Lambda_e(\cdot,v,\bar{t},\bar{s})\right)$ is $HA(\alpha,\beta,\tilde{\gamma},\xi,\eta,\tilde{\rho},\theta)$-V-quasisonvex at $y$;

(iii) $\bar{\rho}(x,y)+\tilde{\rho}(x,y)>0$.

Then $\varphi(x)\nleq\lambda$.

*Proof.* (a): Suppose to the contrary that $\varphi(x)\leqslant\lambda$. Then $f_i(x)-\lambda_i g_i(x)\leqq 0$, $i\in\underline{p}$, with strict inequality holding for at least one index $\ell\in\underline{p}$. Since $u>0$, we see that for each $\tau\in D$,

$$\sum_{i\in I_\tau}u_i[f_i(x)-\lambda_i g_i(x)]\leqq 0, \tag{2.27}$$

with strict inequality holding for at least one index $\tau\in D$.

Now using this inequality, we see that

$$\begin{aligned}
&\Pi_\tau(x,u,v,\lambda,\bar{t},\bar{s})\\
&=\sum_{i\in I_\tau}u_i[f_i(x)-\lambda_i g_i(x)]+\sum_{m\in J_\tau}v_m G_{j_m}(x,t^m)+\sum_{m\in K_\tau}v_m H_{k_m}(x,s^m)\\
&\leqq\sum_{i\in I_\tau}u_i[f_i(x)-\lambda_i g_i(x)]\\
&\quad\text{(by the primal feasibility of } x \text{ and positivity of } v_m,\ m\in\underline{\nu_0})\\
&\leqq 0\quad\text{(by (27))}\\
&\leqq\Pi_\tau(y,u,v,\lambda,\bar{t},\bar{s})-\frac{1}{2}\langle z,\nabla^2\Pi_\tau(y,u,v,\lambda,\bar{t},\bar{s})z\rangle\ \text{(by (25))},
\end{aligned}$$

with strict inequality holding for at least one index $\tau\in D$. Since $\xi_\tau(x,y)>0$ for each $\tau\in D$, the above inequalities yield

$$\begin{aligned}
&\sum_{\tau\in D}\xi_\tau(x,y)\Pi_\tau(x,u,v,\lambda,\bar{t},\bar{s})\\
&<\sum_{\tau\in D}\xi_\tau(x,y)\left[\Pi_\tau(y,u,v,\lambda,\bar{t},\bar{s})-\frac{1}{2}\langle z,\nabla^2\Pi_\tau(y,u,v,\lambda,\bar{t},\bar{s})z\rangle\right],
\end{aligned}$$

and hence

$$\frac{1}{\alpha(x,y)}\bar{\gamma}(x,y)\left(e^{\alpha(x,y)\sum_{\tau\in D}\xi_\tau(x,y)[\Pi_\tau(x,u,v,\lambda,\bar{t},\bar{s})-\Pi_\tau(y,u,v,\lambda,\bar{t},\bar{s})+\frac{1}{2}\langle z,\nabla^2\Pi_\tau(y,u,v,\lambda,\bar{t},\bar{s})z\rangle]}-1\right)<0,$$

which in view of (i) implies that

$$\frac{1}{\beta(x,y)}\Big\langle\sum_{i=1}^{p}u_i[\nabla f_i(y)-\lambda_i\nabla g_i(y)]+\sum_{\tau\in D}\Big[\sum_{m\in J_\tau}v_m\nabla G_{j_m}(y,t^m)+\sum_{m\in K_\tau}v_m\nabla H_{k_m}(y,s^m)\Big]+\Big\{\sum_{i=1}^{p}u_i[\nabla^2 f_i(y)-\lambda_i\nabla^2 g_i(y)]+\sum_{\tau\in D}\Big[\sum_{m\in J_\tau}v_m\nabla^2 G_{j_m}(y,t^m)+\sum_{m\in K_\tau}v_m\nabla^2 H_{k_m}(y,s^m)\Big]\Big\}z,e^{\beta(x,y)\eta(x,y)}-\mathbf{1}\Big\rangle<-\bar{\rho}(x,y)\|\theta(x,y)\|^2. \quad (2.28)$$

As shown in the proof of Theorem 2.11, for each $\tau\in E\backslash D$,

$$\Lambda_\tau(x,v,\bar{t},\bar{s})\leqq\Lambda_\tau(y,v,\bar{t},\bar{s})-\frac{1}{2}\langle z,\nabla^2\Lambda_\tau(y,v,\bar{t},\bar{s})z\rangle,$$

and so

$$\sum_{\tau\in E\backslash D}\pi_\tau(x,y)\Lambda_\tau(x,v,\bar{t},\bar{s})\leqq\sum_{\tau\in E\backslash D}\pi_\tau(x,y)\big[\Lambda_\tau(y,v,\bar{t},\bar{s})-\frac{1}{2}\langle z,\nabla^2\Lambda_\tau(y,v,\bar{t},\bar{s})z\rangle\big],$$

and

$$\frac{1}{\alpha(x,y)}\gamma(x,y)\left(e^{\alpha(x,y)\sum_{\tau\in E\backslash D}\pi_\tau(x,y)[\Lambda_\tau(x,v,\bar{t},\bar{s})-\Lambda_\tau(y,v,\bar{t},\bar{s})+\frac{1}{2}\langle z,\nabla^2\Lambda_\tau(y,v,\bar{t},\bar{s})z\rangle]}-1\right)\leqq 0,$$

which in view of (ii) implies that

$$\frac{1}{\beta(x,y)}\Big\langle \sum_{\tau\in E\setminus D} \Big\{\sum_{m\in J_\tau} v_m\nabla G_{j_m}(y,t^m)+\sum_{m\in K_\tau} v_m\nabla H_{k_m}(y,s^m)+ \Big[\sum_{m\in J_\tau} v_m\nabla^2 G_{j_m}(y,t^m)+\sum_{m\in K_\tau} v_m\nabla^2 H_{k_m}(y,s^m)\Big]z\Big\}, e^{\beta(x,y)\eta(x,y)}-\mathbf{1}\Big\rangle \leqq -\tilde{\rho}(x,y)\|\theta(x,y)\|^2. \quad (2.29)$$

Now combining (31) and (32) and using (iii), we see that

$$\frac{1}{\beta(x,y)}\Big\langle \sum_{i=1}^{p} u_i[\nabla f_i(y)-\lambda_i\nabla g_i(y)]+\sum_{m=1}^{\nu_0} v_m\nabla G_{j_m}(y,t^m) + \sum_{m=\nu_0+1}^{\nu} v_m\nabla H_{k_m}(y,s^m)+\Big\{\sum_{i=1}^{p} u_i[\nabla^2 f_i(y)-\lambda_i\nabla^2 g_i(y)] + \sum_{m=1}^{\nu_0} v_m\nabla^2 G_{j_m}(y,t^m) + \sum_{m=\nu_0+1}^{\nu} v_m\nabla^2 H_{k_m}(y,s^m)\Big\}z, e^{\beta(x,y)\eta(x,y)}-\mathbf{1}\Big\rangle < -[\bar{\rho}(x,y)+\tilde{\rho}(x,y)]\|\theta(x,y)\|^2 \leqq 0,$$

which contradicts (24). Therefore, $\varphi(x)\nleq\lambda$.

(b) and (c) : The proofs are similar to that of part (a).

□

**Theorem 2.18.** (Strong Duality) Let $x^*$ be a normal efficient solution of (P) and assume that for each feasible solution $(y,z,u,v,\lambda,\nu,\nu_0, J_{\nu_0}, K_{\nu\setminus\nu_0}, \bar{t},\bar{s})$ of (DIV), any one of the three sets of conditions specified in Theorem 2.17 is satisfied. Then there exist $u^*, v^*, \lambda^*$, $\nu^*, \nu_0^*, J_{\nu_0^*}, K_{\nu^*\setminus\nu_0^*}, \bar{t}^*$, and $\bar{s}^*$ such that $(x^*, z^*=0, u^*, v^*, \lambda^*, \nu^*, \nu_0^*, J_{\nu_0^*}, K_{\nu^*\setminus\nu_0^*}, \bar{t}^*, \bar{s}^*)$ is an efficient solution of (DIV) and $\varphi(x^*)=\lambda^*$.

*Proof.* The proof is similar to that of Theorem 2.9. □

**Theorem 2.19.** (Strict Converse Duality) Let $x^*$ and $\tilde{w}\equiv(\tilde{x},\tilde{z},\tilde{u},\tilde{v},\tilde{\lambda},\tilde{\nu},\tilde{\nu}_0, J_{\tilde{\nu}_0}, K_{\tilde{\nu}\setminus\tilde{\nu}_0},\bar{\tilde{t}},\bar{\tilde{s}})$ be arbitrary feasible solutions of (P) and (DIV), respectively, such that

$$\sum_{i=1}^{p}\tilde{u}_i[f_i(x^*)-\tilde{\lambda}_i g_i(x^*)]=0. \quad (2.30)$$

Furthermore, assume that any one of the following three sets of hypotheses is satisfied:

(a) The assumptions specified in part (a) of Theorem 2.17 are satisfied for the feasible solution $\tilde{w}$ of (DIV), the function $\big(\Pi_0(\cdot,\tilde{u},\tilde{v},\tilde{\lambda},\bar{\tilde{t}},\bar{\tilde{s}}),\ldots,\Pi_d(\cdot,\tilde{u},\tilde{v},\tilde{\lambda},\bar{\tilde{t}},\bar{\tilde{s}})\big)$ is strictly $HA(\alpha,\beta,\bar{\gamma},\xi,\eta,\bar{\rho},\theta)$-V-pseudosonvex at $\tilde{x}$, and $\xi_0=\xi_1=\ldots=\xi_d=\sigma$.

(b) The assumptions specified in part (b) of Theorem 2.17 are satisfied for the feasible solution $\tilde{w}$ of (DIV), the function $\big(\Pi_0(\cdot,\tilde{u},\tilde{v},\tilde{\lambda},\bar{\tilde{t}},\bar{\tilde{s}}),\ldots,\ \Pi_d(\cdot,\tilde{u},\ \tilde{v},\tilde{\lambda},\bar{\tilde{t}},\bar{\tilde{s}})\big)$ is $HA(\alpha,\beta,\bar{\gamma},\xi,\eta,\bar{\rho},\theta)$-V-quasisonvex at $\tilde{x}$, and $\xi_0=\xi_1=\ldots=\xi_d=\sigma$.

(c) The assumptions specified in part (c) of Theorem 2.17 are satisfied for the feasible solution $\tilde{w}$ of (DIV), the function $\big(\Pi_0(\cdot,\tilde{u},\tilde{v},\tilde{\lambda},\bar{\tilde{t}},\bar{\tilde{s}}),\ldots,\Pi_d(\cdot,\tilde{u},\ \tilde{v},\tilde{\lambda},\bar{\tilde{t}},\bar{\tilde{s}})\big)$ is $HA(\alpha,\beta,\bar{\gamma},\xi,\eta,\bar{\rho},\theta)$-V-quasisonvex at $\tilde{x}$, and $\xi_0=\xi_1=\ldots=\xi_d=\sigma$.

Then $\tilde{x}=x^*$ and $\varphi(x^*)=\tilde{\lambda}$.

*Proof.* (a) : Suppose to the contrary that $\tilde{x}\neq x^*$. It is clear that (24) can be expressed as follows:

$$\sum_{\tau\in D}\Big\{\sum_{i\in I_\tau}\tilde{u}_i[\nabla f_i(\tilde{x})-\tilde{\lambda}_i\nabla g_i(\tilde{x})]+\sum_{m\in J_\tau}\tilde{v}_m\nabla G_{j_m}(\tilde{x},\tilde{t}^m)+\sum_{m\in K_\tau}\tilde{v}_m\nabla H_{k_m}(\tilde{x},\tilde{s}^m)+$$
$$\Big\{\sum_{i\in I_\tau}\tilde{u}_i[\nabla^2 f_i(\tilde{x})-\tilde{\lambda}_i\nabla^2 g_i(\tilde{x})]+\sum_{m\in J_\tau}\tilde{v}_m\nabla^2 G_{j_m}(\tilde{x},\tilde{t}^m)+\sum_{m\in K_\tau}\tilde{v}_m\nabla^2 H_{k_m}(\tilde{x},\tilde{s}^m)\Big\}\tilde{z}\Big\}+$$
$$\sum_{\tau\in E\backslash D}\Big\{\sum_{m\in J_\tau}\tilde{v}_m\nabla G_{j_m}(\tilde{x},\tilde{t}^m)+\sum_{m\in K_\tau}\tilde{v}_m\nabla H_{k_m}(\tilde{x},\tilde{s}^m)+\Big[\sum_{m\in J_\tau}\tilde{v}_m\nabla^2 G_{j_m}(\tilde{x},\tilde{t}^m)$$
$$+\sum_{m\in K_\tau}\tilde{v}_m\nabla^2 H_{k_m}(\tilde{x},\tilde{s}^m)\Big]\tilde{z}\Big\}=0. \quad (2.31)$$

Since $x^*\in\mathbb{F}$, $\tilde{v}_m\geqq 0$, $m\in\underline{\nu}$, and (26) holds, it follows that for each $\tau\in E\backslash D$,

$$\begin{aligned}\Lambda_\tau(x^*,\tilde{v},\bar{\tilde{t}},\bar{\tilde{s}})&=\sum_{m\in J_\tau}\tilde{v}_m G_{j_m}(x^*,\tilde{t}^m)+\sum_{m\in K_\tau}\tilde{v}_m H_{k_m}(x^*,\tilde{s}^m)\\&\leqq 0\\&\leqq\Lambda_\tau(\tilde{x},\tilde{v},\bar{\tilde{t}},\bar{\tilde{s}})-\frac{1}{2}\langle z,\nabla^2\Lambda_\tau(\tilde{x},\tilde{v},\bar{\tilde{t}},\bar{\tilde{s}})\tilde{z}\rangle\ \text{(by (25))},\end{aligned}$$

and hence

$$\begin{aligned}&\sum_{\tau\in E\backslash D}\pi_\tau(x^*,\tilde{x})\Lambda_\tau(x^*,\tilde{v},\bar{\tilde{t}},\bar{\tilde{s}})\\&\leqq\sum_{\tau\in E\backslash D}\pi_\tau(x^*,\tilde{x})\big[\Lambda_\tau(\tilde{x},\tilde{v},\bar{\tilde{t}},\bar{\tilde{s}})-\frac{1}{2}\big\langle z,\nabla^2\Lambda_\tau(\tilde{x},\tilde{v},\bar{\tilde{t}},\bar{\tilde{s}})\tilde{z}\big\rangle\big].\end{aligned}$$

Using this inequality in conjunction with the properties of the functions $\alpha$, $\tilde{\gamma}$, and $\zeta\to e^\zeta$, we get

$$\begin{aligned}&\frac{1}{\alpha(x^*,\tilde{x})}\tilde{\gamma}(x^*,\tilde{x})\\&\Big(e^{\alpha(x^*,\tilde{x})\sum_{\tau\in E\backslash D}\pi_\tau(x^*,\tilde{x})[\Lambda_\tau(x^*,\tilde{v},\bar{\tilde{t}},\bar{\tilde{s}})-\Lambda_\tau(\tilde{x},\tilde{v},\bar{\tilde{t}},\bar{\tilde{s}})+\frac{1}{2}\langle z,\nabla^2\Lambda_\tau(\tilde{x},\tilde{v},\bar{\tilde{t}},\bar{\tilde{s}})\tilde{z}\rangle]}-1\Big)\leqq 0,\end{aligned}$$

which in view of (ii) implies that

$$\begin{aligned}&\frac{1}{\beta(x^*,\tilde{x})}\\&\Big\langle \sum_{\tau\in E\setminus D}\Big\{\sum_{m\in J_\tau}\tilde{v}_m\nabla G_{j_m}(\tilde{x},\tilde{t}^m)+\sum_{m\in K_\tau}\tilde{v}_m\nabla H_{k_m}(\tilde{x},\tilde{s}^m)\\&+\Big[\sum_{m\in J_\tau}\tilde{v}_m\nabla^2 G_{j_m}(\tilde{x},\tilde{t}^m)+\sum_{m\in K_\tau}\tilde{v}_m\nabla^2 H_{k_m}(\tilde{x},\tilde{s}^m)\Big]\tilde{z}\Big\},e^{\beta(x^*,\tilde{x})\eta(x^*,\tilde{x})}-\mathbf{1}\Big\rangle\\&\leqq -\tilde{\rho}(x^*,\tilde{x})\|\theta(x^*,\tilde{x})\|^2.\end{aligned} \tag{2.32}$$

Combining (31) and (32), and using (iii) we get

$$\begin{aligned}&\frac{1}{\beta(x,y)}\Big\langle\sum_{\tau\in D}\Big\{\sum_{i\in I_\tau}\tilde{u}_i[\nabla f_i(\tilde{x})-\tilde{\lambda}_i\nabla g_i(\tilde{x})]+\sum_{m\in J_\tau}\tilde{v}_m\nabla G_{j_m}(\tilde{x},\tilde{t}^m)\\&+\sum_{m\in K_\tau}\tilde{v}_m\nabla H_{k_m}(\tilde{x},\tilde{s}^m)+\Big\{\sum_{i\in I_\tau}\tilde{u}_i[\nabla^2 f_i(\tilde{x})-\tilde{\lambda}\nabla^2 g_i(\tilde{x})]+\sum_{m\in J_\tau}\tilde{v}_m\nabla^2 G_{j_m}(\tilde{x},\tilde{t}^m)\\&+\sum_{m\in K_\tau}\tilde{v}_m\nabla^2 H_{k_m}(\tilde{x},\tilde{s}^m)\Big\}\tilde{z}\Big\},e^{\beta(x^*,\tilde{x})\eta(x^*,\tilde{x})}-\mathbf{1}\Big\rangle\geqq\tilde{\rho}(x^*,\tilde{x})\|\theta(x^*,\tilde{x})\|^2\\&\geqq-\bar{\rho}(x^*,\tilde{x})\|\theta(x^*,\tilde{x})\|^2,\end{aligned} \tag{2.33}$$

which by virtue of our strict $HA(\alpha,\beta,\bar{\gamma},\xi,\eta,\bar{\rho},\theta)$-V-pseudosonvexity hypothesis implies that

$$\begin{aligned}&\frac{1}{\alpha(x^*,\tilde{x})}\bar{\gamma}(x^*,\tilde{x})\\&\Big(e^{\alpha(x^*,\tilde{x})\sigma(x^*,\tilde{x})\sum_{\tau\in D}[\Pi_\tau(x^*,\tilde{u},\tilde{v},\tilde{\lambda},\bar{\tilde{t}},\tilde{s})-\Pi_\tau(\tilde{x},\tilde{u},\tilde{v},\tilde{\lambda},\bar{\tilde{t}},\bar{s})+\frac{1}{2}\langle z,\nabla^2\Pi_\tau(\tilde{x},\tilde{u},\tilde{v},\tilde{\lambda},\bar{\tilde{t}},\bar{s})\tilde{z}\rangle]}-1\Big)\\&>0.\end{aligned}$$

Since $\bar{\gamma}(x^*,\tilde{x})>0$ and $\sigma(x^*,\tilde{x})>0$, we deduce from this inequality that

$$\begin{aligned}&\sum_{\tau\in D}\Pi_\tau(x^*,\tilde{u},\tilde{v},\tilde{\lambda},\bar{\tilde{t}},\bar{\tilde{s}})\\&>\sum_{\tau\in D}\Big[\Pi_\tau(\tilde{x},\tilde{u},\tilde{v},\tilde{\lambda},\bar{\tilde{t}},\bar{\tilde{s}})-\frac{1}{2}\langle z,\nabla^2\Pi_\tau(\tilde{x},\tilde{u},\tilde{v},\tilde{\lambda},\bar{\tilde{t}},\bar{\tilde{s}})\tilde{z}\rangle\Big]\geqq 0,\end{aligned}$$

where the second inequality follows from the dual feasibility of $\tilde{w}$ and (25). Therefore, we have

$$\begin{aligned}0&<\sum_{\tau\in D}\Pi_\tau(x^*,\tilde{u},\tilde{v},\tilde{\lambda},\bar{\tilde{t}},\bar{\tilde{s}})\\&=\sum_{\tau\in D}\sum_{i\in I_\tau}\tilde{u}_i[f_i(x^*)-\tilde{\lambda}_i g_i(x^*)]+\sum_{\tau\in D}\Big[\sum_{m\in J_\tau}\tilde{v}_m G_{j_m}(x^*,\tilde{t}^m)+\sum_{m\in K_\tau}\tilde{v}_m H_{k_m}(x^*,\tilde{s}^m)\Big]\\&\leqq\sum_{i=1}^{p}\tilde{u}_i[f_i(x^*)-\tilde{\lambda}_i g_i(x^*)]\ \text{(by the primal feasibility of } x^*\text{)},\end{aligned}$$

which contradicts (30). Therefore, we conclude that $\tilde{x}=x^*$ and $\varphi(x^*)=\tilde{\lambda}$.

(b) and (c) : The proofs are similar to that of part (a). □

The duality models $(DIV)$ and $(\tilde{D}IV)$ are two families of dual problems for $(P)$ whose members can easily be identified by appropriate choices of the partitioning sets $I_\mu$, $\mu \in D$, and $J_\tau$ and $K_\tau$, $\tau \in E$. These duality models and their special cases contain the semiinfinite analogues of various dual problems proposed and investigated previously for many types of nonlinear programming problems, as well as numerous other duality formulations.

# 7. General Remarks

In this chapter, we have considered eight second-order parametric duality models and established a fairly large number of duality theorems applying various generalized $HA(\alpha,\beta,\gamma,\xi,\eta,\rho,\theta)$-V-sonvexity hypotheses for a semiinfinite multiobjective fractional programming problem. It seems that all these results are very promising to the context of the promoting further research in area of semiinfinite programming and beyond. We have a strong feeling that all the duality results obtained here can be applied in a similar manner to each one of the seven problems designated as $(P1)-(P7)$ in [12], they collectively provide a truly vast number of second-order duality results for various classes of conventional as well as semiinfinite nonlinear programming problems. Furthermore, the approach and techniques employed in this chapter can be utilized to establish similar second-order duality results as well as first-order duality models for some other related classes of optimization problems. For example, one can adopt a similar approach to investigate the second-order duality aspects of the following class of semiinfinite continuous minmax fractional programming problems:

$$\underset{x\in\mathbb{F}}{\text{Minimize}}\ \max_{y\in Y} \frac{f(x,y)}{g(x,y)}.$$

# Bibliography


[1] I. Ahmad and Z. Husain, Second order $(F,\alpha,\rho,d)$-convexity and duality in multiobjective programming, *Inform. Sci.* **176** (2006), 3094 - 3103.

[2] T. Antczak, The notion of V-$r$-invexity in differentiable multiobjective programming, *J. Appl. Anal.* **11** (2005), 63 - 79.

[3] X. Chen, Sufficient conditions and duality for a class of multiobjective fractional programming problems with higher-order $(F,\alpha,\rho,d)$-convexity, *J. Appl. Math. Comput.* **28** (2008), 107 - 121.

[4] M. A. Hanson, Second order invexity and duality in mathematical programming, *Opsearch* **30** (1993), 313 - 320.

[5] O. L. Mangasarian, Second- and higher-order duality theorems in nonlinear programming, *J. Math. Anal. Appl.* **51** (1975), 607 - 620.

[6] B. Mond, Second order duality for nonlinear programs, *Opsearch* **11** (1974), 90 - 99.

[7] B. Mond and T. Weir, Generalized concavity and duality, in *Generalized Concavity in Optimization and Economics* (S. Schaible and W. T. Ziemba, eds.), Academic Press, New York, 1981, pp. 263 - 279.

[8] B. Mond and J. Zhang, Duality for multiobjective programming involving second-order V-invex functions, in *Proceedings of the Optimization Miniconference II* (B. M. Glover and V. Jeyakumar, eds.), University of New South Wales, Sydney, Australia, 1995, pp. 89 - 100.

[9] B. Mond and J. Zhang, Higher order invexity and duality in mathematical programming, in *Generalized Convexity, Generalized Monotonicity : Recent Results* (J. P. Crouzeix, *et al.*, eds.), Kluwer Academic Publishers, printed in the Netherlands, 1998, pp. 357 - 372.

[10] X. M. Yang, Generalized convex duality for multiobjective fractional programs, *Opsearch* **31** (1994), 155 - 163.

[11] G. J. Zalmai, Generalized second-order $(\mathscr{F},\beta,\phi,\rho,\theta)$-univex functions and parametric duality models in semiinfinite discrete minmax fractional programming, *Adv. Nonlin. Var. Ineq.* **15** (2012), 63 - 91.

[12] G. J. Zalmai, Hanson-Antczak-type generalized $(\alpha,\beta,\gamma,\xi,\eta,\rho,\theta)$-V-invex functions in semiinfinite multiobjective fractional programming, Part I: Sufficient efficiency conditions, *Advances in Nonlinear Variational Inequalities* **16** (1) (2013), 91 - 114.

[13] G. J. Zalmai, Hanson-Antczak-type generalized $(\alpha,\beta,\gamma,\xi,\eta,\rho,\theta)$-V-invex functions in semiinfinite multiobjective fractional programming, Part II: First-order parametric duality models, *Advances in Nonlinear Variational Inequalities* **16** (2) (2013), 61 - 90.

[14] G. J. Zalmai, Hanson-Antczak-type generalized $(\alpha,\beta,\gamma,\xi,\eta,\rho,\theta)$-V-invex functions in semiinfinite multiobjective fractional programming, Part III: Second-order parametric duality models, *Advances in Nonlinear Variational Inequalities* **16** (2) (2013), 91 - 126.

[15] G. J. Zalmai and Q. Zhang, Global semiparametric sufficient efficiency conditions for semiinfinite multiobjective fractional programming problems involving generalized $(\alpha,\eta,\rho)$-V-invex functions, *Southeast Asian Bull. Math.* **32** (2008), 573 - 599.

# Chapter 3

# New Parametric Duality Models

First, a significant generalization to the Hanson-Antczak-type $(\alpha, \beta, \gamma, \xi, \eta, \rho, \theta)$- V -invexity (Zalmai [14]), is introduced in the form Hanson-Antczak-type $(\alpha, \beta, \gamma, \xi, \eta, \omega, \rho, \theta)$-V-sonvexities, where $\omega : X \times X \to \mathbb{R}^n$ is a generalizing function. Then several higher-order parametric duality results using new duality models under various generalized Hanson-Antczak-type $(\alpha, \beta, \gamma, \xi, \eta, \omega, \rho, \theta)$- V-sonvexity constraints for a semiinfinite multiobjective fractional programming problem are established. The results thus obtained, applying generalized Hanson-Antczak-type $(\alpha, \beta, \gamma, \xi, \eta, \omega, \rho, \theta)$-V-sonvexities, are new and encompass most of the duality models and duality results in the literature.

## 1. Introduction

To wit, since the plan of the universe is the most perfect, there can be no doubt that all actions in the world can be determined from the observed phenomena and the causes with the aid of the method of maxima and minima.

- Leonhard Euler

True optimization is the revolutionary contribution of modern research to decision processes.

- George Bernhard Dantzig

The Euler "Calculus of Variations" of the year 1744 is one of the most beautiful mathematical works that has ever been written.

- Constantin Caratheodory

Euler truly did not sour his life with limiting value considerations, convergence and continuity criteria and he could not and did not wish to bother about the logical foundation of analysis, but rather he relied - not only on occasion unsuccessfully - on his astonishing certitude of instinct and algorithmic power.

- Emil Alfred Fellman

with different constraint structures which allow for more general HA($\alpha$, $\beta$, $\gamma$, $\xi$, $\eta$, $\omega$, $\rho$, $\theta$) -V-sonvexity conditions under which duality can be established, discussion of second-order duality where partitioning schemes and generalized parametric duality models, and several duality results under various generalized $HA(\alpha,\beta,\gamma,\xi,\eta,\omega,\rho,\theta)$-V-sonvexity assumptions.

Clearly, all the second-order duality results established in this chapter can easily be modified and restated for each one of the seven classes of nonlinear programming problems designated as $(P1)-(P7)$ in [12], which are special cases of $(P)$. In most cases these results can easily be altered and rephrased for each one of the seven problems.

## 2. Preliminaries

In this section we introduce some definitions of certain classes of generalized convex functions which will be needed in the sequel. For a brief discussion of the origins and predecessors of these functions as well as numerous relevant references, the reader is referred to [12].

We first recall the definitions of the second-order versions of $(\alpha,\eta)$-V-invex, $(\alpha,\eta)$-V-pseudoinvex, and $(\alpha,\eta)$-V-quasiinvex functions which were first defined in [8].

Let the function $F=(F_1,\ldots,F_p): X\to\mathbb{R}^p$ be twice differentiable at $x^*$.

**Definition 3.1.** The function $F$ is said to be second-order $(\alpha,\eta)$-V-invex at $x^*$ if there exist functions $\alpha_i : X\times X\to\mathbb{R}_+\backslash\{0\}\equiv(0,\infty)$, $i\in\underline{p}$, and $\eta : X\times X\to\mathbb{R}^n$ such that for each $x\in X$, $z\in\mathbb{R}^n$, and $i\in\underline{p}$,

$$F_i(x)-F_i(x^*)\geqq\langle\alpha_i(x,x^*)[\nabla F_i(x^*)+\nabla^2F_i(x^*)],\eta(x,x^*)\rangle-\frac{1}{2}\langle z,\nabla^2F_i(x^*)z\rangle.$$

**Definition 3.2.** The function $F$ is said to be second-order $(\beta,\eta)$-V-pseudoinvex at $x^*$ if there exist functions $\beta_i : X\times X\to\mathbb{R}_+\backslash\{0\}$, $i\in\underline{p}$, and $\eta : X\times X\to\mathbb{R}^n$ such that for each $x\in X$ and $z\in\mathbb{R}^n$,

$$\Big\langle\sum_{i=1}^{p}[\nabla F_i(x^*)+\nabla^2F_i(x^*)]z,\eta(x,x^*)\Big\rangle\geqq0\ \Rightarrow$$
$$\sum_{i=1}^{p}\beta_i(x,x^*)F_i(x)\geqq\sum_{i=1}^{p}\beta_i(x,x^*)\big[F_i(x^*)-\frac{1}{2}\langle z,\nabla^2F_i(x^*)z\rangle\big].$$

**Definition 3.3.** The function $F$ is said to be second-order $(\gamma,\eta)$-V-quasiinvex at $x^*$ if there exist functions $\gamma_i : X\times X\to\mathbb{R}_+\backslash\{0\}, i\in\underline{p}$, and $\eta : X\times X\to\mathbb{R}^n$ such that for each $x\in X$

and $z \in \mathbb{R}^n$,

$$\sum_{i=1}^{p} \gamma_i(x,x^*)F_i(x) \leqq \sum_{i=1}^{p} \gamma_i(x,x^*)[F_i(x^*) - \frac{1}{2}\langle z, \nabla^2 F_i(x^*)z\rangle] \Rightarrow$$

$$\left\langle \sum_{i=1}^{p} [\nabla F_i(x^*) + \nabla^2 F_i(x^*)], \eta(x,x^*) \right\rangle \leqq 0.$$

The following generalizations of the notions of second-order $(\alpha,\eta)$-V-invexity, $(\alpha,\eta)$-V-pseudoinvexity, and $(\alpha,\eta)$-V-quasiinvexity for vector-valued functions will be used in this chapter. We shall use the word sonvex for second-order invex and call the related generalized invex functions HA($\alpha$, $\beta$, $\gamma$, $\xi$, $\eta$, $\omega$, $\rho$, $\theta$)-V-sonvex, HA($\alpha$, $\beta$, $\gamma$, $\xi$, $\eta$, $\omega$, $\rho$, $\theta$)-V-pseudosonvex, and HA($\alpha$, $\beta$, $\gamma$, $\xi$, $\eta$, $\omega$, $\rho$, $\theta$) -V-quasisonvex functions. They are the second-order counterparts of the classes of generalized $(\alpha,\beta,\gamma,\xi,\eta,\rho,\theta)$-V-invex functions defined in [12].

**Definition 3.4.** The function $F$ is said to be (strictly) $HA(\alpha,\beta,\gamma,\xi,\eta,\omega,\rho,\theta)$-V-sonvex at $x^* \in X$ if there exist functions $\alpha : X \times X \to \mathbb{R}$, $\beta : X \times X \to \mathbb{R}$, $\gamma_i : X \times X \to \mathbb{R}_+$, $\xi_i : X \times X \to \mathbb{R}_+$, $i \in \underline{p}$, $\eta, \omega : X \times X \to \mathbb{R}^n$, $\rho_i : X \times X \to \mathbb{R}$, $i \in \underline{p}$, and $\theta : X \times X \to \mathbb{R}^n$ such that for all $x \in X(x \neq x^*)$, $z \in \mathbb{R}^n$, and $i \in \underline{p}$,

$$\frac{1}{\alpha(x,x^*)}\gamma_i(x,x^*)\left(e^{\alpha(x,x^*)[F_i(x)-F_i(x^*)+\frac{1}{2}\langle\omega(x,x^*),\nabla^2 F_i(x^*)z\rangle]} - 1\right)(>) \geqq$$
$$\frac{1}{\beta(x,x^*)}\langle \xi_i(x,x^*)[\nabla F_i(x^*) + \nabla^2 F_i(x^*)z], e^{\beta(x,x^*)\eta(x,x^*)} - \mathbf{1}\rangle$$
$$+ \rho_i(x,x^*)\|\theta(x,x^*)\|^2$$
$$\text{if } \alpha(x,x^*) \neq 0 \text{ and } \beta(x,x^*) \neq 0 \text{ for all } x \in X,$$

$$\frac{1}{\alpha(x,x^*)}\gamma_i(x,x^*)\left(e^{\alpha(x,x^*)[F_i(x)-F_i(x^*)+\frac{1}{2}\langle\omega(x,x^*),\nabla^2 F_i(x^*)z\rangle]} - 1\right)(>) \geqq$$
$$\langle \xi_i(x,x^*)[\nabla F_i(x^*) + \nabla^2 F_i(x^*)z], \eta(x,x^*)\rangle + \rho_i(x,x^*)\|\theta(x,x^*)\|^2$$
$$\text{if } \alpha(x,x^*) \neq 0 \text{ and } \beta(x,x^*) = 0 \text{ for all } x \in X,$$

$$\gamma_i(x,x^*)[F_i(x) - F_i(x^*) + \frac{1}{2}\langle\omega(x,x^*), \nabla^2 F_i(x^*)z\rangle](>) \geqq$$
$$\frac{1}{\beta(x,x^*)}\langle \xi_i(x,x^*)[\nabla F_i(x^*) + \nabla^2 F_i(x^*)z], e^{\beta(x,x^*)\eta(x,x^*)} - \mathbf{1}\rangle$$
$$+ \rho_i(x,x^*)\|\theta(x,x^*)\|^2$$
$$\text{if } \alpha(x,x^*) = 0 \text{ and } \beta(x,x^*) \neq 0 \text{ for all } x \in X,$$

$$\gamma_i(x,x^*)[F_i(x) - F_i(x^*) + \frac{1}{2}\langle\omega(x,x^*), \nabla^2 F_i(x^*)z\rangle](>) \geqq$$
$$\langle \xi_i(x,x^*)[\nabla F_i(x^*) + \nabla^2 F_i(x^*)z], \eta(x,x^*)\rangle + \rho_i(x,x^*)\|\theta(x,x^*)\|^2$$
$$\text{if } \alpha(x,x^*) = 0 \text{ and } \beta(x,x^*) = 0 \text{ for all } x \in X,$$

where

$$\left(e^{\beta(x,x^*)\eta(x,x^*)} - \mathbf{1}\right) \equiv \left(e^{\beta(x,x^*)\eta_1(x,x^*)} - 1, \ldots, e^{\beta(x,x^*)\eta_n(x,x^*)} - 1\right)$$

and $\|\cdot\|$ is a norm on $\mathbb{R}^n$.

The function $F$ is said to be (strictly) HA($\alpha$, $\beta$, $\gamma$, $\xi$, $\eta$, $\omega$, $\rho$, $\theta$)-V-sonvex on $X$ if it is (strictly) HA($\alpha$, $\beta$, $\gamma$, $\xi$, $\eta$, $\omega$, $\rho$, $\theta$)-V-sonvex at each point $x^* \in X$.

**Definition 3.5.** The function $F$ is said to be (strictly) HA($\alpha$, $\beta$, $\gamma$, $\xi$, $\eta$, $\omega$, $\rho$, $\theta$)-V-pseudosonvex at $x^* \in X$ if there exist functions $\alpha : X \times X \to \mathbb{R}$, $\beta : X \times X \to \mathbb{R}$, $\gamma : X \times X \to \mathbb{R}_+$, $\xi_i : X \times X \to \mathbb{R}_+$, $i \in \underline{p}$, $\eta, \omega : X \times X \to \mathbb{R}^n$, $\rho : X \times X \to \mathbb{R}$, and $\theta : X \times X \to \mathbb{R}^n$ such that for all $x \in X\,(x \neq x^*)$ and $z \in \mathbb{R}^n$,

$$\frac{1}{\beta(x,x^*)}\left\langle \sum_{i=1}^{p} \xi_i(x,x^*)[\nabla F_i(x^*) + \nabla^2 F_i(x^*)z], e^{\beta(x,x^*)\eta(x,x^*)} - \mathbf{1}\right\rangle$$
$$\geqq -\rho(x,x^*)\|\theta(x,x^*)\|^2$$
$$\Rightarrow \frac{1}{\alpha(x,x^*)}\gamma(x,x^*)$$
$$\left(e^{\alpha(x,x^*)\sum_{i=1}^{p}[F_i(x)-F_i(x^*)+\frac{1}{2}\langle\omega(x,x^*),\nabla^2 F_i(x^*)z\rangle]} - 1\right)(>) \geqq 0$$
$$\text{if } \alpha(x,x^*) \neq 0 \text{ and } \beta(x,x^*) \neq 0 \text{ for all } x \in X,$$

$$\left\langle \sum_{i=1}^{p} \xi_i(x,x^*)[\nabla F_i(x^*) + \nabla^2 F_i(x^*)z], \eta(x,x^*)\right\rangle \geqq -\rho(x,x^*)\|\theta(x,x^*)\|^2$$
$$\Rightarrow \frac{1}{\alpha(x,x^*)}\gamma(x,x^*)\left(e^{\alpha(x,x^*)\sum_{i=1}^{p}[F_i(x)-F_i(x^*)+\frac{1}{2}\langle\omega(x,x^*),\nabla^2 F_i(x^*)z\rangle]} - 1\right)(>) \geqq 0$$
$$\text{if } \alpha(x,x^*) \neq 0 \text{ and } \beta(x,x^*) = 0 \text{ for all } x \in X,$$

$$\frac{1}{\beta(x,x^*)}\left\langle \sum_{i=1}^{p} \xi_i(x,x^*)[\nabla F_i(x^*) + \nabla^2 F_i(x^*)z], e^{\beta(x,x^*)\eta(x,x^*)} - \mathbf{1}\right\rangle$$
$$\geqq -\rho(x,x^*)\|\theta(x,x^*)\|^2$$
$$\Rightarrow \gamma(x,x^*)\sum_{i=1}^{p}[F_i(x) - F_i(x^*) + \frac{1}{2}\langle\omega(x,x^*),\nabla^2 F_i(x^*)z\rangle](>) \geqq 0$$
$$\text{if } \alpha(x,x^*) = 0 \text{ and } \beta(x,x^*) \neq 0 \text{ for all } x \in X,$$

$$\left\langle \sum_{i=1}^{p} \xi_i(x,x^*)[\nabla F_i(x^*) + \nabla^2 F_i(x^*)z], \eta(x,x^*)\right\rangle \geqq -\rho(x,x^*)\|\theta(x,x^*)\|^2$$
$$\Rightarrow \gamma(x,x^*)\sum_{i=1}^{p}[F_i(x) - F_i(x^*) + \frac{1}{2}\langle\omega(x,x^*),\nabla^2 F_i(x^*)z\rangle](>) \geqq 0$$
$$\text{if } \alpha(x,x^*) = 0 \text{ and } \beta(x,x^*) = 0 \text{ for all } x \in X.$$

The function $F$ is said to be (strictly) HA($\alpha$, $\beta$, $\gamma$, $\xi$, $\eta$, $\omega$, $\rho$, $\theta$)-V-pseudosonvex on $X$ if it is (strictly) $HA(\alpha,\beta,\gamma,\xi,\eta,\omega,\rho,\theta)$-V-pseudosonvex at each point $x^* \in X$.

**Definition 3.6.** The function $F$ is said to be (prestrictly) HA($\alpha$, $\beta$, $\gamma$, $\xi$, $\eta$, $\omega$, $\rho$, $\theta$)-V-quasisonvex at $x^* \in X$ if there exist functions $\alpha : X \times X \to \mathbb{R}$, $\beta : X \times X \to \mathbb{R}$, $\gamma : X \times X \to \mathbb{R}_+$, $\xi_i : X \times X \to \mathbb{R}_+$, $i \in \underline{p}$, $\eta, \omega : X \times X \to \mathbb{R}^n$, $\rho : X \times X \to \mathbb{R}$, and $\theta : X \times X \to \mathbb{R}^n$ such that for all $x \in X$ and $z \in \mathbb{R}^n$,

$$\frac{1}{\alpha(x,x^*)}\gamma(x,x^*)\left(e^{\alpha(x,x^*)\sum_{i=1}^{p}[F_i(x)-F_i(x^*)+\frac{1}{2}\langle\omega(x,x^*),\nabla^2F_i(x^*)z\rangle]}-1\right)(<)\leqq 0$$
$$\Rightarrow \frac{1}{\beta(x,x^*)}\left\langle\sum_{i=1}^{p}\xi_i(x,x^*)[\nabla F_i(x^*)+\nabla^2F_i(x^*)z], e^{\beta(x,x^*)\eta(x,x^*)}-\mathbf{1}\right\rangle$$
$$\leqq -\rho(x,x^*)\|\theta(x,x^*)\|^2$$
$$\text{if } \alpha(x,x^*)\neq 0 \text{ and } \beta(x,x^*)\neq 0 \text{ for all } x\in X,$$

$$\frac{1}{\alpha(x,x^*)}\gamma(x,x^*)\left(e^{\alpha(x,x^*)\sum_{i=1}^{p}[F_i(x)-F_i(x^*)+\frac{1}{2}\langle\omega(x,x^*),\nabla^2F_i(x^*)z\rangle]}-1\right)(<)\leqq 0$$
$$\Rightarrow \left\langle\sum_{i=1}^{p}\xi_i(x,x^*)[\nabla F_i(x^*)+\nabla^2F_i(x^*)z], \eta(x,x^*)\right\rangle\leqq -\rho(x,x^*)\|\theta(x,x^*)\|^2$$
$$\text{if } \alpha(x,x^*)\neq 0 \text{ and } \beta(x,x^*)= 0 \text{ for all } x\in X,$$

$$\gamma(x,x^*)\sum_{i=1}^{p}[F_i(x)-F_i(x^*)+\frac{1}{2}\langle\omega(x,x^*),\nabla^2F_i(x^*)z\rangle](<)\leqq 0$$
$$\Rightarrow \frac{1}{\beta(x,x^*)}\left\langle\sum_{i=1}^{p}\xi_i(x,x^*)[\nabla F_i(x^*)+\nabla^2F_i(x^*)z], e^{\beta(x,x^*)\eta(x,x^*)}-\mathbf{1}\right\rangle$$
$$\leqq -\rho(x,x^*)\|\theta(x,x^*)\|^2$$
$$\text{if } \alpha(x,x^*)= 0 \text{ and } \beta(x,x^*)\neq 0 \text{ for all } x\in X,$$

$$\gamma(x,x^*)\sum_{i=1}^{p}[F_i(x)-F_i(x^*)+\frac{1}{2}\langle\omega(x,x^*),\nabla^2F_i(x^*)z\rangle](<)\leqq 0$$
$$\Rightarrow \left\langle\sum_{i=1}^{p}\xi_i(x,x^*)[\nabla F_i(x^*)+\nabla^2F_i(x^*)z], \eta(x,x^*)\right\rangle\leqq -\rho(x,x^*)\|\theta(x,x^*)\|^2$$
$$\text{if } \alpha(x,x^*)= 0 \text{ and } \beta(x,x^*)= 0 \text{ for all } x\in X.$$

The function $F$ is said to be (prestrictly) $HA(\alpha,\beta,\gamma,\xi,\eta,\omega,\rho,\theta)$-V-quasisonvex on $X$ if it is (prestrictly) $HA(\alpha,\beta,\gamma,\xi,\eta,\omega,\rho,\theta)$-V-quasisonvex at each point $x^* \in X$.

**Definition 3.7.** [14] The function $F$ is said to be (strictly) $HA(\alpha,\beta,\gamma,\xi,\eta,\rho,\theta)$-V-sonvex at $x^* \in X$ if there exist functions $\alpha : X \times X \to \mathbb{R}$, $\beta : X \times X \to \mathbb{R}$, $\gamma_i : X \times X \to \mathbb{R}_+$, $\xi_i$ :

$X \times X \to \mathbb{R}_+$, $i \in \underline{p}$, $\eta : X \times X \to \mathbb{R}^n$, $\rho_i : X \times X \to \mathbb{R}$, $i \in \underline{p}$, and $\theta : X \times X \to \mathbb{R}^n$ such that for all $x \in X(x \neq x^*)$, $z \in \mathbb{R}^n$, and $i \in \underline{p}$,

$$\begin{aligned}
\frac{1}{\alpha(x,x^*)}\gamma_i(x,x^*)\Big(e^{\alpha(x,x^*)[F_i(x)-F_i(x^*)+\frac{1}{2}\langle z,\nabla^2 F_i(x^*)z\rangle]}-1\Big)(>) \geqq \\
\frac{1}{\beta(x,x^*)}\langle \xi_i(x,x^*)[\nabla F_i(x^*)+\nabla^2 F_i(x^*)z], e^{\beta(x,x^*)\eta(x,x^*)}-\mathbf{1}\rangle \\
+\rho_i(x,x^*)\|\theta(x,x^*)\|^2 \\
\text{if } \alpha(x,x^*) \neq 0 \text{ and } \beta(x,x^*) \neq 0 \text{ for all } x \in X,
\end{aligned}$$

$$\begin{aligned}
\frac{1}{\alpha(x,x^*)}\gamma_i(x,x^*)\Big(e^{\alpha(x,x^*)[F_i(x)-F_i(x^*)+\frac{1}{2}\langle z,\nabla^2 F_i(x^*)z\rangle]}-1\Big)(>) \geqq \\
\big\langle \xi_i(x,x^*)[\nabla F_i(x^*)+\nabla^2 F_i(x^*)z], \eta(x,x^*)\big\rangle + \rho_i(x,x^*)\|\theta(x,x^*)\|^2 \\
\text{if } \alpha(x,x^*) \neq 0 \text{ and } \beta(x,x^*) = 0 \text{ for all } x \in X,
\end{aligned}$$

$$\begin{aligned}
\gamma_i(x,x^*)[F_i(x)-F_i(x^*)+\frac{1}{2}\langle z,\nabla^2 F_i(x^*)z\rangle](>) \geqq \\
\frac{1}{\beta(x,x^*)}\langle \xi_i(x,x^*)[\nabla F_i(x^*)+\nabla^2 F_i(x^*)z], e^{\beta(x,x^*)\eta(x,x^*)}-\mathbf{1}\rangle \\
+\rho_i(x,x^*)\|\theta(x,x^*)\|^2 \\
\text{if } \alpha(x,x^*) = 0 \text{ and } \beta(x,x^*) \neq 0 \text{ for all } x \in X,
\end{aligned}$$

$$\begin{aligned}
\gamma_i(x,x^*)[F_i(x)-F_i(x^*)+\frac{1}{2}\langle z,\nabla^2 F_i(x^*)z\rangle](>) \geqq \\
\big\langle \xi_i(x,x^*)[\nabla F_i(x^*)+\nabla^2 F_i(x^*)z], \eta(x,x^*)\big\rangle + \rho_i(x,x^*)\|\theta(x,x^*)\|^2 \\
\text{if } \alpha(x,x^*) = 0 \text{ and } \beta(x,x^*) = 0 \text{ for all } x \in X,
\end{aligned}$$

where

$$\Big(e^{\beta(x,x^*)\eta(x,x^*)}-\mathbf{1}\Big) \equiv \Big(e^{\beta(x,x^*)\eta_1(x,x^*)}-1,\ldots,e^{\beta(x,x^*)\eta_n(x,x^*)}-1\Big)$$

and $\|\cdot\|$ is a norm on $\mathbb{R}^n$.

The function $F$ is said to be (strictly) $HA(\alpha,\beta,\gamma,\xi,\eta,\rho,\theta)$-V-sonvex on $X$ if it is (strictly) $HA(\alpha,\beta,\gamma,\xi,\eta,\rho,\theta)$-V-sonvex at each point $x^* \in X$.

**Definition 3.8.** [14] The function $F$ is said to be (strictly) $HA(\alpha,\beta,\gamma,\xi,\eta,\rho,\theta)$-V-pseudosonvex at $x^* \in X$ if there exist functions $\alpha : X \times X \to \mathbb{R}$, $\beta : X \times X \to \mathbb{R}$, $\gamma : X \times X \to \mathbb{R}_+$, $\xi_i : X \times X \to \mathbb{R}_+$, $i \in \underline{p}$, $\eta : X \times X \to \mathbb{R}^n$, $\rho : X \times X \to \mathbb{R}$, and $\theta : X \times X \to \mathbb{R}^n$ such that for all $x \in X(x \neq x^*)$ and $z \in \mathbb{R}^n$,

$$\frac{1}{\beta(x,x^*)}\Big\langle \sum_{i=1}^{p} \xi_i(x,x^*)[\nabla F_i(x^*)+\nabla^2 F_i(x^*)z], e^{\beta(x,x^*)\eta(x,x^*)}-\mathbf{1}\Big\rangle$$
$$\geqq -\rho(x,x^*)\|\theta(x,x^*)\|^2$$
$$\Rightarrow \quad \frac{1}{\alpha(x,x^*)}\gamma(x,x^*)\Big(e^{\alpha(x,x^*)\sum_{i=1}^{p}[F_i(x)-F_i(x^*)+\frac{1}{2}\langle z,\nabla^2 F_i(x^*)z\rangle]}-1\Big)(>)\geqq 0$$
$$\text{if } \alpha(x,x^*)\neq 0 \text{ and } \beta(x,x^*)\neq 0 \text{ for all } x\in X,$$

$$\Big\langle \sum_{i=1}^{p} \xi_i(x,x^*)[\nabla F_i(x^*)+\nabla^2 F_i(x^*)z], \eta(x,x^*)\Big\rangle \geqq -\rho(x,x^*)\|\theta(x,x^*)\|^2$$
$$\Rightarrow \quad \frac{1}{\alpha(x,x^*)}\gamma(x,x^*)\Big(e^{\alpha(x,x^*)\sum_{i=1}^{p}[F_i(x)-F_i(x^*)+\frac{1}{2}\langle z,\nabla^2 F_i(x^*)z\rangle]}-1\Big)(>)\geqq 0$$
$$\text{if } \alpha(x,x^*)\neq 0 \text{ and } \beta(x,x^*)=0 \text{ for all } x\in X,$$

$$\frac{1}{\beta(x,x^*)}\Big\langle \sum_{i=1}^{p} \xi_i(x,x^*)[\nabla F_i(x^*)+\nabla^2 F_i(x^*)z], e^{\beta(x,x^*)\eta(x,x^*)}-\mathbf{1}\Big\rangle$$
$$\geqq -\rho(x,x^*)\|\theta(x,x^*)\|^2$$
$$\Rightarrow \quad \gamma(x,x^*)\sum_{i=1}^{p}[F_i(x)-F_i(x^*)+\frac{1}{2}\langle z,\nabla^2 F_i(x^*)z\rangle](>)\geqq 0$$
$$\text{if } \alpha(x,x^*)=0 \text{ and } \beta(x,x^*)\neq 0 \text{ for all } x\in X,$$

$$\Big\langle \sum_{i=1}^{p} \xi_i(x,x^*)[\nabla F_i(x^*)+\nabla^2 F_i(x^*)z], \eta(x,x^*)\Big\rangle \geqq \quad \rho(x,x^*)\|\theta(x,x^*)\|^2$$
$$\Rightarrow \quad \gamma(x,x^*)\sum_{i=1}^{p}[F_i(x)-F_i(x^*)+\frac{1}{2}\langle z,\nabla^2 F_i(x^*)z\rangle](>)\geqq 0$$
$$\text{if } \alpha(x,x^*)=0 \text{ and } \beta(x,x^*)=0 \text{ for all } x\in X.$$

The function $F$ is said to be (strictly) HA($\alpha$, $\beta$, $\gamma$, $\xi$, $\eta$, $\rho$, $\theta$)-V-pseudosonvex on $X$ if it is (strictly) $HA(\alpha,\beta,\gamma,\xi,\eta,\rho,\theta)$-V-pseudosonvex at each point $x^* \in X$.

**Definition 3.9.** [14] The function $F$ is said to be (prestrictly) HA($\alpha$, $\beta$, $\gamma$, $\xi$, $\eta$, $\rho$, $\theta$)-V-quasisonvex at $x^* \in X$ if there exist functions $\alpha : X\times X \to \mathbb{R}$, $\beta : X\times X \to \mathbb{R}$, $\gamma : X\times X \to \mathbb{R}_+$, $\xi_i : X\times X \to \mathbb{R}_+$, $i \in \underline{p}$, $\eta : X\times X \to \mathbb{R}^n$, $\rho : X\times X \to \mathbb{R}$, and $\theta : X\times X \to \mathbb{R}^n$ such that for all $x \in X$ and $z \in \mathbb{R}^n$,

$$\frac{1}{\alpha(x,x^*)}\gamma(x,x^*)\left(e^{\alpha(x,x^*)\sum_{i=1}^{p}[F_i(x)-F_i(x^*)+\frac{1}{2}\langle z,\nabla^2 F_i(x^*)z\rangle]}-1\right)(<)\leqq 0$$
$$\Rightarrow \frac{1}{\beta(x,x^*)}\left\langle \sum_{i=1}^{p}\xi_i(x,x^*)[\nabla F_i(x^*)+\nabla^2 F_i(x^*)z], e^{\beta(x,x^*)\eta(x,x^*)}-\mathbf{1}\right\rangle$$
$$\leqq -\rho(x,x^*)\|\theta(x,x^*)\|^2$$
$$\text{if } \alpha(x,x^*)\neq 0 \text{ and } \beta(x,x^*)\neq 0 \text{ for all } x\in X,$$

$$\frac{1}{\alpha(x,x^*)}\gamma(x,x^*)\left(e^{\alpha(x,x^*)\sum_{i=1}^{p}[F_i(x)-F_i(x^*)+\frac{1}{2}\langle z,\nabla^2 F_i(x^*)z\rangle]}-1\right)(<)\leqq 0$$
$$\Rightarrow \left\langle \sum_{i=1}^{p}\xi_i(x,x^*)[\nabla F_i(x^*)+\nabla^2 F_i(x^*)z], \eta(x,x^*)\right\rangle \leqq -\rho(x,x^*)\|\theta(x,x^*)\|^2$$
$$\text{if } \alpha(x,x^*)\neq 0 \text{ and } \beta(x,x^*)= 0 \text{ for all } x\in X,$$

$$\gamma(x,x^*)\sum_{i=1}^{p}[F_i(x)-F_i(x^*)+\frac{1}{2}\langle z,\nabla^2 F_i(x^*)z\rangle](<)\leqq 0$$
$$\Rightarrow \frac{1}{\beta(x,x^*)}\left\langle \sum_{i=1}^{p}\xi_i(x,x^*)[\nabla F_i(x^*)+\nabla^2 F_i(x^*)z], e^{\beta(x,x^*)\eta(x,x^*)}-\mathbf{1}\right\rangle$$
$$\leqq -\rho(x,x^*)\|\theta(x,x^*)\|^2$$
$$\text{if } \alpha(x,x^*)= 0 \text{ and } \beta(x,x^*)\neq 0 \text{ for all } x\in X,$$

$$\gamma(x,x^*)\sum_{i=1}^{p}[F_i(x)-F_i(x^*)+\frac{1}{2}\langle z,\nabla^2 F_i(x^*)z\rangle](<)\leqq 0$$
$$\Rightarrow \left\langle \sum_{i=1}^{p}\xi_i(x,x^*)[\nabla F_i(x^*)+\nabla^2 F_i(x^*)z], \eta(x,x^*)\right\rangle \leqq -\rho(x,x^*)\|\theta(x,x^*)\|^2$$
$$\text{if } \alpha(x,x^*)= 0 \text{ and } \beta(x,x^*)= 0 \text{ for all } x\in X.$$

The function $F$ is said to be (prestrictly) HA($\alpha$, $\beta$, $\gamma$, $\xi$, $\eta$, $\rho$, $\theta$)-V-quasisonvex on $X$ if it is (prestrictly) $HA(\alpha,\beta,\gamma,\xi,\eta,\rho,\theta)$-V-quasisonvex at each point $x^* \in X$.

**Definition 3.10.** The function $F$ is said to be first-order (strictly) HA($\alpha,\beta,\gamma,\xi,\eta,\rho,\theta$)-V-invex at $x^* \in X$ if there exist functions $\alpha : X\times X \to \mathbb{R}$, $\beta : X\times X \to \mathbb{R}$, $\gamma_i : X\times X \to \mathbb{R}_+$, $\xi_i : X\times X \to \mathbb{R}_+\backslash\{0\}$, $i \in \underline{p}$, $\eta : X\times X \to \mathbb{R}^n$, $\rho_i : X\times X \to \mathbb{R}$, $i \in \underline{p}$, and $\theta : X\times X \to \mathbb{R}^n$ such that for all $x \in X (x \neq x^*)$ and $i \in \underline{p}$,

$$\frac{1}{\alpha(x,x^*)}\gamma_i(x,x^*)\left(e^{\alpha(x,x^*)[F_i(x)-F_i(x^*)]}-1\right)(>)$$
$$\geqq \frac{1}{\beta(x,x^*)}\langle \xi_i(x,x^*)\nabla F_i(x^*), e^{\beta(x,x^*)\eta(x,x^*)}-\mathbf{1}\rangle$$
$$+\rho_i(x,x^*)\|\theta(x,x^*)\|^2 \quad \text{if } \alpha(x,x^*)\neq 0 \text{ and } \beta(x,x^*)\neq 0 \text{ for all } x\in X,$$

$$\frac{1}{\alpha(x,x^*)}\gamma_i(x,x^*)\left(e^{\alpha(x,x^*)[F_i(x)-F_i(x^*)]}-1\right)(>)\geqq\langle\xi_i(x,x^*)\nabla F_i(x^*),\eta(x,x^*)\rangle$$
$$+\rho_i(x,x^*)\|\theta(x,x^*)\|^2 \text{ if } \alpha(x,x^*)\neq 0 \text{ and } \beta(x,x^*)=0 \text{ for all } x\in X,$$

$$\gamma_i(x,x^*)[F_i(x)-F_i(x^*)](>)\geqq\frac{1}{\beta(x,x^*)}\langle\xi_i(x,x^*)\nabla F_i(x^*),e^{\beta(x,x^*)\eta(x,x^*)}-\mathbf{1}\rangle$$
$$+\rho_i(x,x^*)\|\theta(x,x^*)\|^2 \text{ if } \alpha(x,x^*)=0 \text{ and } \beta(x,x^*)\neq 0 \text{ for all } x\in X,$$

$$\gamma_i(x,x^*)[F_i(x)-F_i(x^*)](>)\geqq\langle\xi_i(x,x^*)\nabla F_i(x^*),\eta(x,x^*)\rangle+\rho_i(x,x^*)\|\theta(x,x^*)\|^2$$
$$\text{if } \alpha(x,x^*)=0 \text{ and } \beta(x,x^*)=0 \text{ for all } x\in X,$$

where $\|\cdot\|$ is a norm on $\mathbb{R}^n$ and

$$\left(e^{\beta(x,x^*)\eta(x,x^*)}-\mathbf{1}\right)\equiv\left(e^{\beta(x,x^*)\eta_1(x,x^*)}-1,\ldots,e^{\beta(x,x^*)\eta_n(x,x^*)}-1\right).$$

The function $F$ is said to be first-order (strictly) $HA(\alpha,\beta,\gamma,\xi,\eta,\rho,\theta)$-V-invex on $X$ if it is (strictly) $HA(\alpha,\beta,\gamma,\xi,\eta,\rho,\theta)$-V-invex at each point $x^*\in X$.

**Definition 3.11.** The function $F$ is said to be first-order (strictly) HA($\alpha$, $\beta$, $\gamma$, $\xi$, $\eta$, $\rho$, $\theta$)-V-pseudoinvex at $x^*\in X$ if there exist functions $\alpha: X\times X\to\mathbb{R}$, $\beta: X\times X\to\mathbb{R}$, $\gamma: X\times X\to\mathbb{R}_+$, $\xi_i: X\times X\to\mathbb{R}_+\backslash\{0\}$, $i\in\underline{p}$, $\eta: X\times X\to\mathbb{R}^n$, $\rho: X\times X\to\mathbb{R}$, and $\theta: X\times X\to\mathbb{R}^n$ such that for all $x\in X(x\neq x^*)$,

$$\frac{1}{\beta(x,x^*)}\left\langle\sum_{i=1}^{p}\nabla F_i(x^*),e^{\beta(x,x^*)\eta(x,x^*)}-\mathbf{1}\right\rangle\geqq-\rho(x,x^*)\|\theta(x,x^*)\|^2$$
$$\Rightarrow\ \frac{1}{\alpha(x,x^*)}\gamma(x,x^*)\left(e^{\alpha(x,x^*)\sum_{i=1}^{p}\xi_i(x,x^*)[F_i(x)-F_i(x^*)]}-1\right)(>)\geqq 0$$
$$\text{if } \alpha(x,x^*)\neq 0 \text{ and } \beta(x,x^*)\neq 0 \text{ for all } x\in X,$$

$$\left\langle\sum_{i=1}^{p}\nabla F_i(x^*),\eta(x,x^*)\right\rangle\geq-\rho(x,x^*)\|\theta(x,x^*)\|^2$$
$$\Rightarrow\ \frac{1}{\alpha(x,x^*)}\gamma(x,x^*)\left(e^{\alpha(x,x^*)\sum_{i=1}^{p}\xi_i(x,x^*)[F_i(x)-F_i(x^*)]}-1\right)(>)\geqq 0$$
$$\text{if } \alpha(x,x^*)\neq 0 \text{ and } \beta(x,x^*)=0 \text{ for all } x\in X,$$

$$\frac{1}{\beta(x,x^*)}\left\langle\sum_{i=1}^{p}\nabla F_i(x^*),e^{\beta(x,x^*)\eta(x,x^*)}-\mathbf{1}\right\rangle\geqq-\rho(x,x^*)\|\theta(x,x^*)\|^2$$
$$\Rightarrow\ \gamma(x,x^*)\sum_{i=1}^{p}\xi_i(x,x^*)[F_i(x)-F_i(x^*)](>)\geqq 0$$
$$\text{if } \alpha(x,x^*)=0 \text{ and } \beta(x,x^*)\neq 0 \text{ for all } x\in X,$$

$$\left\langle \sum_{i=1}^{p} \nabla F_i(x^*), \eta(x,x^*) \right\rangle \geqq -\rho(x,x^*)\|\theta(x,x^*)\|^2$$

$$\Rightarrow \gamma(x,x^*) \sum_{i=1}^{p} \xi_i(x,x^*)[F_i(x) - F_i(x^*)](>) \geqq 0$$

$$\text{if } \alpha(x,x^*) = 0 \text{ and } \beta(x,x^*) = 0 \text{ for all } x \in X.$$

The function $F$ is said to be (strictly) HA($\alpha$, $\beta$, $\gamma$, $\xi$, $\eta$, $\rho$, $\theta$)-V-pseudoinvex on $X$ if it is (strictly) $HA(\alpha,\beta,\gamma,\xi,\eta,\rho,\theta)$-V-pseudoinvex at each point $x^* \in X$.

**Definition 3.12.** The function $F$ is said to be first-order (prestrictly) HA($\alpha$, $\beta$, $\gamma$, $\xi$, $\eta$, $\rho$, $\theta$)-V-quasiinvex at $x^* \in X$ if there exist functions $\alpha : X \times X \to \mathbb{R}$, $\beta : X \times X \to \mathbb{R}$, $\gamma : X \times X \to \mathbb{R}_+$, $\xi_i : X \times X \to \mathbb{R}_+ \backslash \{0\}$, $i \in \underline{p}$, $\eta : X \times X \to \mathbb{R}^n$, $\rho : X \times X \to \mathbb{R}$, and $\theta : X \times X \to \mathbb{R}^n$ such that for all $x \in X$,

$$\frac{1}{\alpha(x,x^*)}\gamma(x,x^*)\left(e^{\alpha(x,x^*)\sum_{i=1}^{p}\xi_i(x,x^*)[F_i(x)-F_i(x^*)]} - 1\right)(<) \leqq 0$$

$$\Rightarrow \frac{1}{\beta(x,x^*)}\left\langle \sum_{i=1}^{p} \nabla F_i(x^*), e^{\beta(x,x^*)\eta(x,x^*)} - \mathbf{1} \right\rangle \leqq -\rho(x,x^*)\|\theta(x,x^*)\|^2$$

$$\text{if } \alpha(x,x^*) \neq 0 \text{ and } \beta(x,x^*) \neq 0 \text{ for all } x \in X,$$

$$\frac{1}{\alpha(x,x^*)}\gamma(x,x^*)\left(e^{\alpha(x,x^*)\sum_{i=1}^{p}\xi_i(x,x^*)[F_i(x)-F_i(x^*)]} - 1\right)(<) \leqq 0$$

$$\Rightarrow \left\langle \sum_{i=1}^{p} \nabla F_i(x^*), \eta(x,x^*) \right\rangle \leqq -\rho(x,x^*)\|\theta(x,x^*)\|^2$$

$$\text{if } \alpha(x,x^*) \neq 0 \text{ and } \beta(x,x^*) = 0 \text{ for all } x \in X,$$

$$\gamma(x,x^*) \sum_{i=1}^{p} \xi_i(x,x^*)[F_i(x) - F_i(x^*)](<) \leqq 0$$

$$\Rightarrow \frac{1}{\beta(x,x^*)}\left\langle \sum_{i=1}^{p} \nabla F_i(x^*), e^{\beta(x,x^*)\eta(x,x^*)} - \mathbf{1} \right\rangle \leqq -\rho(x,x^*)\|\theta(x,x^*)\|^2$$

$$\text{if } \alpha(x,x^*) = 0 \text{ and } \beta(x,x^*) \neq 0 \text{ for all } x \in X,$$

$$\gamma(x,x^*) \sum_{i=1}^{p} \xi_i(x,x^*)[F_i(x) - F_i(x^*)](<) \leqq 0 \Rightarrow$$

$$\left\langle \sum_{i=1}^{p} \nabla F_i(x^*), \eta(x,x^*) \right\rangle \leqq -\rho(x,x^*)\|\theta(x,x^*)\|^2$$

$$\text{if } \alpha(x,x^*) = 0 \text{ and } \beta(x,x^*) = 0 \text{ for all } x \in X.$$

The function $F$ is said to be first-order (prestrictly) $HA(\alpha,\beta,\gamma,\xi,\eta,\rho,\theta)$-V-quasiinvex on $X$ if it is (prestrictly) $HA(\alpha,\beta,\gamma,\xi,\eta,\rho,\theta)$-V-quasiinvex at each point $x^* \in X$.

In the proofs of the sufficient efficiency theorems, sometimes may be designed for using certain alternative but equivalent forms of the above definitions for achieving proofs. These are obtained by considering the contrapositive statements. For example, $HA(\alpha,\beta,\gamma,\xi,\eta,\rho,\theta)$-V-pseudoinvexity (when $\alpha(x,x^*)\neq 0$ and $\beta(x,x^*)\neq 0$ for all $x\in X$) can be defined in the following equivalent way:
The function $F$ is said to be first-order $HA(\alpha,\beta,\gamma,\xi,\eta,\rho,\theta)$-V-pseudoinvex at $x^*\in X$ if there exist functions $\alpha: X\times X\to\mathbb{R}$, $\beta: X\times X\to\mathbb{R}$, $\gamma: X\times X\to\mathbb{R}_+$, $\xi_i: X\times X\to\mathbb{R}_+\backslash\{0\}$, $i\in\underline{p}$, $\eta: X\times X\to\mathbb{R}^n$, $\rho: X\times X\to\mathbb{R}$, and $\theta: X\times X\to\mathbb{R}^n$ such that for all $x\in X$,

$$\frac{1}{\alpha(x,x^*)}\gamma(x,x^*)\left(e^{\alpha(x,x^*)\sum_{i=1}^{p}\xi_i(x,x^*)[F_i(x)-F_i(x^*)]}-1\right)<0$$
$$\Rightarrow\ \frac{1}{\beta(x,x^*)}\Big\langle\sum_{i=1}^{p}\nabla F_i(x^*),e^{\beta(x,x^*)\eta(x,x^*)}-\mathbf{1}\Big\rangle$$
$$<-\rho(x,x^*)\|\theta(x,x^*)\|^2.$$

Similary, we observe that during the proofs of the duality theorems, sometimes it may be more appropriate and convenient to use certain alternative but equivalent forms of the above definitions. These are obtained by considering the contrapositive statements. For example, the second-order $HA(\alpha,\beta,\gamma,\xi,\eta,\omega,\rho,\theta)$-V-pseudosonvexity (when $\alpha(x,x^*)\neq 0$ and $\beta(x,x^*)\neq 0$ for all $x\in X$) can be defined in the following equivalent manner:
The function $f$ is said to be second-order $HA(\alpha,\beta,\gamma,\xi,\eta,\omega,\rho,\theta)$-V-pseudosonvex at $x^*\in X$ if there exist functions $\alpha: X\times X\to\mathbb{R}$, $\beta: X\times X\to\mathbb{R}$, $\gamma: X\times X\to\mathbb{R}_+$, $\xi_i: X\times X\to\mathbb{R}_+$, $i\in\underline{p}$, $\eta,\omega: X\times X\to\mathbb{R}^n$, $\rho: X\times X\to\mathbb{R}$, and $\theta: X\times X\to\mathbb{R}^n$ such that for all $x\in X$ and $z\in\mathbb{R}^n$,

$$\frac{1}{\alpha(x,x^*)}\gamma(x,x^*)\left(e^{\alpha(x,x^*)\sum_{i=1}^{p}[F_i(x)-F_i(x^*)+\frac{1}{2}\langle\omega(x,x^*),\nabla^2F_i(x^*)z\rangle]}-1\right)<0$$
$$\Rightarrow\ \frac{1}{\beta(x,x^*)}\Big\langle\sum_{i=1}^{p}\xi_i(x,x^*)[\nabla F_i(x^*)+\nabla^2F_i(x^*)z],e^{\beta(x,x^*)\eta(x,x^*)}-\mathbf{1}\Big\rangle$$
$$<-\rho(x,x^*)\|\theta(x,x^*)\|^2.$$

**Definition 3.13.** The function $F$ is said to be (strictly) first-order HA($\alpha$, $\beta$, $\gamma$, $\eta$, $\rho$, $\theta$) -V-invex at $x^*\in X$ if there exist functions $\alpha: X\times X\to\mathbb{R}$, $\beta: X\times X\to\mathbb{R}$, $\gamma_i: X\times X\to\mathbb{R}_+$, $\eta: X\times X\to\mathbb{R}^n$, $\rho_i: X\times X\to\mathbb{R}$, $i\in\underline{p}$, and $\theta: X\times X\to\mathbb{R}^n$ such that for all $x\in X(x\neq x^*)$ and $i\in\underline{p}$,

$$\frac{1}{\alpha(x,x^*)}\gamma_i(x,x^*)\left(e^{\alpha(x,x^*)[F_i(x)-F_i(x^*)]}-1\right)(>)$$
$$\geqq\frac{1}{\beta(x,x^*)}\langle\nabla F_i(x^*),e^{\beta(x,x^*)\eta(x,x^*)}-\mathbf{1}\rangle$$
$$+\rho_i(x,x^*)\|\theta(x,x^*)\|^2\ \text{ if }\alpha(x,x^*)\neq 0\text{ and }\beta(x,x^*)\neq 0\text{ for all }x\in X,$$

$$\frac{1}{\alpha(x,x^*)}\gamma_i(x,x^*)\left(e^{\alpha(x,x^*)[F_i(x)-F_i(x^*)]}-1\right)(>)\geqq\langle\nabla F_i(x^*),\eta(x,x^*)\rangle$$
$$+\rho_i(x,x^*)\|\theta(x,x^*)\|^2 \text{ if } \alpha(x,x^*)\neq 0 \text{ and } \beta(x,x^*)=0 \text{ for all } x\in X,$$

$$\gamma_i(x,x^*)[F_i(x)-F_i(x^*)](>)\geqq\frac{1}{\beta(x,x^*)}\langle\nabla F_i(x^*),e^{\beta(x,x^*)\eta(x,x^*)}-\mathbf{1}\rangle$$
$$+\rho_i(x,x^*)\|\theta(x,x^*)\|^2 \text{ if } \alpha(x,x^*)=0 \text{ and } \beta(x,x^*)\neq 0 \text{ for all } x\in X,$$

$$\gamma_i(x,x^*)[F_i(x)-F_i(x^*)](>)\geqq\langle\nabla F_i(x^*),\eta(x,x^*)\rangle+\rho_i(x,x^*)\|\theta(x,x^*)\|^2$$
$$\text{if } \alpha(x,x^*)=0 \text{ and } \beta(x,x^*)=0 \text{ for all } x\in X,$$

where $\|\cdot\|$ is a norm on $\mathbb{R}^n$ and

$$\left(e^{\beta(x,x^*)\eta(x,x^*)}-\mathbf{1}\right)\equiv\left(e^{\beta(x,x^*)\eta_1(x,x^*)}-1,\ldots,e^{\beta(x,x^*)\eta_n(x,x^*)}-1\right).$$

The function $F$ is said to be (strictly) $HA(\alpha,\beta,\gamma,\eta,\rho,\theta)$-V-invex on $X$ if it is (strictly) $HA(\alpha,\beta,\gamma,\eta,\rho,\theta)$-V-invex at each point $x^*\in X$.

**Definition 3.14.** The function $F$ is said to be (strictly) first-order HA($\alpha$, $\beta$, $\gamma$, $\eta$, $\rho$, $\theta$) -V-pseudoinvex at $x^*\in X$ if there exist functions $\alpha: X\times X\to\mathbb{R}$, $\beta: X\times X\to\mathbb{R}$, $\gamma: X\times X\to\mathbb{R}_+$, $i\in\underline{p}$, $\eta: X\times X\to\mathbb{R}^n$, $\rho: X\times X\to\mathbb{R}$, and $\theta: X\times X\to\mathbb{R}^n$ such that for all $x\in X(x\neq x^*)$,

$$\frac{1}{\beta(x,x^*)}\left\langle\sum_{i=1}^p\nabla F_i(x^*),e^{\beta(x,x^*)\eta(x,x^*)}-\mathbf{1}\right\rangle\geqq-\rho(x,x^*)\|\theta(x,x^*)\|^2$$
$$\Rightarrow\frac{1}{\alpha(x,x^*)}\gamma(x,x^*)\left(e^{\alpha(x,x^*)\sum_{i=1}^p[F_i(x)-F_i(x^*)]}-1\right)(>)\geqq 0$$
$$\text{if } \alpha(x,x^*)\neq 0 \text{ and } \beta(x,x^*)\neq 0 \text{ for all } x\in X,$$

$$\left\langle\sum_{i=1}^p\nabla F_i(x^*),\eta(x,x^*)\right\rangle\geqq-\rho(x,x^*)\|\theta(x,x^*)\|^2$$
$$\Rightarrow\frac{1}{\alpha(x,x^*)}\gamma(x,x^*)\left(e^{\alpha(x,x^*)\sum_{i=1}^p[F_i(x)-F_i(x^*)]}-1\right)(>)\geqq 0$$
$$\text{if } \alpha(x,x^*)\neq 0 \text{ and } \beta(x,x^*)=0 \text{ for all } x\in X,$$

$$\frac{1}{\beta(x,x^*)}\left\langle\sum_{i=1}^p\nabla F_i(x^*),e^{\beta(x,x^*)\eta(x,x^*)}-\mathbf{1}\right\rangle\geqq-\rho(x,x^*)\|\theta(x,x^*)\|^2$$
$$\Rightarrow\gamma(x,x^*)\sum_{i=1}^p[F_i(x)-F_i(x^*)](>)\geqq 0$$
$$\text{if } \alpha(x,x^*)=0 \text{ and } \beta(x,x^*)\neq 0 \text{ for all } x\in X,$$

$$\left\langle \sum_{i=1}^{p} \nabla F_i(x^*), \eta(x,x^*) \right\rangle \geqq -\rho(x,x^*)\|\theta(x,x^*)\|^2$$
$$\Rightarrow \gamma(x,x^*) \sum_{i=1}^{p} [F_i(x) - F_i(x^*)](>) \geqq 0$$
$$\text{if } \alpha(x,x^*) = 0 \text{ and } \beta(x,x^*) = 0 \text{ for all } x \in X.$$

The function $F$ is said to be (strictly) $HA(\alpha,\beta,\gamma,\eta,\rho,\theta)$-V-pseudoinvex on $X$ if it is (strictly) $HA(\alpha,\beta,\gamma,\eta,\rho,\theta)$-V-pseudoinvex at each point $x^* \in X$.

**Definition 3.15.** The function $F$ is said to be (prestrictly) first-order HA($\alpha$, $\beta$, $\gamma$, $\eta$, $\rho$, $\theta$)-V-quasiinvex at $x^* \in X$ if there exist functions $\alpha : X \times X \to \mathbb{R}$, $\beta : X \times X \to \mathbb{R}$, $\gamma : X \times X \to \mathbb{R}_+$, $i \in \underline{p}$, $\eta : X \times X \to \mathbb{R}^n$, $\rho : X \times X \to \mathbb{R}$, and $\theta : X \times X \to \mathbb{R}^n$ such that for all $x \in X$,

$$\frac{1}{\alpha(x,x^*)}\gamma(x,x^*)\left(e^{\alpha(x,x^*)\sum_{i=1}^{p}[F_i(x)-F_i(x^*)]} - 1\right)(<) \leqq 0$$
$$\Rightarrow \frac{1}{\beta(x,x^*)}\left\langle \sum_{i=1}^{p} \nabla F_i(x^*), e^{\beta(x,x^*)\eta(x,x^*)} - \mathbf{1} \right\rangle \leqq -\rho(x,x^*)\|\theta(x,x^*)\|^2$$
$$\text{if } \alpha(x,x^*) \neq 0 \text{ and } \beta(x,x^*) \neq 0 \text{ for all } x \in X,$$

$$\frac{1}{\alpha(x,x^*)}\gamma(x,x^*)\left(e^{\alpha(x,x^*)\sum_{i=1}^{p}[F_i(x)-F_i(x^*)]} - 1\right)(<) \leqq 0$$
$$\Rightarrow \left\langle \sum_{i=1}^{p} \nabla F_i(x^*), \eta(x,x^*) \right\rangle \leqq -\rho(x,x^*)\|\theta(x,x^*)\|^2$$
$$\text{if } \alpha(x,x^*) \neq 0 \text{ and } \beta(x,x^*) = 0 \text{ for all } x \in X,$$

$$\gamma(x,x^*) \sum_{i=1}^{p} [F_i(x) - F_i(x^*)](<) \leqq 0$$
$$\Rightarrow \frac{1}{\beta(x,x^*)}\left\langle \sum_{i=1}^{p} \nabla F_i(x^*), e^{\beta(x,x^*)\eta(x,x^*)} - \mathbf{1} \right\rangle \leqq -\rho(x,x^*)\|\theta(x,x^*)\|^2$$
$$\text{if } \alpha(x,x^*) = 0 \text{ and } \beta(x,x^*) \neq 0 \text{ for all } x \in X,$$

$$\gamma(x,x^*) \sum_{i=1}^{p} [F_i(x) - F_i(x^*)](<) \leqq 0 \Rightarrow$$
$$\left\langle \sum_{i=1}^{p} \nabla F_i(x^*), \eta(x,x^*) \right\rangle \leqq -\rho(x,x^*)\|\theta(x,x^*)\|^2$$
$$\text{if } \alpha(x,x^*) = 0 \text{ and } \beta(x,x^*) = 0 \text{ for all } x \in X.$$

The function $F$ is said to be (prestrictly) $HA(\alpha,\beta,\gamma,\eta,\rho,\theta)$-V-quasiinvex on $X$ if it is (prestrictly) $HA(\alpha,\beta,\gamma,\eta,\rho,\theta)$-V-quasiinvex at each point $x^* \in X$.

We observe the role of some first-order V-sonvexities in the proofs of the sufficient efficiency theorems is quite useful by considering certain alternative but equivalent forms of the above definitions. These are obtained by considering the contrapositive statements. For example, $HA(\alpha,\beta,\gamma,\eta,\rho,\theta)$-V-pseudoinvexity (when $\alpha(x,x^*)\neq 0$ and $\beta(x,x^*)\neq 0$ for all $x\in X$) can be defined in the following equivalent way:

The function $F$ is said to be first-order $HA(\alpha,\beta,\gamma,\eta,\rho,\theta)$-V-pseudoinvex at $x^*\in X$ if there exist functions $\alpha : X\times X\to\mathbb{R}$, $\beta : X\times X\to\mathbb{R}$, $\gamma : X\times X\to\mathbb{R}_+$, $\eta : X\times X\to\mathbb{R}^n$, $\rho : X\times X\to\mathbb{R}$, and $\theta : X\times X\to\mathbb{R}^n$ such that for all $x\in X$,

$$\frac{1}{\alpha(x,x^*)}\gamma(x,x^*)\Big(e^{\alpha(x,x^*)\sum_{i=1}^p[F_i(x)-F_i(x^*)]}-1\Big)<0$$
$$\Rightarrow\quad \frac{1}{\beta(x,x^*)}\Big\langle\sum_{i=1}^p\nabla F_i(x^*),e^{\beta(x,x^*)\eta(x,x^*)}-\mathbf{1}\Big\rangle<-\rho(x,x^*)\|\theta(x,x^*)\|^2.$$

In the sequel, we shall also develop a consistent notation for vector inequalities. For $a,b\in\mathbb{R}^m$, the following order notation will be used: $a\geqq b$ if and only if $a_i\geqq b_i$ for all $i\in\underline{m}$; $a\geqslant b$ if and only if $a_i\geqq b_i$ for all $i\in\underline{m}$, but $a\neq b$; $a>b$ if and only if $a_i>b_i$ for all $i\in\underline{m}$; and $a\ngeqslant b$ is the negation of $a\geqslant b$.

Consider the multiobjective problem

$$(P^*)\qquad \underset{x\in\mathbb{F}}{\text{Minimize}}\ F(x)=(F_1(x),\ldots,F_p(x)),$$

where $F_i$, $i\in\underline{p}$, are real-valued functions defined on $\mathbb{R}^n$.

An element $x^\circ\in\mathbb{F}$ is said to be an *efficient (Pareto optimal, nondominated, noninferior)* solution of $(P^*)$ if there exists no $x\in\mathbb{F}$ such that $F(x)\leqslant F(x^\circ)$. There are several versions of the concept of efficiency that are currently used in the area of multiobjective programming. However, throughout this section we shall deal exclusively with the efficient solutions of $(P)$ in the sense defined above.

We conclude this section by recalling a set of parametric necessary efficiency conditions for $(P)$ which will be needed for proving strong duality theorems.

**Theorem 3.16.** [14] Let $x^*\in\mathbb{F}$, $\lambda^*=\varphi(x^*)$, for each $i\in\underline{p}$, let $f_i$ and $g_i$ be continuously differentiable at $x^*$, for each $j\in\underline{q}$, let the function $\zeta\to G_j(\zeta,t)$ be continuously differentiable at $x^*$ for all $t\in T_j$, and for each $k\in\underline{r}$, let the function $\zeta\to H_k(\zeta,s)$ be continuously differentiable at $x^*$ for all $s\in S_k$. If $x^*$ is an efficient solution of (P), if the generalized Guignard constraint qualification holds at $x^*$, and if for each $i_0\in\underline{p}$, the set $cone(\{\nabla G_j(x^*,t):t\in\hat{T}_j(x^*),j\in\underline{q}\}\cup\{\nabla f_i(x^*)-\lambda_i^*\nabla g_i(x^*):i\in\underline{p},i\neq i_0\})+span(\{\nabla H_k(x^*,s):s\in S_k,k\in\underline{r}\})$ is closed, then there exist $u^*\in U$ and integers $\nu_0$ and $\nu$, with $0\leqq\nu_0\leqq\nu\leqq n+1$, such that there exist $\nu_0$ indices $j_m$, with $1\leqq j_m\leqq q$, together with $\nu_0$ points $t^m\in\hat{T}_{j_m}(x^*)$, $m\in\underline{\nu_0}$, $\nu-\nu_0$ indices $k_m$, with $1\leqq k_m\leqq r$, together with $\nu-\nu_0$ points $s^m\in S_{k_m}$ for $m\in\underline{\nu}\backslash\underline{\nu_0}$, and $\nu$ real numbers $v_m^*$, with $v_m^* > 0$ for $m \in \underline{\nu_0}$, with the property that

$$\sum_{i=1}^{p} u_i^*[\nabla f_i(x^*) - \lambda_i^* \nabla g_i(x^*)] + \sum_{m=1}^{\nu_0} v_m^* \nabla G_{j_m}(x^*, t^m)$$
$$+ \sum_{m=\nu_0+1}^{\nu} v_m^* \nabla H_{k_m}(x^*, s^m) = 0,$$

where $cone(V)$ is the conic hull of the set $V \subset \mathbb{R}^n$ (i.e., the smallest convex cone containing $V$), $span(V)$ is the linear hull of $V$ (i.e., the smallest subspace containing $V$), $\hat{T}_j(x^*) = \{t \in T_j : G_j(x^*,t) = 0\}$, $U = \{u \in \mathbb{R}^p : u > 0, \sum_{i=1}^{p} u_i = 1\}$, and $\underline{\nu^*} \backslash \underline{\nu_0^*}$ is the complement of the set $\underline{\nu_0^*}$ relative to the set $\underline{\nu^*}$.

For brevity, we shall henceforth refer to an efficient solution $x^* \in \mathbb{F}$ as a *normal* efficient solution of $(P)$ if the generalized Guignard constraint qualification is satisfied at $x^*$ and for each $i_0 \in \underline{p}$, the set $cone(\{\nabla G_j(x^*,t) : t \in \hat{T}_j(x^*), j \in \underline{q}\} \cup \{\nabla f_i(x^*) - \lambda_i^* \nabla g_i(x^*) : i \in \underline{p}, i \neq i_0\}) + span(\{\nabla H_k(x^*,s) : s \in S_k, k \in \underline{r}\})$ is closed, where $\lambda_i^* = f_i(x^*)/g_i(x^*)$, $i \in \underline{p}$.

In the remainder of this chapter, we assume that the functions $f_i$, $g_i$, $i \in \underline{p}$, $\zeta \to G_j(\zeta,t)$, and $\zeta \to H_k(\zeta,s)$, are twice continuously differentiable on $X$ for all $t \in T_j$, $j \in \underline{q}$, and $s \in S_k$, $k \in \underline{r}$.

With regard to the choice of the type of generalized $HA(\alpha,\beta,\gamma,\xi,\eta,\omega,\rho,\theta)$-V-sonvex functions, specified in Definitions 3.4 - 3.6, to be used in the statements and proofs of our second-order duality theorems, we shall consistently use the cases in which the functions $\alpha$ and $\beta$ are nonzero for all $(x,y) \in X \times X$. All the duality results established in this paper can be modified, restated, and proved for the other cases in a straightforward manner.

# 3. Generalized Duality Results I

## 3.1. Duality Models I

In this section, we consider two dual problems with relatively simple constraint structures and prove weak, strong, and strict converse duality theorems under an appropriate $HA(\alpha,\beta,\gamma,\xi,\eta,\omega,\rho,\theta)$-V-pseudosonvexity condition. More general duality models and results for $(P)$ will be discussed in the subsequent sections.

Let

$$\mathbb{H} = \Big\{ (y,z,u,v,\lambda,\nu,\nu_0,J_{\nu_0},K_{\nu\backslash\nu_0},\bar{t},\bar{s}) : y \in X;\ z \in \mathbb{R}^n;\ u \in U;\ \lambda \in \mathbb{R}^p;$$
$$0 \leqq \nu_0 \leqq \nu \leqq n+1;\ v \in \mathbb{R}^{\nu}, v_i > 0, 1 \leqq i \leqq \nu_0;$$
$$J_{\nu_0} = (j_1, j_2, \ldots, j_{\nu_0}), 1 \leqq j_i \leqq q;\ K_{\nu\backslash\nu_0} = (k_{\nu_0+1}, \ldots, k_{\nu}),$$
$$1 \leqq k_i \leqq r;\ \bar{t} = (t^1, t^2, \ldots, t^{\nu_0}), t^i \in T_{j_i};\ \bar{s} = (s^{\nu_0+1}, \ldots, s^{\nu}), s^i \in S_{k_i} \Big\}.$$

Consider the following two problems:

$(DI)$ $$\sup_{(y,z,u,v,\lambda,\nu,\nu_0,J_{\nu_0},K_{\nu\backslash\nu_0},\bar{t},\bar{s})\in\mathbb{H}} \lambda=(\lambda_1,\ldots,\lambda_p)$$

subject to

$$\sum_{i=1}^{p}u_i[\nabla f_i(y)-\lambda_i\nabla g_i(y)]+\sum_{m=1}^{\nu_0}v_m\nabla G_{j_m}(y,t^m)+\sum_{m=\nu_0+1}^{\nu}v_m\nabla H_{k_m}(y,s^m)+ \Big\{\sum_{i=1}^{p}u_i[\nabla^2 f_i(y)-\lambda_i\nabla^2 g_i(y)]+\sum_{m=1}^{\nu_0}v_m\nabla^2 G_{j_m}(y,t^m) + \sum_{m=\nu_0+1}^{\nu}v_m\nabla^2 H_{k_m}(y,s^m)\Big\}z=0, \quad (3.1)$$

$$\sum_{i=1}^{p}u_i[f_i(y)-\lambda_i g_i(y)]+\sum_{m=1}^{\nu_0}v_m G_{j_m}(y,t^m)+\sum_{m=\nu_0+1}^{\nu}v_m H_{k_m}(y,s^m) -\frac{1}{2}\Big\langle\omega(x,y),\Big\{\sum_{i=1}^{p}u_i[\nabla^2 f_i(y) -\lambda_i\nabla^2 g_i(y)]+\sum_{m=1}^{\nu_0}v_m\nabla^2 G_{j_m}(y,t^m)+\sum_{m=\nu_0+1}^{\nu}v_m\nabla^2 H_{k_m}(y,s^m)\Big\}z\Big\rangle \geqq 0 \quad (3.2)$$

for all $x\in\mathbb{F}$;

$(\tilde{D}I)$ $$\sup_{(y,z,u,v,\lambda,\nu,\nu_0,J_{\nu_0},K_{\nu\backslash\nu_0},\bar{t},\bar{s})\in\mathbb{H}} \lambda=(\lambda_1,\ldots,\lambda_p)$$

subject to (2) and

$$\frac{1}{\beta(x,y)}\Big\langle\sum_{i=1}^{p}u_i[\nabla f_i(y)-\lambda_i\nabla g_i(y)]+\sum_{m=1}^{\nu_0}v_m\nabla G_{j_m}(y,t^m) +\sum_{m=\nu_0+1}^{\nu}v_m\nabla H_{k_m}(y,s^m)+ \Big\{\sum_{i=1}^{p}u_i[\nabla^2 f_i(y)-\lambda_i\nabla^2 g_i(y)]+\sum_{m=1}^{\nu_0}v_m\nabla^2 G_{j_m}(y,t^m)+ \sum_{m=\nu_0+1}^{\nu}v_m\nabla^2 H_{k_m}(y,s^m)\Big\}z, e^{\beta(x,y)\eta(x,y)}-\mathbf{1}\Big\rangle \geqq 0 \text{ for all } x\in\mathbb{F}, \quad (3.3)$$

where $\beta$ and $\eta$ are functions from $X\times X$ to $\mathbb{R}$ and $\mathbb{R}^n$, respectively.

Comparing $(DI)$ and $(\tilde{D}I)$, we see that $(\tilde{D}I)$ is relatively more general than $(DI)$ in the sense that any feasible solution of $(DI)$ is also feasible for $(\tilde{D}I)$, but the converse is not necessarily true. Furthermore, we observe that (1) is a system of $n$ equations, whereas (3) is a single inequality. Clearly, from a computational point of view, $(DI)$ is preferable to $(\tilde{D}I)$ because of the dependence of (3) on the feasible set of $(P)$.

Despite these apparent differences, it turns out that the statements and proofs of all the duality theorems for $(P)-(DI)$ and $(P)-(\tilde{D}I)$ are almost identical and, therefore, we shall consider only the pair $(P)-(DI)$.

The next two theorems show that $(DI)$ is a dual problem for $(P)$.

**Theorem 3.17.** Let $x$ and $w \equiv (y, z, u, v, \lambda, \nu, \nu_0, J_{\nu_0}, K_{\nu \backslash \nu_0}, \bar{t}, \bar{s})$ be arbitrary feasible solutions of (P) and (DI), respectively, and assume that the function $L(\cdot, u, v, \lambda, \bar{t}, \bar{s}) : X \to \mathbb{R}^p$ defined by

$$L(\zeta, u, v, \lambda, \bar{t}, \bar{s}) = \big(L_1(\zeta, u, v, \lambda, \bar{t}, \bar{s}), \ldots, L_p(\zeta, u, v, \lambda, \bar{t}, \bar{s})\big)$$

is HA($\alpha$, $\beta$, $\gamma$, $\xi$, $\eta$, $\omega$, 0, $\theta$)-V-pseudosonvex at $y$ for $\gamma(x,y) > 0$, $\xi_i(x,y) > 0$, $i \in \underline{p}$, where

$$L_i(\zeta, u, v, \lambda, \bar{t}, \bar{s}) = u_i\Big[f_i(\zeta) - \lambda_i g_i(\zeta)\Big] + \sum_{m=1}^{\nu_0} v_m G_{j_m}(\zeta, t^m) + \sum_{m=\nu_0+1}^{\nu} v_m H_{k_m}(\zeta, s^m), \quad i \in \underline{p}.$$

Then $\varphi(x) \nleq \lambda$.

*Proof.* Using the $HA(\alpha, \beta, \gamma, \xi, \eta, \omega, 0, \theta)$-V-pseudosonvexity assumption and (1) we conclude that

$$\frac{1}{\alpha(x,y)}\gamma(x,y)\left(e^{\alpha(x,y)\sum_{i=1}^{p}[L_i(x,u,v,\lambda,\bar{t},\bar{s})-L_i(y,u,v,\lambda,\bar{t},\bar{s})+\frac{1}{2}\langle\omega(x,y),\nabla^2 L_i(y,u,v,\lambda,\bar{t},\bar{s})\omega(x,y)\rangle]} - 1\right) \geqq 0.$$

We need to consider two cases: $\alpha(x,y) > 0$ and $\alpha(x,y) < 0$. If we assume that $\alpha(x,y) > 0$ and recall that $\gamma(x,y) > 0$ and $\xi_i(x,y) > 0$, then we see that the above inequality becomes

$$e^{\alpha(x,y)\sum_{i=1}^{p}\left[L_i(x,u,v,\lambda,\bar{t},\bar{s})-L_i(y,u,v,\lambda,\bar{t},\bar{s})+\frac{1}{2}\left\langle\omega(x,y),\nabla^2 L_i(y,u,v,\lambda,\bar{t},\bar{s})z\right\rangle\right]} \geqq 1.$$

This implies that

$$\sum_{i=1}^{p} L_i(x, u, v, \lambda, \bar{t}, \bar{s}) \geqq \sum_{i=1}^{p}\left[L_i(y, u, v, \lambda, \bar{t}, \bar{s}) - \frac{1}{2}\left\langle\omega(x,y), \nabla^2 L_i(y, u, v, \lambda, \bar{t}, \bar{s})z\right\rangle\right] \geqq 0,$$

where the second inequality follows from (2). Inasmuch as $x \in \mathbb{F}$ and $v_m > 0$, $m \in \underline{\nu_0}$, the above inequality reduces to

$$\sum_{i=1}^{p} u_i[f_i(x) - \lambda_i g_i(x)] \geqq 0. \tag{3.4}$$

Since $u > 0$, (4) implies that

$$\big(f_1(x) - \lambda_1 g_1(x), \ldots, f_p(x) - \lambda_p g_p(x)\big) \nleq (0, \ldots, 0),$$

which in turn implies that

$$\varphi(x) = \left(\frac{f_1(x)}{g_1(x)}, \ldots, \frac{f_p(x)}{g_p(x)}\right) \nleq (\lambda_1, \ldots, \lambda_p) = \lambda.$$

If we assume that $\alpha(x,y) < 0$, we arrive at the same conclusion.

□

**Theorem 3.18.** (Strong Duality) Let $x^*$ be a normal efficient solution of (P) and assume that the conditions set forth in Theorem 3.17 are satisfied for all feasible solutions of (DI). Then there exist $u^*, v^*, \lambda^*, \nu^*, \nu_0^*, J_{\nu_0^*}, K_{\nu^*\backslash\nu_0^*}, t^*$, and $\bar{s}^*$ such that $(x^*, z^* = 0, u^*, v^*, \lambda^*, \nu^*, \nu_0^*, J_{\nu_0^*}, K_{\nu^*\backslash\nu_0^*}, \bar{t}^*, \bar{s}^*)$ is an efficient solution of (DI) and $\varphi(x^*) = \lambda^*$.

*Proof.* Since $x^*$ is a normal efficient solution of $(P)$, by Theorem 3.16, there exist $u^*, v^*, \lambda^*, \nu^*, \nu_0^*, J_{\nu_0^*}, K_{\nu^*\backslash\nu_0^*}, \bar{t}^*$, and $\bar{s}^*$ such that ($x^*$, $z^* = 0$, $u^*$, $v^*$, $\lambda^*$, $\nu^*$, $\nu_0^*$, $J_{\nu_0^*}$, $K_{\nu^*\backslash\nu_0^*}$, $\bar{t}^*$, $\bar{s}^*$) is a feasible solution of $(DI)$ and $\varphi(x^*) = \lambda^*$. If it were not efficient, then there would exist a feasible solution $(\tilde{x}, \tilde{z}, \tilde{u}, \tilde{v}, \tilde{\lambda}, \tilde{\nu}, \tilde{\nu}_0, J_{\tilde{\nu}_0}, K_{\tilde{\nu}\backslash\tilde{\nu}_0}, \tilde{\bar{t}}, \tilde{\bar{s}})$ of $(DI)$ such that $\tilde{\lambda} \geqslant \lambda^* = \varphi(x^*)$, contradicting Theorem 3.17. Therefore, ($x^*$, $z^* = 0$, $u^*$, $v^*$, $\lambda^*$, $\nu^*$, $\nu_0^*$, $J_{\nu_0^*}$, $K_{\nu^*\backslash\nu_0^*}$, $\bar{t}^*$, $\bar{s}^*$) is an efficient solution of $(DI)$. □

We also have the following converse duality result for $(P)$ and $(DI)$.

**Theorem 3.19.** (Strict Converse Duality) Let $x^*$ and $\tilde{w} \equiv (\tilde{x}, \tilde{z}, \tilde{u}, \tilde{v}, \tilde{\lambda}, \tilde{\nu}, \tilde{\nu}_0, J_{\tilde{\nu}_0}, K_{\tilde{\nu}\backslash\tilde{\nu}_0}, \tilde{\bar{t}}, \tilde{\bar{s}})$ be arbitrary feasible solutions of (P) and (DI), respectively, such that

$$\sum_{i=1}^{p} \tilde{u}_i \xi_i(x^*, \tilde{x})[f_i(x^*) - \tilde{\lambda}_i g_i(x^*)] = 0, \tag{3.5}$$

where for each $i \in \underline{p}$, $\xi_i(x^*, \tilde{x}) > 0$. Furthermore, assume that the function $\zeta \to L(\zeta, \tilde{u}, \tilde{v}, \tilde{\lambda}, \tilde{\bar{t}}, \tilde{\bar{s}})$ is strictly $HA(\alpha, \beta, \gamma, \xi, \eta, 0, \theta)$-V-pseudosonvex at $\tilde{x}$. Then $\tilde{x} = x^*$ and $\varphi(x^*) = \tilde{\lambda}$.

*Proof.* (a): Suppose to the contrary that $\tilde{x} \neq x^*$. Now proceeding as in the proof of Theorem 3.17 (with $x$ replaced by $x^*$ and $w$ by $\tilde{w}$), we arrive at the strict inequality

$$\sum_{i=1}^{p} \tilde{u}_i \xi_i(x^*, \tilde{x})[f_i(x^*) - \tilde{\lambda}_i g_i(x^*)] > 0,$$

which contradicts (5). Therefore, we conclude that $\tilde{x} = x^*$ and $\varphi(x^*) = \tilde{\lambda}$.

□

**Observation.** Prove Theorem 3.19 without using the method of contradiction to emphasize the role of the strict converse duality for the general understanding.

## 3.2. Duality Model II

In this subsection, we consider certain variants of $(DI)$ and $(\tilde{D}I)$ that allow for more general $HA(\alpha,\beta,\gamma,\xi,\eta,\omega,\rho,\theta)$-V-sonvexity conditions under which duality can be established. These duality models can be stated as follows:

$$(DII) \qquad \sup_{(y,z,u,v,\lambda,\nu,\nu_0,J_{\nu_0},K_{\nu\backslash\nu_0},\bar{t},\bar{s})\in\mathbb{H}} \lambda = (\lambda_1,\ldots,\lambda_p)$$

subject to

$$\sum_{i=1}^{p} u_i[\nabla f_i(y) - \lambda_i \nabla g_i(y)] + \sum_{m=1}^{\nu_0} v_m \nabla G_{j_m}(y,t^m) + \sum_{m=\nu_0+1}^{\nu} v_m \nabla H_{k_m}(y,s^m) + \Big\{\sum_{i=1}^{p} u_i[\nabla^2 f_i(y) - \lambda_i \nabla^2 g_i(y)] + \sum_{m=1}^{\nu_0} v_m \nabla^2 G_{j_m}(y,t^m) + \sum_{m=\nu_0+1}^{\nu} v_m \nabla^2 H_{k_m}(y,s^m)\Big\} z = 0, \quad (3.6)$$

$$f_i(y) - \lambda_i g_i(y) - \frac{1}{2}\langle \omega(x,y), [\nabla^2 f_i(y) - \lambda_i \nabla^2 g_i(y)]z\rangle \geqq 0, \quad i \in \underline{p}, \quad (3.7)$$

$$G_{j_m}(y,t^m) - \frac{1}{2}\langle \omega(x,y), \nabla^2 G_{j_m}(y,t^m)z\rangle \geqq 0, \quad m \in \underline{\nu_0}, \quad (3.8)$$

$$v_m H_{k_m}(y,s^m) - \frac{1}{2}\langle \omega(x,y), v_m \nabla^2 H_{k_m}(y,s^m)z\rangle \geqq 0, \quad m \in \underline{\nu}\backslash\underline{\nu_0}; \quad (3.9)$$

$$(\tilde{D}II) \qquad \sup_{(y,z,u,v,\lambda,\nu,\nu_0,J_{\nu_0},K_{\nu\backslash\nu_0},\bar{t},\bar{s})\in\mathbb{H}} \lambda = (\lambda_1,\ldots,\lambda_p)$$

subject to (3) and (7) - (9).

The remarks and observations made earlier about the relationships between $(DI)$ and $(\tilde{D}I)$ are, of course, also valid for $(DII)$ and $(\tilde{D}II)$. Since the constraint inequalities of $(DII)$ are formed by splitting the inequality (7) into three inequalities (7) - (9), it is clear that Theorems 3.17 - 3.19 are valid for the pair $(P)-(DII)$. Below, we shall establish some duality results in which appropriate generalized $HA(\alpha,\beta,\gamma,\xi,\eta,\omega,\rho,\theta)$-V-sonvexity requirements will be placed on certain vector functions including $(\mathscr{E}_1(\cdot,u,\lambda),\ldots,\mathscr{E}_p(\cdot,u,\lambda))$, where for each $i \in \underline{p}$, the component function $\mathscr{E}_i(\cdot,u,\lambda)$ is defined, for fixed $\lambda$ and $u$, on $X$ by

$$\mathscr{E}_i(\zeta,u,\lambda) = u_i[f_i(\zeta) - \lambda_i g_i(\zeta)].$$

The following theorems show that $(DII)$ is a dual problem for $(P)$.

**Theorem 3.20.** Let $x$ and $w \equiv (y,z,u,v,\lambda,\nu,\nu_0,J_{\nu_0},K_{\nu\backslash\nu_0},\bar{t},\bar{s})$ be arbitrary feasible solutions of (P) and (DII), respectively, and assume that any one of the following four sets of hypotheses is satisfied:

(a) (i) $\left(\mathscr{E}_1(\cdot,u,\lambda),\ldots,\mathscr{E}_p(\cdot,u,\lambda)\right)$ is HA($\alpha$, $\beta$, $\bar{\gamma}$, $\xi$, $\eta$, $\omega$, $\bar{\rho}$, $\theta$)- V -pseudosonvex at $y$ for $\bar{\gamma}(x,y) > 0$ and $\xi_i(x,y) > 0$, $i \in \underline{p}$;

(ii) $\left(v_1 G_{j_1}(\cdot,t^1),\ldots,v_{v_0} G_{j_{v_0}}(\cdot,t^{v_0})\right)$ is HA($\alpha$, $\beta$, $\tilde{\gamma}$, $\pi$, $\eta$, $\omega$, $\tilde{\rho}$, $\theta$)- V-quasisonvex at $y$ for $\tilde{\gamma}(x,y) > 0$ and $\pi_m(x,y) > 0$ for each $m \in \underline{v_0}$;

(iii) $\left(v_{v_0+1} H_{k_{v_0+1}}(\cdot,s^{v_0+1}),\ldots,v_v H_{k_v}(\cdot,s^v)\right)$ is HA($\alpha$, $\beta$, $\breve{\gamma}$, $\delta$, $\eta$, $\omega$, $\breve{\rho}$, $\theta$)-V-quasisonvex at $y$ for $\breve{\gamma}(x,y) > 0$ and $\delta_m(x,y) > 0$ for each $m \in \underline{v}$ ;

(iv) $\bar{\rho}(x,y) + \tilde{\rho}(x,y) + \breve{\rho}(x,y) \geqq 0$;

(b) (i) $\left(\mathscr{E}_1(\cdot,u,\lambda),\ldots,\mathscr{E}_p(\cdot,u,\lambda)\right)$ is prestrictly HA($\alpha$, $\beta$, $\gamma$, $\xi$, $\eta$, $\omega$, $\rho$, $\theta$)-V-quasisonvex at $y$ for $\bar{\gamma}(x,y) > 0$ and $\xi(x,y) > 0$;

(ii) $\left(v_1 G_{j_1}(\cdot,t^1),\ldots,v_{v_0} G_{j_{v_0}}(\cdot,t^{v_0})\right)$ is HA($\alpha$, $\beta$, $\tilde{\gamma}$, $\pi$, $\eta$, $\omega$, $\tilde{\rho}$, $\theta$) -V-quasisonvex at $y$;

(iii) $\left(v_{v_0+1} H_{k_{v_0+1}}(\cdot,s^{v_0+1}),\ldots,v_v H_{k_v}(\cdot,s^v)\right)$ is $HA(\alpha,\beta,\breve{\gamma},\delta,\eta,\omega,\breve{\rho},\theta)$-V-quasisonvex at $y$;

(iv) $\bar{\rho}(x,y) + \tilde{\rho}(x,y) + \breve{\rho}(x,y) > 0$;

(c) (i) $\left(\mathscr{E}_1(\cdot,u,\lambda),\ldots,\mathscr{E}_p(\cdot,u,\lambda)\right)$ is prestrictly HA($\alpha$, $\beta$, $\bar{\gamma}$, $\xi$, $\eta$, $\omega$, $\bar{\rho}$, $\theta$)-V-quasisonvex at $y$ for $\bar{\gamma}(x,y) > 0$ and $\xi(x,y) > 0$;

(ii) $\left(v_1 G_{j_1}(\cdot,t^1),\ldots,v_{v_0} G_{j_{v_0}}(\cdot,t^{v_0})\right)$ is strictly $HA(\alpha,\beta,\tilde{\gamma},\pi,\eta,\omega,\tilde{\rho},\theta)$-V-pseudosonvex at $y$;

(iii) $\left(v_{v_0+1} H_{k_{v_0+1}}(\cdot,s^{v_0+1}),\ldots,v_v H_{k_v}(\cdot,s^v)\right)$ is $HA(\alpha,\beta,\breve{\gamma},\delta,\eta,\omega,\breve{\rho},\theta)$-V-quasisonvex at $y$;

(iv) $\bar{\rho}(x,y) + \tilde{\rho}(x,y) + \breve{\rho}(x,y) \geqq 0$;

(d) (i) $\left(\mathscr{E}_1(\cdot,u,\lambda),\ldots,\mathscr{E}_p(\cdot,u,\lambda)\right)$ is prestrictly HA($\alpha$, $\beta$, $\bar{\gamma}$, $\xi$, $\eta$, $\omega$, $\bar{\rho}$, $\theta$)-V-quasisonvex at $y$ for $\bar{\gamma}(x,y) > 0$ and $\xi(x,y) > 0$;

(ii) $\left(v_1 G_{j_1}(\cdot,t^1),\ldots,v_{v_0} G_{j_{v_0}}(\cdot,t^{v_0})\right)$ is HA($\alpha$, $\beta$, $\tilde{\gamma}$, $\pi$, $\eta$, $\omega$, $\tilde{\rho}$, $\theta$) - V-quasisonvex at $y$;

(iii) $\left(v_{v_0+1} H_{k_{v_0+1}}(\cdot,s^{v_0+1}),\ldots,v_v H_{k_v}(\cdot,s^v)\right)$ is strictly HA($\alpha$, $\beta$, $\breve{\gamma}$, $\delta$, $\eta$, $\omega$, $\breve{\rho}$, $\theta$)-V-
pseudosonvex at $y$;

(iv) $\bar{\rho}(x,y) + \tilde{\rho}(x,y) + \breve{\rho}(x,y) \geqq 0$.

Then $\varphi(x) \nleq \lambda$.

*Proof.* (a): Applying (8) and the primal feasibility of $x$, we have

$$G_{j_m}(x,t^m) \leqq 0 \leqq G_{j_m}(y,t^m) - \frac{1}{2}\langle \omega(x,y), \nabla^2 G_{j_m}(y,t^m) z\rangle \text{ for each } m \in \underline{v_0}.$$

(a) The assumptions specified in part (a) of Theorem 3.20 are satisfied for the feasible solution $\tilde{w}$ of (DII), and the function $\big(\mathscr{E}_1(\cdot,\tilde{u},\tilde{\lambda}),\ldots,\mathscr{E}_p(\cdot,\tilde{u},\tilde{\lambda})\big)$ is strictly $HA(\alpha,\beta,\bar{\gamma},\xi,\eta,\omega,\bar{\rho},\theta)$-V-pseudosonvex at $\tilde{x}$.

(b) The assumptions specified in part (b) of Theorem 3.20 are satisfied for the feasible solution $\tilde{w}$ of (DII), and the function $\big(\mathscr{E}_1(\cdot,\tilde{u},\tilde{\lambda}),\ldots,\mathscr{E}_p(\cdot,\tilde{u},\tilde{\lambda})\big)$ is $HA(\alpha,\beta,\bar{\gamma},\xi,\eta,\omega,\bar{\rho},\theta)$-V-quasisonvex at $\tilde{x}$.

(c) The assumptions specified in part (c) of Theorem 3.20 are satisfied for the feasible solution $\tilde{w}$ of (DII), and the function $\big(\mathscr{E}_1(\cdot,\tilde{u},\tilde{\lambda}),\ldots,\mathscr{E}_p(\cdot,\tilde{u},\tilde{\lambda})\big)$ is $HA(\alpha,\beta,\bar{\gamma},\xi,\eta,\bar{\rho},\theta)$-V-quasisonvex at $\tilde{x}$.

(d) The assumptions specified in part (d) of Theorem 3.20 are satisfied for the feasible solution $\tilde{w}$ of (DII), and the function $\big(\mathscr{E}_1(\cdot,\tilde{u},\tilde{\lambda}),\ldots,\mathscr{E}_p(\cdot,\tilde{u},\tilde{\lambda})\big)$ is HA($\alpha$, $\beta$, $\bar{\gamma}$, $\xi$, $\eta$, $\bar{\rho}$, $\theta$)-V-quasisonvex at $\tilde{x}$.

Then $\tilde{x}=x^*$ and $\varphi(x^*)=\tilde{\lambda}$.

*Proof.* (a) and (b) : The proofs are similar to that of Theorem 3.19.
(c) : Suppose to the contrary that $\tilde{x}\neq x^*$. Now proceeding as in the proof of part (c) of Theorem 3.20 (with $x$ replaced by $x^*$ and $w$ by $\tilde{w}$), we arrive at (13) and (14) which can be combined as follows:

$$-\frac{1}{\beta(x^*,\tilde{x})}\Big\langle\sum_{m=1}^{\tilde{v}_0}\tilde{v}_m\big[\nabla G_{j_m}(\tilde{x},\tilde{t}^m)+\nabla^2 G_{j_m}(\tilde{x},\tilde{t}^m)\tilde{z}\big]+\sum_{m=\tilde{v}_0+1}^{\tilde{v}}\tilde{v}_m\big[\nabla H_{k_m}(\tilde{x},\tilde{s}^m)+\nabla^2 H_{k_m}(\tilde{x},\tilde{s}^m)\tilde{z}\big],e^{\beta(x^*,\tilde{x})\eta(x^*,\tilde{x})}-\mathbf{1}\Big\rangle$$
$$>[\bar{\rho}(x^*,\tilde{x})+\breve{\rho}(x^*,\tilde{x})]\|\theta(x^*,\tilde{x})\|^2.$$

Combining this inequality with (6) and using (iv), we obtain

$$\frac{1}{\beta(x^*,\tilde{x})}\Big\langle\sum_{i=1}^{p}\big[\nabla\mathscr{E}_i(\tilde{x},\tilde{u},\tilde{\lambda})+\nabla^2\mathscr{E}_i(\tilde{x},\tilde{u},\tilde{\lambda})\tilde{z}\big],e^{\beta(x^*,\tilde{x})\eta(x^*,\tilde{x})}-\mathbf{1}\Big\rangle$$
$$>-\bar{\rho}(x^*,\tilde{x})\|\theta(x^*,\tilde{x})\|^2,$$

which by virtue of our $HA(\alpha,\beta,\bar{\gamma},\xi,\eta,\omega,\bar{\rho},\theta)$-V-quasisonvexity hypothesis implies that

$$\frac{1}{\alpha(x^*,\tilde{x})}\bar{\gamma}(x^*,\tilde{x})\Big(e^{\alpha(x^*,\tilde{x})\sum_{i=1}^{p}\xi_i(x^*,\tilde{x})\{\mathscr{E}_i(x^*,\tilde{u},\tilde{\lambda})-[\mathscr{E}_i(\tilde{x},\tilde{u},\tilde{\lambda})-\frac{1}{2}\langle\omega(x^*,\tilde{x}),\nabla^2\mathscr{E}_i(\tilde{x},\tilde{u},\tilde{\lambda})\tilde{z}\rangle]\}}-1\Big)>0.$$

Since $\bar{\gamma}(x^*,\tilde{x})>0$, this inequality implies that

$$\sum_{i=1}^{p}\xi_i(x^*,\tilde{x})\mathscr{E}_i(x^*,\tilde{u},\tilde{\lambda})>\sum_{i=1}^{p}\xi_i(x^*,\tilde{x})\big[\mathscr{E}_i(\tilde{x},\tilde{u},\tilde{\lambda})-\frac{1}{2}\langle\tilde{z},\nabla^2\mathscr{E}_i(\tilde{x},\tilde{u},\tilde{\lambda})\tilde{z}\rangle\big]\geqq 0,$$



Since $\pi_m(x,y)>0$ for each $m\in\underline{v_0}$, it follows that

$$\sum_{m=1}^{v_0}v_m G_{j_m}(x,t^m)\leqq\sum_{m=1}^{v_0}v_m\big[G_{j_m}(y,t^m)-\frac{1}{2}\langle\omega(x,y),\nabla^2 G_{j_m}(y,t^m)z\rangle\big],$$

and hence

$$\frac{1}{\alpha(x,y)}\tilde{\gamma}(x,y)\Big(e^{\alpha(x,y)\sum_{m=1}^{v_0}v_m\{G_{j_m}(x,t^m)-[G_{j_m}(y,t^m)-\frac{1}{2}\langle\omega(x,y),\nabla^2 G_{j_m}(y,t^m)z\rangle]\}}-1\Big)\leqq 0,$$

which in view of (ii) implies that

$$\Big\langle\sum_{m=1}^{v_0}\pi_m(x,y)\big[v_m\nabla G_{j_m}(y,t^m)+v_m\nabla^2 G_{j_m}(y,t^m)z\big],e^{\beta(x,y)\eta(x,y)}-\mathbf{1}\Big\rangle$$
$$\leqq-\tilde{\rho}(x,y)\|\theta(x,y)\|^2. \tag{3.10}$$

Similarly, we can show that our assumptions in (iii) combined with the primal feasibility of $x$ and (9) lead to the following inequality:

$$\Big\langle\sum_{m=v_0+1}^{v}\delta_m(x,y)\big[v_m\nabla H_{k_m}(y,s^m)+v_m\nabla^2 H_{k_m}(y,s^m)z\big],e^{\beta(x,y)\eta(x,y)}-\mathbf{1}\Big\rangle$$
$$\leqq-\breve{\rho}(x,y)\|\theta(x,y)\|^2. \tag{3.11}$$

Now because of (10), (11), and (iv), (6) yields

$$\frac{1}{\beta(x,y)}\Big\langle\sum_{i=1}^{p}\xi_i(x,y)\big[\nabla\mathscr{E}_i(y,u,\lambda)+\nabla^2\mathscr{E}_i(y,u,\lambda)z\big],e^{\beta(x,y)\eta(x,y)}-\mathbf{1}\Big\rangle$$
$$\geqq-\bar{\rho}(x,y)\|\theta(x,y)\|^2,$$

which in view of (i) implies that

$$\frac{1}{\alpha(x,y)}\bar{\gamma}(x,y)\Big(e^{\alpha(x,y)\sum_{i=1}^{p}\{\mathscr{E}_i(x,u,\lambda)-[\mathscr{E}_i(y,u,\lambda)-\frac{1}{2}\langle\omega(x,y),\nabla^2\mathscr{E}_i(y,u,\lambda)z\rangle]\}}-1\Big)\geq 0.$$

Since $\bar{\gamma}(x,y)>0$, this inequality implies that

$$\sum_{i=1}^{p}u_i\big[f_i(x)-\lambda_i g_i(x)\big]$$
$$\geqq\sum_{i=1}^{p}u_i\xi_i(x,y)\big\{f_i(y)-\lambda_i g_i(y)-\frac{1}{2}\langle\omega(x,y),[\nabla^2 f_i(y)-\lambda_i\nabla^2 g_i(y)]z\rangle\big\}\geqq 0,$$

where the last inequality follows from (7). Since for each $i\in\underline{p}$, $u_i\xi_i(x,y)>0$, this inequality implies that

$$\big(f_1(x)-\lambda_1 g_1(x),\ldots,f_p(x)-\lambda_p g_p(x)\big)\nleq(0,\ldots,0),$$

which in turn implies that

$$\varphi(x) = \left(\frac{f_1(x)}{g_1(x)}, \ldots, \frac{f_p(x)}{g_p(x)}\right) \nleq (\lambda_1, \ldots, \lambda_p) = \lambda.$$

(b) : The proof is similar to that of part (a).

(c) : Suppose to the contrary that $\varphi(x) \leqslant \lambda$. Then $f_i(x) - \lambda_i g_i(x) \leqq 0,\ i \in \underline{p}$, with strict inequality holding for at least one index $\ell \in \underline{p}$. Since $u > 0,\ \xi_i(x,y) > 0,\ i \in \underline{p}$, and (4.2) holds, we see that

$$\sum_{i=1}^{p} u_i[f_i(x) - \lambda_i g_i(x)]$$
$$< 0 \leqq \sum_{i=1}^{p} \xi_i(x,y) u_i \{f_i(y) - \lambda_i g_i(y) - \frac{1}{2}\langle \omega(x,y), [\nabla^2 f_i(y) - \lambda_i \nabla^2 g_i(y)]z\rangle\},$$

and hence

$$\frac{1}{\alpha(x,y)}\bar{\gamma}(x,y)\left(e^{\alpha(x,y)\sum_{i=1}^{p}\{\mathscr{E}_i(x,u,\lambda) - [\mathscr{E}_i(y,u,\lambda) - \frac{1}{2}\langle\omega(x,y), \nabla^2\mathscr{E}_i(y,u,\lambda)z\rangle]\}} - 1\right) < 0,$$

which in view of (i) implies that

$$\frac{1}{\beta(x,y)}\left\langle \sum_{i=1}^{p} \xi_i(x,y)\left[\nabla\mathscr{E}_i(y,u,\lambda) + \nabla^2\mathscr{E}_i(y,u,\lambda)z\right], e^{\beta(x,y)\eta(x,y)} - \mathbf{1}\right\rangle$$
$$\leqq -\bar{\rho}(x,y)\|\theta(x,y)\|^2. \tag{3.12}$$

Because of (8) and the primal feasibility of $x$, we have

$$G_{j_m}(x,t^m) \leqq 0 \leqq G_{j_m}(y,t^m) - \frac{1}{2}\langle\omega(x,y), \nabla^2 G_{j_m}(y,t^m)z\rangle \ \text{ for each } m \in \underline{\nu_0},$$

and hence

$$\sum_{m=1}^{\nu_0} v_m \pi_m(x,y) G_{j_m}(x,t^m)$$
$$\leqq \sum_{m=1}^{\nu_0} v_m \pi_m(x,y)\left[G_{j_m}(y,t^m) - \frac{1}{2}\langle\omega(x,y), \nabla^2 G_{j_m}(y,t^m)z\rangle\right].$$

From this we see that

$$\frac{1}{\alpha(x,y)}\tilde{\gamma}(x,y)$$
$$\left(e^{\alpha(x,y)\sum_{m=1}^{\nu_0} v_m\{G_{jm}(x,t^m) - [G_{jm}(y,t^m) - \frac{1}{2}\langle\omega(x,y), \nabla^2 G_{jm}(y,t^m)z\rangle]\}} - 1\right) \leqq 0,$$

which in view of (ii) implies that

$$\frac{1}{\beta(x,y)}$$
$$\left\langle \sum_{m=1}^{\nu_0} \pi_m(x,y)\left[v_m \nabla G_{j_m}(y,t^m) + v_m \nabla^2 G_{j_m}(y,t^m)z\right], e^{\beta(x,y)\eta(x,y)} - \mathbf{1}\right\rangle$$
$$< -\tilde{\rho}(x,y)\|\theta(x,y)\|^2. \tag{3.13}$$



Similarly, we can show that our assumptions in (iii) combined with the primal $x$ and (9) lead to the following inequality:

$$\frac{1}{\beta(x,y)}$$
$$\left\langle \sum_{m=\nu_0+1}^{\nu} \delta_m(x,y)\left[v_m \nabla H_{k_m}(y,s^m) + v_m \nabla^2 H_{k_m}(y,s^m)z\right], e^{\beta(x,y)\eta(x,y)} - \right.$$
$$\leqq -\breve{\rho}(x,y)\|\theta(x,y)\|^2.$$

Now combining (12) - (14) and using (iv), we get

$$\frac{1}{\beta(x,y)}\left\langle \sum_{i=1}^{p} u_i[\nabla f_i(y) - \lambda_i \nabla g_i(y)]\right.$$
$$+ \sum_{m=1}^{\nu_0} v_m \nabla G_{j_m}(y,t^m) + \sum_{m=\nu_0+1}^{\nu} v_m \nabla H_{k_m}(y,s^m) +$$
$$\left\{\sum_{i=1}^{p} u_i[\nabla^2 f_i(y) - \lambda_i \nabla^2 g_i(y)] + \sum_{m=1}^{\nu_0} v_m \nabla^2 G_{j_m}(y,t^m) + \right.$$
$$\left.\left.\sum_{m=\nu_0+1}^{\nu} v_m \nabla^2 H_{k_m}(y,s^m)\right\}z, e^{\beta(x,y)\eta(x,y)} - \mathbf{1}\right\rangle$$
$$< -[\bar{\rho}(x,y) + \tilde{\rho}(x,y) + \breve{\rho}(x,y)]\|\theta(x,y)\|^2 \leqq 0$$

which contradicts (6). Therefore, we conclude that $\varphi(x) \nleq \lambda$.

(d) : The proof is similar to that of part (c).

It would be a good idea to prove this theorem without the method of contradictio compare the methods.

**Theorem 3.21.** (Strong Duality) Let $x^*$ be a normal efficient solution of (P) and assur for each feasible solution $(y, z, u, v, \lambda, \nu, \nu_0, J_{\nu_0}, K_{\nu\backslash\nu_0}, \bar{t}, \bar{s})$ of (DII), any one of the fo of conditions specified in Theorem 3.20 is satisfied. Then there exist $u^*, v^*, \lambda^*, \nu^*,$ $K_{\nu^*\backslash\nu_0^*}, \bar{t}^*$, and $\bar{s}^*$ such that $(x^*, z^* = 0, u^*, v^*, \lambda^*, \nu^*, \nu_0^*, J_{\nu_0^*}, K_{\nu^*\backslash\nu_0^*}, \bar{t}^*, \bar{s}^*)$ is an e solution of (DII) and $\varphi(x^*) = \lambda^*$.

*Proof.* The proof is similar to that of Theorem 3.18.

**Theorem 3.22.** (Strict Converse Duality) Let $x^*$ and $\tilde{w} \equiv (\tilde{x}, \tilde{z}, \tilde{u}, \tilde{v}, \tilde{\lambda}, \tilde{\nu}, \tilde{\nu}_0, J_{\tilde{\nu}_0}, K$ be arbitrary feasible solutions of (P) and (DII), respectively, such that

$$\sum_{i=1}^{p} \tilde{u}_i \xi_i(x^*, \tilde{x})[f_i(x^*) - \tilde{\lambda}_i g_i(x^*)] = 0,$$

where $\xi_i,\ i \in \underline{p}$, are given real-valued functions from $X \times X$ to $\mathbb{R}_+\backslash\{0\}$. Fı assume that any one of the following four sets of hypotheses is satisfied:

where the second inequality follows from (7). This inequality contradicts (15) and, therefore, we must have $\tilde{x} = x^*$ and $\varphi(x^*) = \tilde{\lambda}$.
(d) : The proof is similar to that of part (c).

□

## 3.3. Duality Model III

In this section, we discuss several families of duality results under various generalized $HA(\alpha, \beta, \gamma, \xi, \eta, \omega, \rho, \theta)$-V-sonvexity hypotheses imposed on certain vector functions whose components are formed by considering different combinations of the problem functions. This is accomplished by employing a partitioning scheme which was originally proposed in [7] for the purpose of constructing generalized dual problems for nonlinear programming problems. Now we intoduce some additional notations.
Let $\nu_0$ and $\nu$ be integers, with $1 \leqq \nu_0 \leqq \nu \leqq n+1$, and let $\{J_0, J_1, \ldots, J_M\}$ and $\{K_0, K_1, \ldots, K_M\}$ be partitions of the sets $\underline{\nu_0}$ and $\underline{\nu} \backslash \underline{\nu_0}$, respectively; thus, $J_i \subseteq \underline{\nu_0}$ for each $i \in \underline{M} \cup \{0\}$, $J_i \cap J_j = \emptyset$ for each $i, j \in \underline{M} \cup \{0\}$ with $i \neq j$, and $\cup_{i=0}^{M} J_i = \underline{\nu_0}$. Obviously, similar properties hold for $\{K_0, K_1, \ldots, K_M\}$. Moreover, if $m_1$ and $m_2$ are the numbers of the partitioning sets of $\underline{\nu_0}$ and $\underline{\nu} \backslash \underline{\nu_0}$, respectively, then $M = \max\{m_1, m_2\}$ and $J_i = \emptyset$ or $K_i = \emptyset$ for $i > \min\{m_1, m_2\}$. In addition, we use the real-valued functions $\Phi_i(\cdot, u, v, \lambda, \bar{t}, \bar{s})$ and $\Lambda_\tau(\cdot, v, \bar{t}, \bar{s})$ defined, for fixed $u, v, \lambda, \bar{t} \equiv (t^1, t^2, \ldots, t^{\nu_0})$, and $\bar{s} \equiv (s^{\nu_0+1}, s^{\nu_0+2}, \ldots, s^{\nu})$, on $X$ as follows:

$$\Phi_i(\zeta, u, v, \lambda, \bar{t}, \bar{s}) = u_i \Big[ f_i(\zeta) - \lambda_i g_i(\zeta) + \sum_{m \in J_0} v_m G_{j_m}(\zeta, t^m) + \sum_{m \in K_0} v_m H_{k_m}(\zeta, s^m) \Big], \quad i \in \underline{p},$$

$$\Lambda_\tau(\zeta, v, \bar{t}, \bar{s}) = \sum_{m \in J_\tau} v_m G_{j_m}(\zeta, t^m) + \sum_{m \in K_\tau} v_m H_{k_m}(\zeta, s^m), \quad \tau \in \underline{M}.$$

Making use of the sets and functions defined above, we can state our general duality models as follows:

$$(DIII) \qquad \sup_{(y,z,u,v,\lambda,\nu,\nu_0,J_{\nu_0},K_{\nu\backslash\nu_0},\bar{t},\bar{s}) \in \mathbb{H}} \lambda = (\lambda_1, \cdots, \lambda_p)$$

subject to

$$\sum_{i=1}^{p} u_i [\nabla f_i(y) - \lambda_i \nabla g_i(y)] + \sum_{m=1}^{\nu_0} v_m \nabla G_{j_m}(y, t^m) + \sum_{m=\nu_0+1}^{\nu} v_m \nabla H_{k_m}(y, s^m) + \Big\{ \sum_{i=1}^{p} u_i [\nabla^2 f_i(y) - \lambda_i \nabla^2 g_i(y)] + \sum_{m=1}^{\nu_0} v_m \nabla^2 G_{j_m}(y, t^m) + \sum_{m=\nu_0+1}^{\nu} v_m \nabla^2 H_{k_m}(y, s^m) \Big\} z = 0, \quad (3.16)$$

$$f_i(y) - \lambda_i g_i(y) + \sum_{m \in J_0} v_m G_{j_m}(y,t^m) + \sum_{m \in K_0} v_m H_{k_m}(y,s^m)$$
$$-\frac{1}{2}\Big\langle \omega(x,y), \Big[\nabla^2 f_i(y) - \lambda_i \nabla^2 g_i(y) + \sum_{m \in J_0} v_m \nabla^2 G_{j_m}(y,t^m) +$$
$$\sum_{m \in K_0} v_m \nabla^2 H_{k_m}(y,s^m)\Big] z\Big\rangle \geqq 0, \ i \in \underline{p}, \quad (3.17)$$

$$\sum_{m \in J_\tau} v_m G_{j_m}(y,t^m) + \sum_{m \in K_\tau} v_m H_{k_m}(y,s^m) - \frac{1}{2}\Big\langle \omega(x,y), \Big[\sum_{m \in J_\tau} v_m \nabla^2 G_{j_m}(y,t^m) +$$
$$\sum_{m \in K_\tau} v_m \nabla^2 H_{k_m}(y,s^m)\Big] z\Big\rangle \geqq 0, \ \tau \in \underline{M}, \quad (3.18)$$

for all $x \in \mathbb{F}$;

$$(\tilde{D}III) \qquad \sup_{(y,z,u,v,\lambda,\nu,\nu_0,J_{\nu_0},K_{\nu\backslash\nu_0},\bar{t},\bar{s}) \in \mathbb{H}} \lambda = (\lambda_1, \ldots, \lambda_p)$$

subject to (3), (17) and (18).

The remarks and observations made earlier about the relationships between $(DI)$ and $(\tilde{D}I)$ are, of course, also valid for $(DIII)$ and $(\tilde{D}III)$.

The next two theorems show that $(DIII)$ is a dual problem for $(P)$.

**Theorem 3.23.** Let $x$ and $(y,z,u,v,\lambda,\nu,\nu_0,J_{\nu_0},K_{\nu\backslash\nu_0},\bar{t},\bar{s})$ be arbitrary feasible solutions of (P) and (DIII), respectively, and assume that any one of the following three sets of hypotheses is satisfied:

(a) (i) $(\Phi_1(\cdot,u,v,\lambda,\bar{t},\bar{s}),\ldots,\Phi_p(\cdot,u,v,\lambda,\bar{t},\bar{s}))$ is $HA(\alpha,\beta,\bar{\gamma},\xi,\eta,\omega,\bar{\rho},\theta)$-V-pseudosonvex at $y$ for $\bar{\gamma}(x,y) > 0$ and $\xi(x,y) > 0$;

(ii) $(\Lambda_1(\cdot,v,\bar{t},\bar{s}),\ldots,\Lambda_M(\cdot,v,\bar{t},\bar{s}))$ is HA($\alpha$, $\beta$, $\tilde{\gamma}$, $\pi$, $\eta$, $\omega$, $\tilde{\rho}$, $\theta$) - V-quasisonvex at $y$;

(iii) $\bar{\rho}(x,y) + \tilde{\rho}(x,y) \geqq 0$;

(b) (i) $(\Phi_1(\cdot,u,v,\lambda,\bar{t},\bar{s}),\ldots,\Phi_p(\cdot,u,v,\lambda,\bar{t},\bar{s}))$ is prestrictly HA($\alpha$, $\beta$, $\bar{\gamma}$, $\xi$, $\eta$, $\omega$, $\bar{\rho}$, $\theta$)-V-
quasisonvex at $y$ for $\bar{\gamma}(x,y) > 0$ and $\xi(x,y) > 0$;

(ii) $(\Lambda_1(\cdot,v,\bar{t},\bar{s}),\ldots,\Lambda_M(\cdot,v,\bar{t},\bar{s}))$ is HA($\alpha$, $\beta$, $\tilde{\gamma}$, $\pi$, $\eta$, $\omega$, $\tilde{\rho}$, $\theta$)-V-quasisonvex at $y$;

(iii) $\bar{\rho}(x,y) + \tilde{\rho}(x,y) > 0$;

(c) (i) $(\Phi_1(\cdot,u,v,\lambda,\bar{t},\bar{s}),\ldots,\Phi_p(\cdot,u,v,\lambda,\bar{t},\bar{s}))$ is prestrictly HA($\alpha$, $\beta$, $\bar{\gamma}$, $\xi$, $\eta$, $\omega$, $\bar{\rho}$, $\theta$)-V-
quasisonvex at $y$ for $\bar{\gamma}(x,y) > 0$ and $\xi(x,y) > 0$;

(ii) $(\Lambda_1(\cdot,v,\bar{t},\bar{s}),\ldots,\Lambda_M(\cdot,v,\bar{t},\bar{s}))$ is strictly HA($\alpha$, $\beta$, $\tilde{\gamma}$, $\pi$, $\eta$, $\omega$, $\tilde{\rho}$, $\theta$) -V-pseudosonvex at $y$;

Since $\pi_m(x,y) > 0$ for each $m \in \underline{\nu_0}$, it follows that

$$\sum_{m=1}^{\nu_0} v_m G_{j_m}(x,t^m) \leqq \sum_{m=1}^{\nu_0} v_m \left[G_{j_m}(y,t^m) - \frac{1}{2}\langle \omega(x,y), \nabla^2 G_{j_m}(y,t^m)z\rangle\right],$$

and hence

$$\frac{1}{\alpha(x,y)}\tilde{\gamma}(x,y)$$
$$\left(e^{\alpha(x,y)\sum_{m=1}^{\nu_0} v_m\{G_{jm}(x,t^m)-[G_{jm}(y,t^m)-\frac{1}{2}\langle\omega(x,y),\nabla^2 G_{jm}(y,t^m)z\rangle]\}} - 1\right) \leqq 0,$$

which in view of (ii) implies that

$$\left\langle \sum_{m=1}^{\nu_0} \pi_m(x,y)\left[v_m \nabla G_{j_m}(y,t^m) + v_m \nabla^2 G_{j_m}(y,t^m)z\right], e^{\beta(x,y)\eta(x,y)} - \mathbf{1}\right\rangle$$
$$\leqq -\tilde{\rho}(x,y)\|\theta(x,y)\|^2. \tag{3.10}$$

Similarly, we can show that our assumptions in (iii) combined with the primal feasibility of $x$ and (9) lead to the following inequality:

$$\left\langle \sum_{m=\nu_0+1}^{\nu} \delta_m(x,y)\left[v_m \nabla H_{k_m}(y,s^m) + v_m \nabla^2 H_{k_m}(y,s^m)z\right], e^{\beta(x,y)\eta(x,y)} - \mathbf{1}\right\rangle$$
$$\leqq -\breve{\rho}(x,y)\|\theta(x,y)\|^2. \tag{3.11}$$

Now because of (10), (11), and (iv), (6) yields

$$\frac{1}{\beta(x,y)}\left\langle \sum_{i=1}^{p} \xi_i(x,y)\left[\nabla \mathscr{E}_i(y,u,\lambda) + \nabla^2 \mathscr{E}_i(y,u,\lambda)z\right], e^{\beta(x,y)\eta(x,y)} - \mathbf{1}\right\rangle$$
$$\geqq -\bar{\rho}(x,y)\|\theta(x,y)\|^2,$$

which in view of (i) implies that

$$\frac{1}{\alpha(x,y)}\bar{\gamma}(x,y)\left(e^{\alpha(x,y)\sum_{i=1}^{p}\{\mathscr{E}_i(x,u,\lambda)-[\mathscr{E}_i(y,u,\lambda)-\frac{1}{2}\langle\omega(x,y),\nabla^2\mathscr{E}_i(y,u,\lambda)z\rangle]\}} - 1\right) \geqq 0.$$

Since $\bar{\gamma}(x,y) > 0$, this inequality implies that

$$\sum_{i=1}^{p} u_i\left[f_i(x) - \lambda_i g_i(x)\right]$$
$$\geqq \sum_{i=1}^{p} u_i\xi_i(x,y)\left\{f_i(y) - \lambda_i g_i(y) - \frac{1}{2}\langle \omega(x,y), [\nabla^2 f_i(y) - \lambda_i \nabla^2 g_i(y)]z\rangle\right\} \geqq 0,$$

where the last inequality follows from (7). Since for each $i \in \underline{p}$, $u_i\xi_i(x,y) > 0$, this inequality implies that

$$\left(f_1(x) - \lambda_1 g_1(x), \ldots, f_p(x) - \lambda_p g_p(x)\right) \nleq (0,\ldots,0),$$

which in turn implies that

$$\varphi(x) = \left(\frac{f_1(x)}{g_1(x)}, \ldots, \frac{f_p(x)}{g_p(x)}\right) \nleq (\lambda_1, \ldots, \lambda_p) = \lambda.$$

(b) : The proof is similar to that of part (a).

(c) : Suppose to the contrary that $\varphi(x) \leqslant \lambda$. Then $f_i(x) - \lambda_i g_i(x) \leqq 0$, $i \in \underline{p}$, with strict inequality holding for at least one index $\ell \in \underline{p}$. Since $u > 0$, $\xi_i(x,y) > 0$, $i \in \underline{p}$, and (4.2) holds, we see that

$$\sum_{i=1}^{p} u_i[f_i(x) - \lambda_i g_i(x)]$$
$$< 0 \leqq \sum_{i=1}^{p} \xi_i(x,y) u_i \{ f_i(y) - \lambda_i g_i(y) - \frac{1}{2}\langle \omega(x,y), [\nabla^2 f_i(y) - \lambda_i \nabla^2 g_i(y)] z\rangle \},$$

and hence

$$\frac{1}{\alpha(x,y)} \bar{\gamma}(x,y) \left(e^{\alpha(x,y)\sum_{i=1}^{p}\{\mathscr{E}_i(x,u,\lambda) - [\mathscr{E}_i(y,u,\lambda) - \frac{1}{2}\langle \omega(x,y), \nabla^2 \mathscr{E}_i(y,u,\lambda) z\rangle]\}} - 1\right) < 0,$$

which in view of (i) implies that

$$\frac{1}{\beta(x,y)} \left\langle \sum_{i=1}^{p} \xi_i(x,y) \left[\nabla \mathscr{E}_i(y,u,\lambda) + \nabla^2 \mathscr{E}_i(y,u,\lambda) z\right], e^{\beta(x,y)\eta(x,y)} - \mathbf{1} \right\rangle$$
$$\leqq -\bar{\rho}(x,y)\|\theta(x,y)\|^2. \tag{3.12}$$

Because of (8) and the primal feasibility of $x$, we have

$$G_{j_m}(x,t^m) \leqq 0 \leqq G_{j_m}(y,t^m) - \frac{1}{2}\langle \omega(x,y), \nabla^2 G_{j_m}(y,t^m) z\rangle \text{ for each } m \in \underline{v_0},$$

and hence

$$\sum_{m=1}^{v_0} v_m \pi_m(x,y) G_{j_m}(x,t^m)$$
$$\leqq \sum_{m=1}^{v_0} v_m \pi_m(x,y) \left[G_{j_m}(y,t^m) - \frac{1}{2}\langle \omega(x,y), \nabla^2 G_{j_m}(y,t^m) z\rangle\right].$$

From this we see that

$$\frac{1}{\alpha(x,y)} \tilde{\gamma}(x,y)$$
$$\left(e^{\alpha(x,y)\sum_{m=1}^{v_0} v_m\{G_{j_m}(x,t^m) - [G_{j_m}(y,t^m) - \frac{1}{2}\langle \omega(x,y), \nabla^2 G_{j_m}(y,t^m) z\rangle]\}} - 1\right) \leqq 0,$$

which in view of (ii) implies that

$$\frac{1}{\beta(x,y)}$$
$$\left\langle \sum_{m=1}^{v_0} \pi_m(x,y) \left[v_m \nabla G_{j_m}(y,t^m) + v_m \nabla^2 G_{j_m}(y,t^m) z\right], e^{\beta(x,y)\eta(x,y)} - \mathbf{1} \right\rangle$$
$$< -\tilde{\rho}(x,y)\|\theta(x,y)\|^2. \tag{3.13}$$

Similarly, we can show that our assumptions in (iii) combined with the primal feasibility of $x$ and (9) lead to the following inequality:

$$\frac{1}{\beta(x,y)} \Big\langle \sum_{m=\nu_0+1}^{\nu} \delta_m(x,y)\left[v_m \nabla H_{k_m}(y,s^m) + v_m \nabla^2 H_{k_m}(y,s^m) z\right], e^{\beta(x,y)\eta(x,y)} - \mathbf{1} \Big\rangle \leqq -\breve{\rho}(x,y)\|\theta(x,y)\|^2. \tag{3.14}$$

Now combining (12) - (14) and using (iv), we get

$$\frac{1}{\beta(x,y)} \Big\langle \sum_{i=1}^{p} u_i[\nabla f_i(y) - \lambda_i \nabla g_i(y)] + \sum_{m=1}^{\nu_0} v_m \nabla G_{j_m}(y,t^m) + \sum_{m=\nu_0+1}^{\nu} v_m \nabla H_{k_m}(y,s^m) + \Big\{ \sum_{i=1}^{p} u_i[\nabla^2 f_i(y) - \lambda_i \nabla^2 g_i(y)] + \sum_{m=1}^{\nu_0} v_m \nabla^2 G_{j_m}(y,t^m) + \sum_{m=\nu_0+1}^{\nu} v_m \nabla^2 H_{k_m}(y,s^m) \Big\} z, e^{\beta(x,y)\eta(x,y)} - \mathbf{1} \Big\rangle < -[\bar{\rho}(x,y) + \tilde{\rho}(x,y) + \breve{\rho}(x,y)]\|\theta(x,y)\|^2 \leqq 0,$$

which contradicts (6). Therefore, we conclude that $\varphi(x) \nleq \lambda$.
(d) : The proof is similar to that of part (c).
It would be a good idea to prove this theorem without the method of contradiction and compare the methods. □

**Theorem 3.21.** (Strong Duality) Let $x^*$ be a normal efficient solution of (P) and assume that for each feasible solution $(y,z,u,v,\lambda,\nu,\nu_0,J_{\nu_0},K_{\nu\backslash\nu_0},\bar{t},\bar{s})$ of (DII), any one of the four sets of conditions specified in Theorem 3.20 is satisfied. Then there exist $u^*, v^*, \lambda^*, \nu^*, \nu_0^*, J_{\nu_0^*}, K_{\nu^*\backslash\nu_0^*}, \bar{t}^*$, and $\bar{s}^*$ such that $(x^*, z^* = 0, u^*, v^*, \lambda^*, \nu^*, \nu_0^*, J_{\nu_0^*}, K_{\nu^*\backslash\nu_0^*}, \bar{t}^*, \bar{s}^*)$ is an efficient solution of (DII) and $\varphi(x^*) = \lambda^*$.

*Proof.* The proof is similar to that of Theorem 3.18. □

**Theorem 3.22.** (Strict Converse Duality) Let $x^*$ and $\tilde{w} \equiv (\tilde{x},\tilde{z},\tilde{u},\tilde{v},\tilde{\lambda},\tilde{\nu},\tilde{\nu}_0,J_{\tilde{\nu}_0},K_{\tilde{\nu}\backslash\tilde{\nu}_0},\bar{\tilde{t}},\bar{\tilde{s}})$ be arbitrary feasible solutions of (P) and (DII), respectively, such that

$$\sum_{i=1}^{p} \tilde{u}_i \xi_i(x^*,\tilde{x})[f_i(x^*) - \tilde{\lambda}_i g_i(x^*)] = 0, \tag{3.15}$$

where $\xi_i$, $i \in \underline{p}$, are given real-valued functions from $X \times X$ to $\mathbb{R}_+ \backslash \{0\}$. Furthermore, assume that any one of the following four sets of hypotheses is satisfied:

(a) The assumptions specified in part (a) of Theorem 3.20 are satisfied for the feasible solution $\tilde{w}$ of (DII), and the function $\big(\mathscr{E}_1(\cdot,\tilde{u},\tilde{\lambda}),\ldots,$ $\mathscr{E}_p(\cdot,\tilde{u},\tilde{\lambda})\big)$ is strictly $HA(\alpha,\beta,\bar{\gamma},\xi,$ $\eta,\omega,\bar{\rho},\theta)$-V-pseudosonvex at $\tilde{x}$.

(b) The assumptions specified in part (b) of Theorem 3.20 are satisfied for the feasible solution $\tilde{w}$ of (DII), and the function $\big(\mathscr{E}_1(\cdot,\tilde{u},\tilde{\lambda}),\ldots,$ $\mathscr{E}_p(\cdot,\tilde{u},\tilde{\lambda})\big)$ is $HA(\alpha,\beta,\bar{\gamma},\xi,\eta,\omega,\bar{\rho},\theta)$-V-quasisonvex at $\tilde{x}$.

(c) The assumptions specified in part (c) of Theorem 3.20 are satisfied for the feasible solution $\tilde{w}$ of (DII), and the function $\big(\mathscr{E}_1(\cdot,\tilde{u},\tilde{\lambda}),\ldots,$ $\mathscr{E}_p(\cdot,\tilde{u},\tilde{\lambda})\big)$ is $HA(\alpha,\beta,\bar{\gamma},\xi,\eta,\bar{\rho},\theta)$-V-quasisonvex at $\tilde{x}$.

(d) The assumptions specified in part (d) of Theorem 3.20 are satisfied for the feasible solution $\tilde{w}$ of (DII), and the function $\big(\mathscr{E}_1(\cdot,\tilde{u},\tilde{\lambda}),\ldots,$ $\mathscr{E}_p(\cdot,\tilde{u},\tilde{\lambda})\big)$ is HA($\alpha$, $\beta$, $\bar{\gamma}$, $\xi$, $\eta$, $\bar{\rho}$, $\theta$)-V-quasisonvex at $\tilde{x}$.

Then $\tilde{x}=x^*$ and $\varphi(x^*)=\tilde{\lambda}$.

*Proof.* (a) and (b) : The proofs are similar to that of Theorem 3.19.
(c) : Suppose to the contrary that $\tilde{x}\neq x^*$. Now proceeding as in the proof of part (c) of Theorem 3.20 (with $x$ replaced by $x^*$ and $w$ by $\tilde{w}$), we arrive at (13) and (14) which can be combined as follows:

$$-\frac{1}{\beta(x^*,\tilde{x})}\Big\langle\sum_{m=1}^{\tilde{v}_0}\tilde{v}_m\big[\nabla G_{j_m}(\tilde{x},\tilde{t}^m)+\nabla^2 G_{j_m}(\tilde{x},\tilde{t}^m)\tilde{z}\big]+\sum_{m=\tilde{v}_0+1}^{\tilde{v}}\tilde{v}_m\big[\nabla H_{k_m}(\tilde{x},\tilde{s}^m)+\nabla^2 H_{k_m}(\tilde{x},\tilde{s}^m)\tilde{z}\big],e^{\beta(x^*,\tilde{x})\eta(x^*,\tilde{x})}-\mathbf{1}\Big\rangle$$
$$>[\tilde{\rho}(x^*,\tilde{x})+\check{\rho}(x^*,\tilde{x})]\|\theta(x^*,\tilde{x})\|^2.$$

Combining this inequality with (6) and using (iv), we obtain

$$\frac{1}{\beta(x^*,\tilde{x})}\Big\langle\sum_{i=1}^{p}\big[\nabla\mathscr{E}_i(\tilde{x},\tilde{u},\tilde{\lambda})+\nabla^2\mathscr{E}_i(\tilde{x},\tilde{u},\tilde{\lambda})\tilde{z}\big],e^{\beta(x^*,\tilde{x})\eta(x^*,\tilde{x})}-\mathbf{1}\Big\rangle$$
$$>-\bar{\rho}(x^*,\tilde{x})\|\theta(x^*,\tilde{x})\|^2,$$

which by virtue of our $HA(\alpha,\beta,\bar{\gamma},\xi,\eta,\omega,\bar{\rho},\theta)$-V-quasisonvexity hypothesis implies that

$$\frac{1}{\alpha(x^*,\tilde{x})}\bar{\gamma}(x^*,\tilde{x})\Big(e^{\alpha(x^*,\tilde{x})\sum_{i=1}^{p}\xi_i(x^*,\tilde{x})\{\mathscr{E}_i(x^*,\tilde{u},\tilde{\lambda})-[\mathscr{E}_i(\tilde{x},\tilde{u},\tilde{\lambda})-\frac{1}{2}\langle\omega(x^*,\tilde{x}),\nabla^2\mathscr{E}_i(\tilde{x},\tilde{u},\tilde{\lambda})\tilde{z}\rangle]\}}-1\Big)>0.$$

Since $\bar{\gamma}(x^*,\tilde{x})>0$, this inequality implies that

$$\sum_{i=1}^{p}\xi_i(x^*,\tilde{x})\mathscr{E}_i(x^*,\tilde{u},\tilde{\lambda})>\sum_{i=1}^{p}\xi_i(x^*,\tilde{x})\big[\mathscr{E}_i(\tilde{x},\tilde{u},\tilde{\lambda})-\frac{1}{2}\langle\tilde{z},\nabla^2\mathscr{E}_i(\tilde{x},\tilde{u},\tilde{\lambda})\tilde{z}\rangle\big]\geqq 0,$$

(iii) $\bar{\rho}(x,y)+\tilde{\rho}(x,y) \geqq 0$.

Then $\varphi(x) \nleq \lambda$.

*Proof.* (a): It is clear that (16) can be expressed as follows:

$$\begin{aligned}
\frac{1}{\beta(x,y)}\Big\langle & \sum_{i=1}^{p} u_i[\nabla f_i(y)-\lambda_i \nabla g_i(y)] + \sum_{m\in J_0} v_m \nabla G_{j_m}(y,t^m) \\
&+ \sum_{m\in K_0} v_m \nabla H_{k_m}(y,s^m) + \\
&\Big\{\sum_{i=1}^{p} u_i[\nabla^2 f_i(y)-\lambda_i \nabla^2 g_i(y)] \\
&+ \sum_{m\in J_0} v_m \nabla^2 G_{j_m}(y,t^m) + \sum_{m\in K_0} v_m \nabla^2 H_{k_m}(y,s^m)\Big\} z \\
&+ \sum_{\tau=1}^{M}\Big\{\sum_{m\in J_\tau} v_m \nabla G_{j_m}(y,t^m) \\
&+ \sum_{m\in K_\tau} v_m \nabla H_{k_m}(y,s^m) \\
&+ \Big[\sum_{m\in J_\tau} v_m \nabla^2 G_{j_m}(y,t^m) + \\
&\sum_{m\in K_\tau} v_m \nabla^2 H_{k_m}(y,s^m)\Big] z\Big\}, e^{\beta(x,y)\eta(x,y)} - \mathbf{1}\Big\rangle = 0. \qquad (3.19)
\end{aligned}$$

Since $x\in\mathbb{F}$, $v_m>0$, $m\in \underline{v_0}$, and (18) holds, it follows that for each $\tau\in \underline{M}$,

$$\begin{aligned}
\Lambda_\tau(x,v,\bar{t},\bar{s}) &= \sum_{m\in J_\tau} v_m G_{j_m}(x,t^m) + \sum_{m\in K_\tau} v_m H_{k_m}(x,s^m) \\
&\leqq 0 \\
&\leqq \Lambda_\tau(y,v,\bar{t},\bar{s}) - \frac{1}{2}\Big\langle \omega(x,y), \nabla^2\Lambda_\tau(y,v,\bar{t},\bar{s})z\Big\rangle.
\end{aligned}$$

As $\pi_\tau(x,y)>0$ for each $\tau\in M$, the above inequalities yield

$$\begin{aligned}
&\sum_{\tau=1}^{M} \pi_\tau(x,y)\Lambda_\tau(x,v,\bar{t},\bar{s}) \\
&\leqq \sum_{\tau=1}^{M} \pi_\tau(x,y)\big[\Lambda_\tau(y,v,\bar{t},\bar{s}) - \frac{1}{2}\langle \omega(x,y), \nabla^2\Lambda_\tau(y,v,\bar{t},\bar{s})z\rangle\big],
\end{aligned}$$

and hence

$$\begin{aligned}
&\frac{1}{\alpha(x,y)}\tilde{\gamma}(x,y) \\
&\Big(e^{\alpha(x,y)\sum_{\tau=1}^{M}\pi_\tau(x,y)[\Lambda_\tau(x,v,\bar{t},\bar{s})-\Lambda_\tau(y,v,\bar{t},\bar{s})+\frac{1}{2}\langle z,\nabla^2\Lambda_\tau(y,v,\bar{t},\bar{s})z\rangle]} - 1\Big) \leqq 0,
\end{aligned}$$

which in view of (ii) implies that

$$\frac{1}{\beta(x,y)}\Big\langle \sum_{\tau=1}^{M}\Big\{\sum_{m\in J_\tau} v_m\nabla G_{j_m}(y,t^m)+\sum_{m\in K_\tau} v_m\nabla H_{k_m}(y,s^m) + \Big[\sum_{m\in J_\tau} v_m\nabla^2 G_{j_m}(y,t^m) + \sum_{m\in K_\tau} v_m\nabla^2 H_{k_m}(y,s^m)\Big]z\Big\}, e^{\beta(x,y)\eta(x,y)}-\mathbf{1}\Big\rangle \leqq -\tilde{\rho}(x,y)\|\theta(x,y)\|^2. \quad (3.20)$$

Combining (19) and (20), and using (iii) we get

$$\frac{1}{\beta(x,y)}\Big\langle \sum_{i=1}^{p} u_i[\nabla f_i(y)-\lambda_i\nabla g_i(y)]+\sum_{m\in J_0} v_m\nabla G_{j_m}(y,t^m) + \sum_{m\in K_0} v_m\nabla H_{k_m}(y,s^m) + \Big\{\sum_{i=1}^{p} u_i[\nabla^2 f_i(y)-\lambda_i\nabla^2 g_i(y)]+\sum_{m\in J_0} v_m\nabla^2 G_{j_m}(y,t^m) + \sum_{m\in K_0} v_m\nabla^2 H_{k_m}(y,s^m\Big\}z, e^{\beta(x,y)\eta(x,y)}-\mathbf{1}\Big\rangle \geqq \tilde{\rho}(x,y)\|\theta(x,y)\|^2 \geqq -\bar{\rho}(x,y)\|\theta(x,y)\|^2,$$

which by virtue of (i) implies that

$$\frac{1}{\alpha(x,y)}\bar{\gamma}(x,y) \Big(e^{\alpha(x,y)\sum_{i=1}^{p}\xi_i(x,y)[\Phi_i(x,u,v,\lambda,\bar{t},\bar{s})-\Phi_i(y,u,v,\lambda,\bar{t},\bar{s})+\frac{1}{2}\langle\omega(x,y),\nabla^2\Phi(y,u,v,\lambda,\bar{t},\bar{s})z\rangle]} -1\Big) \geqq 0.$$

Since $\bar{\gamma}(x,y)>0$, we can deduce from this inequality that

$$\sum_{i=1}^{p}\xi_i(x,y)\Phi_i(x,u,v,\lambda,\bar{t},\bar{s}) \geqq \sum_{i=1}^{p}\xi_i(x,y)\big[\Phi_i(y,u,v,\lambda,\bar{t},\bar{s})-\frac{1}{2}\langle\omega(x,y),\nabla^2\Phi(y,u,v,\lambda,\bar{t},\bar{s})z\big]\geqq 0,$$

where the second inequality follows from (17). Therefore, we have

$$0\leqq \sum_{i=1}^{p}\xi_i(x,y)\Phi_i(x,u,v,\lambda,\bar{t},\bar{s}) = \sum_{i=1}^{p} u_i\xi_i(x,y)\Big[f_i(x)-\lambda_i g_i(x)]+\sum_{m\in J_0} v_m G_{j_m}(x,t^m)+\sum_{m\in K_0} v_m H_{k_m}(x,s^m)\Big] \leqq \sum_{i=1}^{p} u_i\xi_i(x,y)[f_i(x)-\lambda_i g_i(x)] \text{ (by the primal feasibility of } x).$$

As shown in the proof of Theorem 3.20, this inequality leads to the conclusion that $\varphi(x) \nleq \lambda$.

(b): The proof is similar to that of part (a).

(c) : Suppose to the contrary that $\varphi(x) \leqslant \lambda$. This implies that for each $i \in \underline{p}$,

$$f_i(x) - \lambda_i g_i(x) \leqq 0, \tag{3.21}$$

with strict inequality holding for at least one index $\ell \in \underline{p}$. Keeping in mind that $v_m > 0$ for each $m \in \underline{v_0}$, we have

$$\begin{aligned}
&\Phi_i(x,u,v,\lambda,\bar{t},\bar{s}) \\
&\leqq \sum_{i=1}^{p} u_i[f_i(x) - \lambda_i g_i(x)] \text{ (by the primal feasibility of } x) \\
&< 0 \text{ (by (21) and the fact that } u > 0) \\
&\leqq \Phi_i(y,u,v,\lambda,\bar{t},\bar{s}) - \frac{1}{2}\langle \omega(x,y), \nabla^2\Phi_i(y,u,v,\lambda,\bar{t},\bar{s})z\rangle \text{ (by (17))}
\end{aligned}$$

and hence

$$\begin{aligned}
&\sum_{i=1}^{p} \xi_i(x,y)\Phi_i(x,u,v,\lambda,\bar{t},\bar{s}) \\
&< \sum_{i=1}^{p} \xi_i(x,y)\Big[\Phi_i(y,u,v,\lambda,\bar{t},\bar{s}) - \frac{1}{2}\big\langle \omega(x,y), \nabla^2\Phi_i(y,u,v,\lambda,\bar{t},\bar{s})z\big\rangle\Big],
\end{aligned}$$

and

$$\begin{aligned}
&\frac{1}{\alpha(x,y)}\bar{\gamma}(x,y) \\
&\Big(e^{\alpha(x,y)\sum_{i=1}^{p}\xi_i(x,y)[\Phi_i(x,u,v,\lambda,\bar{t},\bar{s}) - \Phi_i(y,u,v,\lambda,\bar{t},\bar{s}) + \frac{1}{2}\langle\omega(x,y),\nabla^2\Phi(y,u,v,\lambda,\bar{t},\bar{s})z\rangle]} \\
&-1\Big) < 0,
\end{aligned}$$

which in view of (i) implies that

$$\begin{aligned}
\frac{1}{\beta(x,y)}\Big\langle \sum_{i=1}^{p} u_i[\nabla f_i(y) - \lambda_i \nabla g_i(y)] &+ \sum_{m\in J_0} v_m \nabla G_{j_m}(y,t^m) \\
&+ \sum_{m\in K_0} v_m \nabla H_{k_m}(y,s^m) \\
+\Big\{\sum_{i=1}^{p} u_i[\nabla^2 f_i(y) - \lambda_i \nabla^2 g_i(y)] &+ \sum_{m\in J_0} v_m \nabla^2 G_{j_m}(y,t^m) \\
+ \sum_{m\in K_0} v_m \nabla^2 H_{k_m}(y,s^m\Big\}z, e^{\beta(x,y)\eta(x,y)} - \mathbf{1}\Big\rangle &\leqq -\bar{\rho}(x,y)\|\theta(x,y)\|^2. \qquad (3.22)
\end{aligned}$$

As shown in the proof of part (a), for each $\tau \in \underline{M}$, we have

$$\Lambda_\tau(x,v,\bar{t},\bar{s}) \leqq \Lambda_\tau(y,v,\bar{t},\bar{s}) - \frac{1}{2}\langle \omega(x,y), \nabla^2\Lambda_\tau(y,v,\bar{t},\bar{s})z\rangle,$$

Making use of the sets and functions defined above, we consider the following second-order duality models:

$$(DIV) \qquad \sup_{(y,z,u,v,\lambda,\nu,\nu_0,J_{\nu_0},K_{\nu\backslash\nu_0},\bar{t},\bar{s})\in\mathbb{H}} \lambda=(\lambda_1,\cdots,\lambda_p)$$

subject to

$$\sum_{i=1}^{p}u_i[\nabla f_i(y)-\lambda_i\nabla g_i(y)]+\sum_{m=1}^{\nu_0}v_m\nabla G_{j_m}(y,t^m)+\sum_{m=\nu_0+1}^{\nu}v_m\nabla H_{k_m}(y,s^m)$$
$$+\Big\{\sum_{i=1}^{p}u_i[\nabla^2 f_i(y)-\lambda_i\nabla^2 g_i(y)]$$
$$+\sum_{m=1}^{\nu_0}v_m\nabla^2 G_{j_m}(y,t^m)+\sum_{m=\nu_0+1}^{\nu}v_m\nabla^2 H_{k_m}(y,s^m)\Big\}z=0, \quad (3.24)$$

$$\sum_{i\in I_\tau}u_i[f_i(y)-\lambda_i g_i(y)]+\sum_{m\in J_\tau}v_m G_{j_m}(y,t^m)+\sum_{m\in K_\tau}v_m H_{k_m}(y,s^m)$$
$$-\frac{1}{2}\Big\langle\omega(x,y),\Big\{\sum_{i\in I_\tau}u_i[\nabla^2 f_i(y)-\lambda_i\nabla^2 g_i(y)]+\sum_{m\in J_\tau}v_m\nabla^2 G_{j_m}(y,t^m)+$$
$$\sum_{m\in K_\tau}v_m\nabla^2 H_{k_m}(y,s^m)\Big\}z\Big\rangle\geqq 0, \quad \tau\in D, and\ x\in\mathbb{F}, \quad (3.25)$$

$$\sum_{m\in J_\tau}v_m G_{j_m}(y,t^m)$$
$$+\sum_{m\in K_\tau}v_m H_{k_m}(y,s^m)-\frac{1}{2}\Big\langle\omega(x,y),\Big[\sum_{m\in J_\tau}v_m\nabla^2 G_{j_m}(y,t^m)+$$
$$\sum_{m\in K_\tau}v_m\nabla^2 H_{k_m}(y,s^m)\Big]z\Big\rangle$$
$$\geqq 0, \quad \tau\in E\backslash D, x\in\mathbb{F}; \quad (3.26)$$

$$(\tilde{D}IV) \qquad \sup_{(y,z,u,v,\lambda,\nu,\nu_0,J_{\nu_0},K_{\nu\backslash\nu_0},\bar{t},\bar{s})\in\mathbb{H}} \lambda=(\lambda_1,\ldots,\lambda_p)$$

subject to (3), (25) and (26).

The remarks and observations made earlier about the relationships between $(DI)$ and $(\tilde{D}I)$ are, of course, also valid for $(DIV)$ and $(\tilde{D}IV)$.

The next two theorems show that $(DIV)$ is a dual problem for $(P)$.

**Theorem 3.26.** (Weak Duality) Let $x$ and $(y,z,u,v,\lambda,\nu,\nu_0,J_{\nu_0},K_{\nu\backslash\nu_0},\bar{t},\bar{s})$ be arbitrary feasible solutions of (P) and (DIV), respectively, and assume that any one of the following three sets of hypotheses is satisfied:

(a) (i) $\big(\Pi_0(\cdot,u,v,\lambda,\bar{t},\bar{s}),\ldots,\Pi_d(\cdot,u,v,\lambda,\bar{t},\bar{s})\big)$ is $HA(\alpha,\beta,\bar{\gamma},\xi,\eta,\omega,\bar{\rho},\theta)$-V-pseudosonvex at $y$ for $\bar{\gamma}(x,y)>0$ and $\xi(x,y)>0$;

**Theorem 3.25.** (Strict Converse Duality) Let $x^*$ and $\tilde{w} \equiv (\tilde{x}, \tilde{z}, \tilde{u}, \tilde{v}, \tilde{\lambda}, \tilde{v}, \tilde{v}_0, J_{\tilde{v}_0}, K_{\tilde{v}\backslash\tilde{v}_0}, \bar{\tilde{t}}, \bar{\tilde{s}})$ be arbitrary feasible solutions of (P) and (DIII), respectively, such that

$$\sum_{i=1}^{p} \tilde{u}_i \xi_i(x^*, \tilde{x})[f_i(x^*) - \tilde{\lambda}_i g_i(x^*)] = 0, \tag{3.23}$$

where $\xi_i$, $i \in \underline{p}$, are given real-valued functions from $X \times X$ to $\mathbb{R}_+ \backslash \{0\}$. Furthermore, assume that any one of the following three sets of hypotheses is satisfied:

(a) The assumptions specified in part (a) of Theorem 3.23 are satisfied for the feasible solution $\tilde{w}$ of (DIII), and the function $\big(\Phi_1(\cdot, \tilde{u}, \tilde{v}, \tilde{\lambda}, \bar{\tilde{t}}, \bar{\tilde{s}}), \ldots, \Phi_p(\cdot, \tilde{u}, \tilde{v}, \tilde{\lambda}, \bar{\tilde{t}}, \bar{\tilde{s}})\big)$ is strictly $HA(\alpha, \beta, \bar{\gamma}, \xi, \eta, \omega, \bar{\rho}, \theta)$-V-pseudosonvex at $\tilde{x}$.

(b) The assumptions specified in part (b) of Theorem 3.23 are satisfied for the feasible solution $\tilde{w}$ of (DIII), and the function $\big(\Phi_1(\cdot, \tilde{u}, \tilde{v}, \tilde{\lambda}, \bar{\tilde{t}}, \bar{\tilde{s}}), \ldots, \Phi_p(\cdot, \tilde{u}, \tilde{v}, \tilde{\lambda}, \bar{\tilde{t}}, \bar{\tilde{s}})\big)$ is $HA(\alpha, \beta, \bar{\gamma}, \xi, \eta\omega, barp\rho, \theta)$-V-quasisonvex $\tilde{x}$.

(c) The assumptions specified in part (c) of Theorem 3.23 are satisfied for the feasible solution $\tilde{w}$ of (DIII), and the function $\big(\Phi_1(\cdot, \tilde{u}, \tilde{v}, \tilde{\lambda}, \bar{\tilde{t}}, \bar{\tilde{s}}), \ldots, \Phi_p(\cdot, \tilde{u}, \tilde{v}, \tilde{\lambda}, \bar{\tilde{t}}, \bar{\tilde{s}})\big)$ is $HA(\alpha, \beta, \bar{\gamma}, \xi, \eta, \bar{\rho}, \theta)$-V-quasisonvex $\tilde{x}$.

Then $\tilde{x} = x^*$ and $\varphi(x^*) = \tilde{\lambda}$.

*Proof.* The proof is similar to that of Theorem 3.22. □

Each one of the three sets of results given in Theorems 3.23 - 3.25 can be viewed as a family of duality results whose members can easily be identified by appropriate choices of the partitioning sets $J_\mu$ and $K_\mu$, $\mu \in \underline{M} \cup \{0\}$.

## 3.4. Duality Model IV

In this section we discuss another collection of duality results for $(P)$ which are different from those stated in Theorems 3.23 - 3.25. In formulating these duality results, we utilize a partition of $\underline{p}$ in addition to those of $\underline{v_0}$ and $\underline{v}\backslash\underline{v_0}$. It appears that this partitioning scheme was first proposed in [10] for a multiobjective fractional programming problem with a finite number of constraints. In our theorems, we impose appropriate generalized $HA(\alpha, \beta, \gamma, \xi, \eta, \omega, \rho, \theta)$-V-sonvexity requirements on certain vector functions whose components comprise some combinations of the functions $f_i$, $g_i$, $i \in \underline{p}$, $G_j(\cdot, t)$, $j \in \underline{q}$, and $H_k(\cdot, s)$, $k \in \underline{r}$.

Let $\{I_0, I_1, \ldots, I_d\}$, $\{J_0, J_1, \ldots, J_e\}$, and $\{K_0, K_1, \ldots, K_e\}$ be partitions of $\underline{p}$, $\underline{v_0}$, and $\underline{v}\backslash\underline{v_0}$, respectively, such that $D = \{0, 1, 2, \ldots, d\} \subset E = \{0, 1, \ldots, e\}$, and let the function $\Pi_\tau(\cdot, u, v, \lambda, \bar{t}, \bar{s}) : X \to \mathbb{R}$ be defined, for fixed $u, v, \lambda, \bar{t}$, and $\bar{s}$ by

$$\Pi_\tau(\zeta, u, v, \lambda, \bar{t}, \bar{s}) = \sum_{i \in I_\tau} u_i[f_i(\zeta) - \lambda_i g_i(\zeta)] + \sum_{m \in J_\tau} v_m G_{j_m}(\zeta, t^m) + \sum_{m \in K_\tau} v_m H_{k_m}(\zeta, s^m), \quad \tau \in D.$$

Making use of the sets and functions defined above, we consider the following second-order duality models:

$$(DIV) \qquad \sup_{(y,z,u,v,\lambda,\nu,\nu_0,J_{\nu_0},K_{\nu\backslash\nu_0},\bar{t},\bar{s})\in\mathbb{H}} \lambda=(\lambda_1,\cdots,\lambda_p)$$

subject to

$$\sum_{i=1}^{p}u_i[\nabla f_i(y)-\lambda_i\nabla g_i(y)]+\sum_{m=1}^{\nu_0}v_m\nabla G_{j_m}(y,t^m)+\sum_{m=\nu_0+1}^{\nu}v_m\nabla H_{k_m}(y,s^m)$$
$$+\Big\{\sum_{i=1}^{p}u_i[\nabla^2 f_i(y)-\lambda_i\nabla^2 g_i(y)]$$
$$+\sum_{m=1}^{\nu_0}v_m\nabla^2 G_{j_m}(y,t^m)+\sum_{m=\nu_0+1}^{\nu}v_m\nabla^2 H_{k_m}(y,s^m)\Big\}z=0, \quad (3.24)$$

$$\sum_{i\in I_\tau}u_i[f_i(y)-\lambda_i g_i(y)]+\sum_{m\in J_\tau}v_m G_{j_m}(y,t^m)+\sum_{m\in K_\tau}v_m H_{k_m}(y,s^m)$$
$$-\frac{1}{2}\Big\langle\omega(x,y),\Big\{\sum_{i\in I_\tau}u_i[\nabla^2 f_i(y)-\lambda_i\nabla^2 g_i(y)]+\sum_{m\in J_\tau}v_m\nabla^2 G_{j_m}(y,t^m)+$$
$$\sum_{m\in K_\tau}v_m\nabla^2 H_{k_m}(y,s^m)\Big\}z\Big\rangle\geqq 0, \quad \tau\in D, and\ x\in\mathbb{F}, \quad (3.25)$$

$$\sum_{m\in J_\tau}v_m G_{j_m}(y,t^m)$$
$$+\sum_{m\in K_\tau}v_m H_{k_m}(y,s^m)-\frac{1}{2}\Big\langle\omega(x,y),\Big[\sum_{m\in J_\tau}v_m\nabla^2 G_{j_m}(y,t^m)+$$
$$\sum_{m\in K_\tau}v_m\nabla^2 H_{k_m}(y,s^m)\Big]z\Big\rangle$$
$$\geqq 0, \quad \tau\in E\backslash D, x\in\mathbb{F}; \quad (3.26)$$

$$(\tilde{D}IV) \qquad \sup_{(y,z,u,v,\lambda,\nu,\nu_0,J_{\nu_0},K_{\nu\backslash\nu_0},\bar{t},\bar{s})\in\mathbb{H}} \lambda=(\lambda_1,\ldots,\lambda_p)$$

subject to (3), (25) and (26).

The remarks and observations made earlier about the relationships between $(DI)$ and $(\tilde{D}I)$ are, of course, also valid for $(DIV)$ and $(\tilde{D}IV)$.

The next two theorems show that $(DIV)$ is a dual problem for $(P)$.

**Theorem 3.26.** (Weak Duality) Let $x$ and $(y,z,u,v,\lambda,\nu,\nu_0,J_{\nu_0},K_{\nu\backslash\nu_0},\bar{t},\bar{s})$ be arbitrary feasible solutions of (P) and (DIV), respectively, and assume that any one of the following three sets of hypotheses is satisfied:

(a) (i) $\big(\Pi_0(\cdot,u,v,\lambda,\bar{t},\bar{s}),\ldots,\Pi_d(\cdot,u,v,\lambda,\bar{t},\bar{s})\big)$ is $HA(\alpha,\beta,\bar{\gamma},\xi,\eta,\omega,\bar{\rho},\theta)$-V-pseudosonvex at $y$ for $\bar{\gamma}(x,y)>0$ and $\xi(x,y)>0$;

(ii) $\left(\Lambda_{d+1}(\cdot,v,\bar{t},\bar{s}),\ldots,\Lambda_e(\cdot,v,\bar{t},\bar{s})\right)$ is HA($\alpha$, $\beta$, $\tilde{\gamma}$, $\pi$, $\eta$, $\omega$, $\tilde{\rho}$, $\theta$)-V-quasisonvex at $y$;

(iii) $\bar{\rho}(x,y)+\tilde{\rho}(x,y) \geqq 0$;

(b) (i) $\left(\Pi_0(\cdot,u,v,\lambda,\bar{t},\bar{s}),\ldots,\Pi_d(\cdot,u,v,\lambda,\bar{t},\bar{s})\right)$ is prestrictly HA($\alpha$, $\beta$, $\bar{\gamma}$, $\xi$, $\eta$, $\omega$, $\bar{\rho}$, $\theta$)-V-
quasisonvex at $y$ for $\bar{\gamma}(x,y) > 0$ and $\xi(x,y) > 0$;

(ii) $\left(\Lambda_{d+1}(\cdot,v,\bar{t},\bar{s}),\ldots,\Lambda_e(\cdot,v,\bar{t},\bar{s})\right)$ is strictly $HA(\alpha,\beta,\tilde{\gamma},\pi,\eta,\omega,\tilde{\rho},\theta)$-V-pseudosonvex at $y$;

(iii) $\bar{\rho}(x,y)+\tilde{\rho}(x,y) \geqq 0$;

(c) (i) $\left(\Pi_0(\cdot,u,v,\lambda,\bar{t},\bar{s}),\ldots,\Pi_d(\cdot,u,v,\lambda,\bar{t},\bar{s})\right)$ is prestrictly HA($\alpha$, $\beta$, $\bar{\gamma}$, $\xi$, $\eta$, $\omega$, $\bar{\rho}$, $\theta$)-V-
quasisonvex at $y$ for $\bar{\gamma}(x,y) > 0$ and $\xi(x,y) > 0$;

(ii) $\left(\Lambda_{d+1}(\cdot,v,\bar{t},\bar{s}),\ldots,\Lambda_e(\cdot,v,\bar{t},\bar{s})\right)$ is HA($\alpha$, $\beta$, $\tilde{\gamma}$, $\xi$, $\eta$, $\omega$, $\tilde{\rho}$, $\theta$) -V-quasisonvex at $y$;

(iii) $\bar{\rho}(x,y)+\tilde{\rho}(x,y) > 0$.

Then $\varphi(x) \nleq \lambda$.

*Proof.* (a): Suppose to the contrary that $\varphi(x) \leqslant \lambda$. Then $f_i(x)-\lambda_i g_i(x) \leqq 0$, $i \in \underline{p}$, with strict inequality holding for at least one index $\ell \in \underline{p}$. Since $u > 0$, we see that for each $\tau \in D$,

$$\sum_{i\in I_\tau} u_i[f_i(x)-\lambda_i g_i(x)] \leqq 0, \tag{3.27}$$

with strict inequality holding for at least one index $\tau \in D$.

Now using this inequality, we see that

$$\begin{aligned}
&\Pi_\tau(x,u,v,\lambda,\bar{t},\bar{s})\\
&= \sum_{i\in I_\tau} u_i[f_i(x)-\lambda_i g_i(x)] + \sum_{m\in J_\tau} v_m G_{j_m}(x,t^m) + \sum_{m\in K_\tau} v_m H_{k_m}(x,s^m)\\
&\leqq \sum_{i\in I_\tau} u_i[f_i(x)-\lambda_i g_i(x)]\\
&\quad \text{(by the primal feasibility of } x \text{ and positivity of } v_m,\ m \in \underline{v_0})\\
&\leqq 0 \quad \text{(by (27))}\\
&\leqq \Pi_\tau(y,u,v,\lambda,\bar{t},\bar{s}) - \frac{1}{2}\langle \omega(x,y), \nabla^2\Pi_\tau(y,u,v,\lambda,\bar{t},\bar{s})z\rangle \quad \text{(by (25))},
\end{aligned}$$

with strict inequality holding for at least one index $\tau \in D$. Since $\xi_\iota(x,y) > 0$ for each $\tau \in D$, the above inequalities yield

$$\begin{aligned}
&\sum_{\tau\in D} \xi_\tau(x,y)\Pi_\tau(x,u,v,\lambda,\bar{t},\bar{s})\\
&< \sum_{\tau\in D} \xi_\tau(x,y)\left[\Pi_\tau(y,u,v,\lambda,\bar{t},\bar{s}) - \frac{1}{2}\langle \omega(x,y), \nabla^2\Pi_\tau(y,u,v,\lambda,\bar{t},\bar{s})z\rangle\right],
\end{aligned}$$

and hence

$$\frac{1}{\alpha(x,y)}\bar{\gamma}(x,y)\left(e^{\alpha(x,y)\sum_{\tau\in D}\xi_\tau(x,y)[\Pi_\tau(x,u,v,\lambda,\bar{t},\bar{s})-\Pi_\tau(y,u,v,\lambda,\bar{t},\bar{s})+\frac{1}{2}\langle\omega(x,y),\nabla^2\Pi_\tau(y,u,v,\lambda,\bar{t},\bar{s})z\rangle]}-1\right)<0,$$

which in view of (i) implies that

$$\frac{1}{\beta(x,y)}\Big\langle\sum_{i=1}^{p}u_i[\nabla f_i(y)-\lambda_i\nabla g_i(y)]+\sum_{\tau\in D}\Big[\sum_{m\in J_\tau}v_m\nabla G_{j_m}(y,t^m)$$
$$+\sum_{m\in K_\tau}v_m\nabla H_{k_m}(y,s^m)\Big]+\Big\{\sum_{i=1}^{p}u_i[\nabla^2 f_i(y)-\lambda_i\nabla^2 g_i(y)]+\sum_{\tau\in D}\Big[\sum_{m\in J_\tau}v_m\nabla^2 G_{j_m}(y,t^m)$$
$$+\sum_{m\in K_\tau}v_m\nabla^2 H_{k_m}(y,s^m)\Big]\Big\}z,e^{\beta(x,y)\eta(x,y)}-\mathbf{1}\Big\rangle<-\bar{\rho}(x,y)\|\theta(x,y)\|^2. \quad (3.28)$$

As shown in the proof of Theorem 3.20, for each $\tau\in E\backslash D$,

$$\Lambda_\tau(x,v,\bar{t},\bar{s})$$
$$\leqq\Lambda_\tau(y,v,\bar{t},\bar{s})-\frac{1}{2}\langle\omega(x,y),\nabla^2\Lambda_\tau(y,v,\bar{t},\bar{s})z\rangle,$$

and so

$$\sum_{\tau\in E\backslash D}\pi_\tau(x,y)\Lambda_\tau(x,v,\bar{t},\bar{s})$$
$$\leqq\sum_{\tau\in E\backslash D}\pi_\tau(x,y)\big[\Lambda_\tau(y,v,\bar{t},\bar{s})-\frac{1}{2}\langle\omega(x,y),\nabla^2\Lambda_\tau(y,v,\bar{t},\bar{s})z\rangle\big],$$

and

$$\frac{1}{\alpha(x,y)}\gamma(x,y)\left(e^{\alpha(x,y)\sum_{\tau\in E\backslash D}\pi_\tau(x,y)[\Lambda_\tau(x,v,\bar{t},\bar{s})-\Lambda_\tau(y,v,\bar{t},\bar{s})+\frac{1}{2}\langle\omega(x,y),\nabla^2\Lambda_\tau(y,v,\bar{t},\bar{s})z\rangle]}-1\right)\leqq 0,$$

which in view of (ii) implies that

$$\frac{1}{\beta(x,y)}\Big\langle\sum_{\tau\in E\backslash D}\Big\{\sum_{m\in J_\tau}v_m\nabla G_{j_m}(y,t^m)+\sum_{m\in K_\tau}v_m\nabla H_{k_m}(y,s^m)+$$
$$\Big[\sum_{m\in J_\tau}v_m\nabla^2 G_{j_m}(y,t^m)+\sum_{m\in K_\tau}v_m\nabla^2 H_{k_m}(y,s^m)\Big]z\Big\},e^{\beta(x,y)\eta(x,y)}-\mathbf{1}\Big\rangle$$
$$\leqq-\tilde{\rho}(x,y)\|\theta(x,y)\|^2. \quad (3.29)$$

Now combining (28) and (29) and using (iii), we see that

$$\frac{1}{\beta(x,y)}\Big\langle \sum_{i=1}^{p} u_i[\nabla f_i(y) - \lambda_i \nabla g_i(y)] + \sum_{m=1}^{\nu_0} v_m \nabla G_{j_m}(y,t^m)$$
$$+ \sum_{m=\nu_0+1}^{\nu} v_m \nabla H_{k_m}(y,s^m) + \Big\{ \sum_{i=1}^{p} u_i[\nabla^2 f_i(y) - \lambda_i \nabla^2 g_i(y)] + \sum_{m=1}^{\nu_0} v_m \nabla^2 G_{j_m}(y,t^m)$$
$$+ \sum_{m=\nu_0+1}^{\nu} v_m \nabla^2 H_{k_m}(y,s^m) \Big\} z, e^{\beta(x,y)\eta(x,y)} - \mathbf{1} \Big\rangle < -[\bar{\rho}(x,y) + \tilde{\rho}(x,y)] \|\theta(x,y)\|^2 \leqq 0,$$

which contradicts (24). Therefore, $\varphi(x) \nleq \lambda$.

(b) and (c) : The proofs are similar to that of part (a).

□

**Theorem 3.27.** (Strong Duality) Let $x^*$ be a normal efficient solution of (P) and assume that for each feasible solution $(y,z,u,v,\lambda,\nu,\nu_0, J_{\nu_0}, K_{\nu\backslash\nu_0}, \bar{t},\bar{s})$ of (DIV), any one of the three sets of conditions specified in Theorem 6.1 is satisfied. Then there exist $u^*, v^*, \lambda^*$, $\nu^*, \nu_0^*, J_{\nu_0^*}, K_{\nu^*\backslash\nu_0^*}, \bar{t}^*$, and $\bar{s}^*$ such that $(x^*, z^* = 0, u^*, v^*, \lambda^*, \nu^*, \nu_0^*, J_{\nu_0^*}, K_{\nu^*\backslash\nu_0^*}, \bar{t}^*, \bar{s}^*)$ is an efficient solution of (DIV) and $\varphi(x^*) = \lambda^*$.

*Proof.* The proof is similar to that of Theorem 3.18.

□

**Theorem 3.28.** (Strict Converse Duality) Let $x^*$ and $\tilde{w} \equiv (\tilde{x},\tilde{z},\tilde{u},\tilde{v},\tilde{\lambda},\tilde{\nu},\tilde{\nu}_0, J_{\tilde{\nu}_0}, K_{\tilde{\nu}\backslash\tilde{\nu}_0}, \bar{\tilde{t}},\bar{\tilde{s}})$ be arbitrary feasible solutions of (P) and (DIV), respectively, such that

$$\sum_{i=1}^{p} \tilde{u}_i[f_i(x^*) - \tilde{\lambda}_i g_i(x^*)] = 0. \tag{3.30}$$

Furthermore, assume that any one of the following three sets of hypotheses is satisfied:

(a) The assumptions specified in part (a) of Theorem 3.26 are satisfied for the feasible solution $\tilde{w}$ of (DIV), the function $\big(\Pi_0(\cdot,\tilde{u},\tilde{v},\tilde{\lambda},\bar{\tilde{t}},\bar{\tilde{s}}),\ldots,\Pi_d(\cdot,\tilde{u},\tilde{v},\tilde{\lambda},\bar{\tilde{t}},\bar{\tilde{s}})\big)$ is strictly $HA(\alpha,\beta,\bar{\gamma},\xi,\eta,\omega,\bar{\rho},\theta)$-V-pseudosonvex at $\tilde{x}$, and $\xi_0 = \xi_1 = \ldots = \xi_d = \sigma$.

(b) The assumptions specified in part (b) of Theorem 3.26 are satisfied for the feasible solution $\tilde{w}$ of (DIV), the function $\big(\Pi_0(\cdot,\tilde{u},\tilde{v},\tilde{\lambda},\bar{\tilde{t}},\bar{\tilde{s}}),\ldots,\Pi_d(\cdot,\tilde{u},\tilde{v},\tilde{\lambda},\bar{\tilde{t}},\bar{\tilde{s}})\big)$ is $HA(\alpha,\beta,\bar{\gamma},\xi,\eta,\omega,\bar{\rho},\theta)$-V-quasisonvex at $\tilde{x}$, and $\xi_0 = \xi_1 = \ldots = \xi_d = \sigma$.

(c) The assumptions specified in part (c) of Theorem 3.26 are satisfied for the feasible solution $\tilde{w}$ of (DIV), the function $\big(\Pi_0(\cdot,\tilde{u},\tilde{v},\tilde{\lambda},\bar{\tilde{t}},\bar{\tilde{s}}),\ldots,\Pi_d(\cdot,\tilde{u},\tilde{v},\tilde{\lambda},\bar{\tilde{t}},\bar{\tilde{s}})\big)$ is $HA(\alpha,\beta,\bar{\gamma},\xi,\eta,\omega,\bar{\rho},\theta)$-V-quasisonvex at $\tilde{x}$, and $\xi_0 = \xi_1 = \ldots = \xi_d = \sigma$.

Then $\tilde{x} = x^*$ and $\varphi(x^*) = \tilde{\lambda}$.

*Proof.* (a) : Suppose to the contrary that $\tilde{x} \neq x^*$. It is clear that (24) can be expressed as follows:

$$\sum_{\tau\in D}\Big\{\sum_{i\in I_\tau}\tilde{u}_i[\nabla f_i(\tilde{x})-\tilde{\lambda}_i\nabla g_i(\tilde{x})]+\sum_{m\in J_\tau}\tilde{v}_m\nabla G_{j_m}(\tilde{x},\tilde{t}^m)+\sum_{m\in K_\tau}\tilde{v}_m\nabla H_{k_m}(\tilde{x},\tilde{s}^m)+$$
$$\Big\{\sum_{i\in I_\tau}\tilde{u}_i[\nabla^2 f_i(\tilde{x})-\tilde{\lambda}_i\nabla^2 g_i(\tilde{x})]+\sum_{m\in J_\tau}\tilde{v}_m\nabla^2 G_{j_m}(\tilde{x},\tilde{t}^m)+\sum_{m\in K_\tau}\tilde{v}_m\nabla^2 H_{k_m}(\tilde{x},\tilde{s}^m)\Big\}\tilde{z}\Big\}+$$
$$\sum_{\tau\in E\backslash D}\Big\{\sum_{m\in J_\tau}\tilde{v}_m\nabla G_{j_m}(\tilde{x},\tilde{t}^m)+\sum_{m\in K_\tau}\tilde{v}_m\nabla H_{k_m}(\tilde{x},\tilde{s}^m)+\Big[\sum_{m\in J_\tau}\tilde{v}_m\nabla^2 G_{j_m}(\tilde{x},\tilde{t}^m)$$
$$+\sum_{m\in K_\tau}\tilde{v}_m\nabla^2 H_{k_m}(\tilde{x},\tilde{s}^m)\Big]\tilde{z}\Big\}=0. \quad (3.31)$$

Since $x^* \in \mathbb{F}$, $\tilde{v}_m \geqq 0$, $m \in \underline{\nu}$, and (26) holds, it follows that for each $\tau \in E\backslash D$,

$$\Lambda_\tau(x^*,\tilde{v},\bar{\tilde{t}},\bar{\tilde{s}})=\sum_{m\in J_\tau}\tilde{v}_m G_{j_m}(x^*,\tilde{t}^m)+\sum_{m\in K_\tau}\tilde{v}_m H_{k_m}(x^*,\tilde{s}^m)$$
$$\leqq 0$$
$$\leqq \Lambda_\tau(\tilde{x},\tilde{v},\bar{\tilde{t}},\bar{\tilde{s}})-\frac{1}{2}\langle\omega(x^*,\tilde{x}),\nabla^2\Lambda_\tau(\tilde{x},\tilde{v},\bar{\tilde{t}},\bar{\tilde{s}})\tilde{z}\rangle \text{ (by (25))},$$

and hence

$$\sum_{\tau\in E\backslash D}\pi_\tau(x^*,\tilde{x})\Lambda_\tau(x^*,\tilde{v},\bar{\tilde{t}},\bar{\tilde{s}})$$
$$\leqq\sum_{\tau\in E\backslash D}\pi_\tau(x^*,\tilde{x})\big[\Lambda_\tau(\tilde{x},\tilde{v},\bar{\tilde{t}},\bar{\tilde{s}})-\frac{1}{2}\big\langle\omega(x^*,\tilde{x}),\nabla^2\Lambda_\tau(\tilde{x},\tilde{v},\bar{\tilde{t}},\bar{\tilde{s}})\tilde{z}\big\rangle\big].$$

Using this inequality in conjunction with the properties of the functions $\alpha$, $\tilde{\gamma}$, and $\zeta \to e^{\zeta}$, we get

$$\frac{1}{\alpha(x^*,\tilde{x})}\tilde{\gamma}(x^*,\tilde{x})$$
$$\Big(e^{\alpha(x^*,\tilde{x})\sum_{\tau\in E\backslash D}\pi_\tau(x^*,\tilde{x})[\Lambda_\tau(x^*,\tilde{v},\bar{\tilde{t}},\bar{\tilde{s}})-\Lambda_\tau(\tilde{x},\tilde{v},\bar{\tilde{t}},\bar{\tilde{s}})+\frac{1}{2}\langle\omega(x^*,\tilde{x}),\nabla^2\Lambda_\tau(\tilde{x},\tilde{v},\bar{\tilde{t}},\bar{\tilde{s}})\tilde{z}\rangle]}$$
$$-1\Big)\leqq 0,$$

which in view of (ii) implies that

$$\frac{1}{\beta(x^*,\tilde{x})}\Big\langle\sum_{\tau\in E\backslash D}\Big\{\sum_{m\in J_\tau}\tilde{v}_m\nabla G_{j_m}(\tilde{x},\tilde{t}^m)+\sum_{m\in K_\tau}\tilde{v}_m\nabla H_{k_m}(\tilde{x},\tilde{s}^m)$$
$$+\Big[\sum_{m\in J_\tau}\tilde{v}_m\nabla^2 G_{j_m}(\tilde{x},\tilde{t}^m)$$
$$+\sum_{m\in K_\tau}\tilde{v}_m\nabla^2 H_{k_m}(\tilde{x},\tilde{s}^m)\Big]\tilde{z}\Big\},e^{\beta(x^*,\tilde{x})\eta(x^*,\tilde{x})}-\mathbf{1}\Big\rangle\leqq-\tilde{\rho}(x^*,\tilde{x})\|\theta(x^*,\tilde{x})\|^2. \quad (3.32)$$

Combining (31) and (32), and using (iii) we get

$$\frac{1}{\beta(x,y)}\Big\langle \sum_{\tau\in D}\Big\{\sum_{i\in I_\tau}\tilde{u}_i[\nabla f_i(\tilde{x})-\tilde{\lambda}_i\nabla g_i(\tilde{x})]+\sum_{m\in J_\tau}\tilde{v}_m\nabla G_{j_m}(\tilde{x},\tilde{t}^m)$$

$$+\sum_{m\in K_\tau}\tilde{v}_m\nabla H_{k_m}(\tilde{x},\tilde{s}^m)+\Big\{\sum_{i\in I_\tau}\tilde{u}_i[\nabla^2 f_i(\tilde{x})-\tilde{\lambda}\nabla^2 g_i(\tilde{x})]+\sum_{m\in J_\tau}\tilde{v}_m\nabla^2 G_{j_m}(\tilde{x},\tilde{t}^m)$$

$$+\sum_{m\in K_\tau}\tilde{v}_m\nabla^2 H_{k_m}(\tilde{x},\tilde{s}^m)\Big\}\tilde{z}\Big\},e^{\beta(x^*,\tilde{x})\eta(x^*,\tilde{x})}-\mathbf{1}\Big\rangle$$

$$\geqq \tilde{\rho}(x^*,\tilde{x})\|\theta(x^*,\tilde{x})\|^2\geqq -\bar{\rho}(x^*,\tilde{x})\|\theta(x^*,\tilde{x})\|^2, \quad (3.33)$$

which by virtue of our strict $HA(\alpha,\beta,\bar{\gamma},\xi,\eta,\omega,\bar{\rho},\theta)$-V-pseudosonvexity hypothesis implies that

$$\frac{1}{\alpha(x^*,\tilde{x})}\bar{\gamma}(x^*,\tilde{x})$$
$$\Big(e^{\alpha(x^*,\tilde{x})\sigma(x^*,\tilde{x})\sum_{\tau\in D}[\Pi_\tau(x^*,\tilde{u},\tilde{v},\tilde{\lambda},\bar{\tilde{t}},\bar{\tilde{s}})-\Pi_\tau(\tilde{x},\tilde{u},\tilde{v},\tilde{\lambda},\bar{\tilde{t}},\bar{\tilde{s}})+\frac{1}{2}\langle\omega(x^*,\tilde{x}),\nabla^2\Pi_\tau(\tilde{x},\tilde{u},\tilde{v},\tilde{\lambda},\bar{\tilde{t}},\bar{\tilde{s}})\tilde{z}\rangle]}$$
$$-1\Big)>0.$$

Since $\bar{\gamma}(x^*,\tilde{x})>0$ and $\sigma(x^*,\tilde{x})>0$, we deduce from this inequality that

$$\sum_{\tau\in D}\Pi_\tau(x^*,\tilde{u},\tilde{v},\tilde{\lambda},\bar{\tilde{t}},\bar{\tilde{s}})$$
$$>\sum_{\tau\in D}\big[\Pi_\tau(\tilde{x},\tilde{u},\tilde{v},\tilde{\lambda},\bar{\tilde{t}},\bar{\tilde{s}})-\frac{1}{2}\langle\omega(x^*,\tilde{x}),\nabla^2\Pi_\tau(\tilde{x},\tilde{u},\tilde{v},\tilde{\lambda},\bar{\tilde{t}},\bar{\tilde{s}})\tilde{z}\rangle\big]\geqq 0,$$

where the second inequality follows from the dual feasibility of $\tilde{w}$ and (25). Therefore, we have

$$\begin{aligned}0&<\sum_{\tau\in D}\Pi_\tau(x^*,\tilde{u},\tilde{v},\tilde{\lambda},\bar{\tilde{t}},\bar{\tilde{s}})\\&=\sum_{\tau\in D}\sum_{i\in I_\tau}\tilde{u}_i[f_i(x^*)-\tilde{\lambda}_i g_i(x^*)]\\&+\sum_{\tau\in D}\Big[\sum_{m\in J_\tau}\tilde{v}_m G_{j_m}(x^*,\tilde{t}^m)+\sum_{m\in K_\tau}\tilde{v}_m H_{k_m}(x^*,\tilde{s}^m)\Big]\\&\leqq\sum_{i=1}^{p}\tilde{u}_i[f_i(x^*)-\tilde{\lambda}_i g_i(x^*)]\ \text{(by the primal feasibility of } x^*\text{)},\end{aligned}$$

which contradicts (30). Therefore, we conclude that $\tilde{x}=x^*$ and $\varphi(x^*)=\tilde{\lambda}$.

(b) and (c) : The proofs are similar to that of part (a). □

The duality models $(DIV)$ and $(\tilde{D}IV)$ are two families of dual problems for $(P)$ whose members can easily be identified by appropriate choices of the partitioning sets $I_\mu$, $\mu\in D$, and $J_\tau$ and $K_\tau$, $\tau\in E$. These duality models and their special cases contain the semiinfinite analogues of various dual problems proposed and investigated previously for many types of nonlinear programming problems, as well as numerous other duality formulations.

# 4. Generalized Duality Results II

This section is essentially based on work initiated in the publication [4] where some information about multiobjective fractional programming is presented, the current status of semiinfinite programming is briefly discussed and numerous key references are cited, further extensions of the variants of the V-invex functions defined recently by Antczak [1] are proposed, and a fairly large number of sets of global parametric sufficient efficiency results under various generalized $HA(\alpha,\beta,\gamma,\xi,\eta,\rho,\theta)$-V-invexity by Zalmai [12 - 14]. Here we shall make use of the sufficient efficiency results developed in [4] and a set of necessary efficiency conditions, which will be recalled in the next section, and construct several first-order parametric duality models for $(P)$ and prove appropriate duality theorems. The second-order counterparts of these duality models and results are discussed in [5].

Although presently numerous duality results exist in the related literature for several classes of multiobjective fractional optimization problems involving certain classes of generalized convex functions, so far no such results are available for any kind of semiinfinite multiobjective programming problems utilizing any type of generalized $HA(\alpha,\beta,\gamma,\xi,\eta,\rho,\theta)$-V-invexity concepts. To the best of our knowledge, all the duality results established in this subsection are new in the area of semiinfinite programming, and are suitable for further research endeavors. The rest of this subsection is about considering two parametric duality models with somewhat limited constraint structures and proving weak, strong, and strict converse duality theorems under appropriate generalized $HA(\alpha,\beta,\gamma,\xi,\eta,\rho,\theta)$-V-invexity hypotheses. This is followed by two other duality models with different constraint structures which allow for more general types of $HA(\alpha,\beta,\gamma,\xi,\eta,\rho,\theta)$-V-invexity conditions under which duality can be established. We continue further subsections of duality where we use two partitioning schemes and construct four generalized parametric duality models and obtain several duality results using various generalized $HA(\alpha,\beta,\gamma,\xi,\eta,\rho,\theta)$-V-invexity assumptions. Finally, we point out some further research opportunities arising from certain modifications of the principal problem model considered in this section. Evidently, all the duality results established in this section can easily be modified and restated for each one of the seven classes of nonlinear programming problems designated as $(P1)-(P7)$ in [4].

Let the function $F=(F_1,F_2,\ldots,F_p):X\to\mathbb{R}^p$ be differentiable at $x^*$. The following notions of V-invexity, V-pseudoinvexity, and V-quasiinvexity for vector-valued functions will be used in this section. They are substantial generalizations of the class of V-r-invex functions recently introduced by Antczak [1].

**Definition 3.29.** The function $F$ is said to be (strictly) first-order HA($\alpha$, $\beta$, $\gamma$, $\xi$, $\eta$, $\rho$, $\theta$) - V-invex at $x^*\in X$ if there exist functions $\alpha:X\times X\to\mathbb{R}$, $\beta:X\times X\to\mathbb{R}$, $\gamma_i:X\times X\to\mathbb{R}_+\equiv[0,\infty)$, $\xi_i:X\times X\to\mathbb{R}_+\backslash\{0\}$, $i\in\underline{p}$, $\eta:X\times X\to\mathbb{R}^n$, $\rho_i:X\times X\to\mathbb{R}$, $i\in\underline{p}$, and $\theta:X\times X\to\mathbb{R}^n$ such that for all $x\in X(x\neq x^*)$ and $i\in\underline{p}$,

$$\frac{1}{\alpha(x,x^*)}\gamma_i(x,x^*)\left(e^{\alpha(x,x^*)[F_i(x)-F_i(x^*)]}-1\right)$$

$$(>)\geqq\frac{1}{\beta(x,x^*)}\langle\xi_i(x,x^*)\nabla F_i(x^*),e^{\beta(x,x^*)\eta(x,x^*)}-\mathbf{1}\rangle+\rho_i(x,x^*)\|\theta(x,x^*)\|^2$$

if $\alpha(x,x^*)\neq0$ and $\beta(x,x^*)\neq0$ for all $x\in X$,

$$\frac{1}{\alpha(x,x^*)}\gamma_i(x,x^*)\left(e^{\alpha(x,x^*)[F_i(x)-F_i(x^*)]}-1\right)$$
$$(>) \geqq \left\langle \xi_i(x,x^*)\nabla F_i(x^*), \eta(x,x^*)\right\rangle$$
$$+\rho_i(x,x^*)\|\theta(x,x^*)\|^2$$
$$\text{if } \alpha(x,x^*) \neq 0 \text{ and } \beta(x,x^*) = 0 \text{ for all } x \in X,$$

$$\gamma_i(x,x^*)[F_i(x)-F_i(x^*)]$$
$$(>) \geqq \frac{1}{\beta(x,x^*)}\left\langle \xi_i(x,x^*)\nabla F_i(x^*), e^{\beta(x,x^*)\eta(x,x^*)}-\mathbf{1}\right\rangle$$
$$+\rho_i(x,x^*)\|\theta(x,x^*)\|^2$$
$$\text{if } \alpha(x,x^*) = 0 \text{ and } \beta(x,x^*) \neq 0 \text{ for all } x \in X,$$

$$\gamma_i(x,x^*)[F_i(x)-F_i(x^*)](>) \geqq \langle \xi_i(x,x^*)\nabla F_i(x^*), \eta(x,x^*)\rangle$$
$$+\rho_i(x,x^*)\|\theta(x,x^*)\|^2$$
$$\text{if } \alpha(x,x^*) = 0 \text{ and } \beta(x,x^*) = 0 \text{ for all } x \in X,$$

where
$$\left(e^{\beta(x,x^*)\eta(x,x^*)}-\mathbf{1}\right) \equiv \left(e^{\beta(x,x^*)\eta_1(x,x^*)}-1,\ldots,e^{\beta(x,x^*)\eta_n(x,x^*)}-1\right)$$
and $\|\cdot\|$ is a norm on $\mathbb{R}^n$.

The function $F$ is said to be (strictly) $HA(\alpha,\beta,\gamma,\xi,\eta,\rho,\theta)$-V-invex on $X$ if it is (strictly) $HA(\alpha,\beta,\gamma,\xi,\eta,\rho,\theta)$-V-invex at each $x^* \in X$.

**Definition 3.30.** The function $F$ is said to be (strictly) first-order HA($\alpha$, $\beta$, $\gamma$, $\xi$, $\eta$, $\rho$, $\theta$)-V-pseudoinvex at $x^* \in X$ if there exist functions $\alpha : X \times X \to \mathbb{R}$, $\beta : X \times X \to \mathbb{R}$, $\gamma : X \times X \to \mathbb{R}_+$, $\xi_i : X \times X \to \mathbb{R}_+\backslash\{0\}$, $i \in \underline{p}$, $\eta : X \times X \to \mathbb{R}^n$, $\rho : X \times X \to \mathbb{R}$, and $\theta : X \times X \to \mathbb{R}^n$ such that for all $x \in X (x \neq x^*)$,

$$\frac{1}{\beta(x,x^*)}\left\langle \sum_{i=1}^{p}\nabla F_i(x^*), e^{\beta(x,x^*)\eta(x,x^*)}-\mathbf{1}\right\rangle \geqq -\rho(x,x^*)\|\theta(x,x^*)\|^2$$
$$\Rightarrow \frac{1}{\alpha(x,x^*)}\gamma(x,x^*)\left(e^{\alpha(x,x^*)\sum_{i=1}^{p}\xi_i(x,x^*)[F_i(x)-F_i(x^*)]}-1\right)(>) \geqq 0$$
$$\text{if } \alpha(x,x^*) \neq 0 \text{ and } \beta(x,x^*) \neq 0 \text{ for all } x \in X,$$

$$\left\langle \sum_{i=1}^{p}\nabla F_i(x^*), \eta(x,x^*)\right\rangle \geqq -\rho(x,x^*)\|\theta(x,x^*)\|^2$$
$$\Rightarrow \frac{1}{\alpha(x,x^*)}\gamma(x,x^*)\left(e^{\alpha(x,x^*)\sum_{i=1}^{p}\xi_i(x,x^*)[F_i(x)-F_i(x^*)]}-1\right)(>) \geqq 0$$
$$\text{if } \alpha(x,x^*) \neq 0 \text{ and } \beta(x,x^*) = 0 \text{ for all } x \in X,$$

$$\frac{1}{\beta(x,x^*)}\left\langle \sum_{i=1}^{p} \nabla F_i(x^*), e^{\beta(x,x^*)\eta(x,x^*)} - \mathbf{1} \right\rangle \geqq -\rho(x,x^*)\|\theta(x,x^*)\|^2$$
$$\Rightarrow \gamma(x,x^*)\sum_{i=1}^{p} \xi_i(x,x^*)[F_i(x) - F_i(x^*)](>) \geqq 0$$
$$\text{if } \alpha(x,x^*) = 0 \text{ and } \beta(x,x^*) \neq 0 \text{ for all } x \in X,$$

$$\left\langle \sum_{i=1}^{p} \nabla F_i(x^*), \eta(x,x^*) \right\rangle \geqq -\rho(x,x^*)\|\theta(x,x^*)\|^2$$
$$\Rightarrow \gamma(x,x^*)\sum_{i=1}^{p} \xi_i(x,x^*)[F_i(x) - F_i(x^*)](>) \geqq 0$$
$$\text{if } \alpha(x,x^*) = 0 \text{ and } \beta(x,x^*) = 0 \text{ for all } x \in X.$$

The function $F$ is said to be (strictly) $HA(\alpha,\beta,\gamma,\xi,\eta,\rho,\theta)$-V-pseudoinvex on $X$ if it is (strictly) $HA(\alpha,\beta,\gamma,\xi,\eta,\rho,\theta)$-V-pseudoinvex at each $x^* \in X$.

**Definition 3.31.** The function $F$ is said to be (prestrictly) first-order HA($\alpha$, $\beta$, $\gamma$, $\xi$, $\eta$, $\rho$, $\theta$)-V-quasiinvex at $x^* \in X$ if there exist functions $\alpha : X \times X \to \mathbb{R}$, $\beta : X \times X \to \mathbb{R}$, $\gamma : X \times X \to \mathbb{R}_+$, $\xi_i : X \times X \to \mathbb{R}_+ \backslash \{0\}$, $i \in \underline{p}$, $\eta : X \times X \to \mathbb{R}^n$, $\rho : X \times X \to \mathbb{R}$, and $\theta : X \times X \to \mathbb{R}^n$ such that for all $x \in X$,

$$\frac{1}{\alpha(x,x^*)}\gamma(x,x^*)\left(e^{\alpha(x,x^*)\sum_{i=1}^{p}\xi_i(x,x^*)[F_i(x)-F_i(x^*)]} - 1\right)(<) \leqq 0$$
$$\Rightarrow \frac{1}{\beta(x,x^*)}\left\langle \sum_{i=1}^{p} \nabla F_i(x^*), e^{\beta(x,x^*)\eta(x,x^*)} - \mathbf{1} \right\rangle \leqq -\rho(x,x^*)\|\theta(x,x^*)\|^2$$
$$\text{if } \alpha(x,x^*) \neq 0 \text{ and } \beta(x,x^*) \neq 0 \text{ for all } x \in X,$$

$$\frac{1}{\alpha(x,x^*)}\gamma(x,x^*)\left(e^{\alpha(x,x^*)\sum_{i=1}^{p}\xi_i(x,x^*)[F_i(x)-F_i(x^*)]} - 1\right)(<) \leqq 0$$
$$\Rightarrow \left\langle \sum_{i=1}^{p} \nabla F_i(x^*), \eta(x,x^*) \right\rangle \leqq -\rho(x,x^*)\|\theta(x,x^*)\|^2$$
$$\text{if } \alpha(x,x^*) \neq 0 \text{ and } \beta(x,x^*) = 0 \text{ for all } x \in X,$$

$$\gamma(x,x^*)\sum_{i=1}^{p} \xi_i(x,x^*)[F_i(x) - F_i(x^*)](<) \leqq 0$$
$$\Rightarrow \frac{1}{\beta(x,x^*)}\left\langle \sum_{i=1}^{p} \nabla F_i(x^*), e^{\beta(x,x^*)\eta(x,x^*)} - \mathbf{1} \right\rangle \leqq -\rho(x,x^*)\|\theta(x,x^*)\|^2$$
$$\text{if } \alpha(x,x^*) = 0 \text{ and } \beta(x,x^*) \neq 0 \text{ for all } x \in X,$$

$$\gamma(x,x^*)\sum_{i=1}^{p}\xi_i(x,x^*)[F_i(x)-F_i(x^*)](<)$$
$$\leqq 0 \;\Rightarrow\; \langle\nabla F_i(x^*),\eta(x,x^*)\rangle \leqq -\rho(x,x^*)\|\theta(x,x^*)\|^2$$
$$\text{if } \alpha(x,x^*)=0 \text{ and } \beta(x,x^*)=0 \text{ for all } x\in X.$$

The function $F$ is said to be (prestrictly) $HA(\alpha,\beta,\gamma,\xi,\eta,\rho,\theta)$-V-quasiinvex on $X$ if it is (prestrictly) $HA(\alpha,\beta,\gamma,\xi,\eta,\rho,\theta)$-V-quasiinvex at each $x^*\in X$.
In certain situations, it is an interesting idea for proving the duality theorems to use some alternative but equivalent forms of the above definitions. These are obtained by considering the contrapositive statements. For example, $HA(\alpha,\beta,\gamma,\xi,\eta,\rho,\theta)$-V-pseudoinvexity (when $\alpha(x,x^*)\neq 0$ and $\beta(x,x^*)\neq 0$ for all $x\in X$) can be defined in the following equivalent way:
The function $f$ is said to be HA($\alpha$, $\beta$, $\gamma$, $\xi$, $\eta$, $\rho$, $\theta$)-V-pseudoinvex at $x^*\in X$ if there exist functions $\alpha: X\times X\to\mathbb{R}$, $\beta: X\times X\to\mathbb{R}$, $\gamma: X\times X\to\mathbb{R}_+$, $\xi_i: X\times X\to\mathbb{R}_+\backslash\{0\}$, $i\in\underline{p}$, $\eta: X\times X\to\mathbb{R}^n$, $\rho: X\times X\to\mathbb{R}$, and $\theta: X\times X\to\mathbb{R}^n$ such that for all $x\in X$,

$$\frac{1}{\alpha(x,x^*)}\gamma(x,x^*)\left(e^{\alpha(x,x^*)\sum_{i=1}^{p}\xi_i(x,x^*)[F_i(x)-F_i(x^*)]}-1\right)<0$$
$$\Rightarrow\quad \frac{1}{\beta(x,x^*)}\langle\sum_{i=1}^{p}\nabla F_i(x^*),e^{\beta(x,x^*)\eta(x,x^*)}-\mathbf{1}\rangle<-\rho(x,x^*)\|\theta(x,x^*)\|^2.$$

In the sequel, we shall also need a consistent notation for vector inequalities. For $a,b\in\mathbb{R}^m$, the following order notation will be used: $a\geqq b$ if and only if $a_i\geqq b_i$ for all $i\in\underline{m}$; $a\geqslant b$ if and only if $a_i\geqq b_i$ for all $i\in\underline{m}$, but $a\neq b$; $a>b$ if and only if $a_i>b_i$ for all $i\in\underline{m}$; and $a\ngeqslant b$ is the negation of $a\geqslant b$.

Consider the multiobjective problem

$$(P^*)\qquad \underset{x\in\mathbb{F}}{\text{Minimize}}\; F(x)=(F_1(x),\ldots,F_p(x)),$$

where $F_i$, $i\in\underline{p}$, are real-valued functions defined on $X$.

An element $x^\circ\in\mathbb{F}$ is said to be an *efficient (Pareto optimal, nondominated, noninferior)* solution of $(P^*)$ if there exists no $x\in\mathbb{F}$ such that $F(x)\leqslant F(x^\circ)$. There are several versions of the concept of efficiency that are currently used in the area of multiobjective programming. However, throughout this chapter we shall deal exclusively with the efficient solutions of $(P)$ in the sense defined above. Now, we recall a set of parametric necessary efficiency conditions for $(P)$ which will be needed for proving strong duality theorems.

**Theorem 3.32.** [6] Let $x^*\in\mathbb{F}$, $\lambda^*=\varphi(x^*)$, for each $i\in\underline{p}$, let $f_i$ and $g_i$ be continuously differentiable at $x^*$, for each $j\in\underline{q}$, let the function $G_j(\cdot,t)$ be continuously differentiable at $x^*$ for all $t\in T_j$, and for each $k\in\underline{r}$, let the function $H_k(\cdot,s)$ be continuously differentiable at $x^*$ for all $s\in S_k$. If $x^*$ is an efficient solution of (P), if the generalized Guignard constraint qualification holds at $x^*$, and if for each $i_0\in\underline{p}$, the set $cone(\{\nabla G_j(x^*,t): t\in\hat{T}_j(x^*), j\in\underline{q}\}\cup\{\nabla f_i(x^*)-\lambda_i^*\nabla g_i(x^*): i\in\underline{p}, i\neq i_0\})+span(\{\nabla H_k(x^*,s): s\in S_k,\, k\in\underline{r}\})$ is closed, then there exist $u^* \in U$ and integers $\nu_0$ and $\nu$, with $0 \leqq \nu_0 \leqq \nu \leqq n+1$, such that there exist $\nu_0$ indices $j_m$, with $1 \leqq j_m \leqq q$, together with $\nu_0$ points $t^m \in \hat{T}_{j_m}(x^*)$, $m \in \underline{\nu_0}$, $\nu - \nu_0$ indices $k_m$, with $1 \leqq k_m \leqq r$, together with $\nu - \nu_0$ points $s^m \in S_{k_m}$ for $m \in \underline{\nu} \backslash \underline{\nu_0}$, and $\nu$ real numbers $v_m^*$, with $v_m^* > 0$ for $m \in \underline{\nu_0}$, with the property that

$$\sum_{i=1}^{p} u_i^* [\nabla f_i(x^*) - \lambda_i^* \nabla g_i(x^*)] + \sum_{m=1}^{\nu_0} v_m^* \nabla G_{j_m}(x^*, t^m)$$
$$+ \sum_{m=\nu_0+1}^{\nu} v_m^* \nabla H_{k_m}(x^*, s^m) = 0,$$

where $cone(V)$ is the conic hull of the set $V \subset \mathbb{R}^n$ (i.e., the smallest convex cone containing $V$), $span(V)$ is the linear hull of $V$ (i.e., the smallest subspace containing $V$), $\hat{T}_j(x^*) = \{t \in T_j : G_j(x^*, t) = 0\}$, $U = \{u \in \mathbb{R}^p : u > 0, \sum_{i=1}^{p} u_i = 1\}$, and $\underline{\nu^*} \backslash \underline{\nu_0^*}$ is the complement of the set $\underline{\nu_0^*}$ relative to the set $\underline{\nu^*}$.

For brevity, we shall henceforth refer to an efficient solution $x^* \in \mathbb{F}$ as a *normal* efficient solution of $(P)$ if the generalized Guignard constraint qualification is satisfied at $x^*$ and for each $i_0 \in \underline{p}$, the set $cone(\{\nabla G_j(x^*, t) : t \in \hat{T}_j(x^*), j \in \underline{q}\} \cup \{\nabla f_i(x^*) - \lambda_i^* \nabla g_i(x^*) : i \in \underline{p}, i \neq i_0\}) + span(\{\nabla H_k(x^*, s) : s \in S_k, k \in \underline{r}\})$ is closed, where $\lambda_i^* = f_i(x^*)/g_i(x^*)$, $i \in \underline{p}$.

We further assume that the functions $f_i$, $g_i$, $i \in \underline{p}$, $G_j(\cdot, t)$, and $H_k(\cdot, s)$, are continuously differentiable on $X$ for all $t \in T_j$, $j \in \underline{q}$, and $s \in S_k$, $k \in \underline{r}$.

With regard to the choice of the type of generalized $HA(\alpha, \beta, \gamma, \xi, \eta, \rho, \theta)$-V-invex functions, specified in Definitions 3.10 - 3.12, to be used in the statements and proofs of our duality theorems, we shall consistently use the cases in which the functions $\alpha$ and $\beta$ are nonzero for all $(x, y) \in X \times X$. All the duality results established in this section can be modified, restated, and proved for the other cases in a natural manner.

## 5. Duality Model I

In this section, we consider two dual problems with relatively simple constraint structures and prove weak, strong, and strict converse duality theorems under appropriate $HA(\alpha, \beta, \gamma, \xi, \eta, \rho, \theta)$-V-invexity conditions. More general duality models and results for $(P)$ will be discussed in the subsequent sections.

Let

$$\mathbb{H} = \Big\{ (y, u, v, \lambda, \nu, \nu_0, J_{\nu_0}, K_{\nu \backslash \nu_0}, \bar{t}, \bar{s}) : y \in X;\ u \in U;\ \lambda \in \mathbb{R}^p;$$
$$0 \leqq \nu_0 \leqq \nu \leqq n+1;$$
$$v \in \mathbb{R}^{\nu}, v_i > 0, 1 \leqq i \leqq \nu_0;\ J_{\nu_0} = (j_1, j_2, \ldots, j_{\nu_0}), 1 \leqq j_i \leqq q;$$
$$K_{\nu \backslash \nu_0} = (k_{\nu_0+1}, \ldots, k_{\nu}),$$
$$1 \leqq k_i \leqq r;\ \bar{t} = (t^1, t^2, \ldots, t^{\nu_0}), t^i \in T_{j_i};\ \bar{s} = (s^{\nu_0+1}, \ldots, s^{\nu}), s^i \in S_{k_i} \Big\}.$$

Consider the following two problems:

$$(DI) \qquad \sup_{(y,u,v,\lambda,\nu,\nu_0,J_{\nu_0},K_{\nu\backslash\nu_0},\bar{t},\bar{s})\in\mathbb{H}} \lambda = (\lambda_1,\cdots,\lambda_p)$$

subject to

$$\sum_{i=1}^{p} u_i[\nabla f_i(y) - \lambda_i \nabla g_i(y)] + \sum_{m=1}^{\nu_0} v_m \nabla G_{j_m}(y,t^m) + \sum_{m=\nu_0+1}^{\nu} v_m \nabla H_{k_m}(y,s^m) = 0, \tag{3.34}$$

$$u_i[f_i(y) - \lambda_i g_i(y)] + \sum_{m=1}^{\nu_0} v_m G_{j_m}(y,t^m) + \sum_{m=\nu_0+1}^{\nu} v_m H_{k_m}(y,s^m) \geqq 0, \;\; i \in \underline{p}; \tag{3.35}$$

$$(\tilde{D}I) \qquad \sup_{(y,u,v,\lambda,\nu,\nu_0,J_{\nu_0},K_{\nu\backslash\nu_0},\bar{t},\bar{s})\in\mathbb{H}} \lambda = (\lambda_1,\cdots,\lambda_p)$$

subject to (35) and

$$\frac{1}{\beta(x,y)}\Big\langle \sum_{i=1}^{p} u_i[\nabla f_i(y) - \lambda_i \nabla g_i(y)] + \sum_{m=1}^{\nu_0} v_m \nabla G_{j_m}(y,t^m) + \sum_{m=\nu_0+1}^{\nu} v_m \nabla H_{k_m}(y,s^m), e^{\beta(x,y)\eta(x,y)} - \mathbf{1}\Big\rangle \geqq 0 \text{ for all } x \in \mathbb{F}, \tag{3.36}$$

where $\beta : X \times X \to \mathbb{R}$ and $\eta : X \times X \to \mathbb{R}^n$ are given functions.

The structures of the two problems designated above as $(DI)$ and $(\tilde{D}I)$, which can be proved under appropriate $HA(\alpha,\beta,\gamma,\xi,\eta,\rho,\theta)$-V-invexity hypotheses to be dual problems for $(P)$, are based directly on the form and contents of the necessary efficiency conditions of Theorem 3.16. This is, of course, the standard method for constructing Wolfe-type dual problems. Comparing $(DI)$ and $(\tilde{D}I)$, we see that $(\tilde{D}I)$ is relatively more general than $(DI)$ in the sense that any feasible solution of $(DI)$ is also feasible for $(\tilde{D}I)$, but the converse is not necessarily true. Furthermore, we observe that (1) is a system of $n$ equations, whereas (3) is a single inequality. Clearly, from a computational point of view, $(DI)$ is preferable to $(\tilde{D}I)$ because of the dependence of (3) on the feasible set of $(P)$. Despite these apparent differences, it turns out that the statements and proofs of all the duality theorems for $(P)-(DI)$ and $(P)-(\tilde{D}I)$ are almost identical and, therefore, we shall consider only the pair $(P)-(DI)$.

The next two theorems show that $(DI)$ is a dual problem for $(P)$.

**Theorem 3.33.** Let $x$ and $(y,u,v,\lambda,\nu,\nu_0,J_{\nu_0},K_{\nu\backslash\nu_0},\bar{t},\bar{s})$ be arbitrary feasible solutions of (P) and (DI), respectively, and assume that the Lagrangian-type function $\big(L_1(\cdot,u,v,\lambda,\bar{t},\bar{s}),\ldots,L_p(\cdot,u,v,\lambda,\bar{t},\bar{s})\big) : X \to \mathbb{R}^p$ is HA($\alpha$, $\beta$, $\gamma$, $\xi$, $\eta$, 0, $\theta$) - V-pseudoinvex and $\gamma(x,y) > 0$, where

$$L_i(z,u,v,\lambda,\bar{t},\bar{s}) = u_i\Big[f_i(z) - \lambda_i g_i(z) + \sum_{m=1}^{\nu_0} v_m G_{j_m}(z,t^m) + \sum_{m=\nu_0+1}^{\nu} v_m H_{k_m}(z,s^m)\Big], \quad i \in \underline{p}.$$

Then $\varphi(x) \nleq \lambda$.

*Proof.* From (34) we observe that

$$\frac{1}{\beta(x,y)}\Big\langle \sum_{i=1}^{p} u_i[\nabla f_i(y)-\lambda_i\nabla g_i(y)]+\sum_{m=1}^{\nu_0} v_m\nabla G_{j_m}(y,t^m) + \sum_{m=\nu_0+1}^{\nu} v_m\nabla H_{k_m}(y,s^m), e^{\beta(x,y)\eta(x,y)}-\mathbf{1}\Big\rangle = 0,$$

which in view of our $HA(\alpha,\beta,\gamma,\xi,\eta,0,\theta)$-V-pseudoinvexity assumption implies that

$$\frac{1}{\alpha(x,y)}\gamma(x,y)\left(e^{\alpha(x,y)\sum_{i=1}^{p}\xi_i(x,y)[L_i(x,u,v,\lambda,\bar{t},\bar{s})-L_i(y,u,v,\lambda,\bar{t},\bar{s})]}-1\right)\geqq 0.$$

We need to consider two cases: $\alpha(x,y)>0$ and $\alpha(x,y)<0$. If we assume that $\alpha(x,y)>0$ and recall that $\gamma(x,y)>0$, then the above inequality becomes

$$e^{\alpha(x,y)\sum_{i=1}^{p}\xi_i(x,y)[L_i(x,u,v,\lambda,\bar{t},\bar{s})-L_i(y,u,v,\lambda,\bar{t},\bar{s})]}\geqq 1,$$

which implies that

$$\sum_{i=1}^{p}\xi_i(x,y)L_i(x,u,v,\lambda,\bar{t},\bar{s})\geqq\sum_{i=1}^{p}\xi_i(x,y)L_i(y,u,v,\lambda,\bar{t},\bar{s}).$$

Because of the dual feasibility of $(y,u,v,\lambda,\nu,\nu_0,J_{\nu_0},K_{\nu\backslash\nu_0},\bar{t},\bar{s})$, it follows the right-hand side of the above inequality is greater than or equal to zero, and hence we have $L(x,u,v,\lambda,\bar{t},\bar{s})\geqq 0$. Inasmuch as $x\in\mathbb{F}$, and $v_m>0$, $m\in\underline{\nu_0}$, this inequality simplifies to

$$\sum_{i=1}^{q}u_i\xi_i(x,y)[f_i(x)-\lambda_i g_i(x)]\geqq 0. \tag{3.37}$$

Since $u>0$ and $\xi_i(x,y)>0$, $i\in\underline{p}$, the above inequality implies that

$$\big(f_1(x)-\lambda_1 g_1(x),\ldots,f_p(x)-\lambda_p g_p(x)\big)\nleq(0,\ldots,0),$$

which in turn implies that

$$\varphi(x)=\left(\frac{f_1(x)}{g_1(x)},\ldots,\frac{f_p(x)}{g_p(x)}\right)\nleq(\lambda_1,\ldots,\lambda_p)=\lambda.$$

If we assume that $\alpha(x,x^*)<0$, we arrive at the same conclusion.

□

**Theorem 3.34.** (Strong Duality) Let $x^*$ be a normal efficient solution of (P) and assume that for each feasible solution $(y,u,v,\lambda,\nu,\nu_0,J_{\nu_0},K_{\nu\backslash\nu_0},\bar{t},\bar{s})$ of (DI), the conditions specified in Theorem 3.33 are satisfied. Then there exist $u^*,v^*,\lambda^*$, $\nu^*,\nu_0^*,J_{\nu_0^*}$, $K_{\nu^*\backslash\nu_0^*},\bar{t}^*$, and $\bar{s}^*$ such that $(x^*,u^*,v^*,\lambda^*,\nu^*,\nu_0^*,J_{\nu_0^*},K_{\nu^*\backslash\nu_0^*},\bar{t}^*,\bar{s}^*)$ is an efficient solution of (DI) and $\varphi(x^*)=\lambda^*$.

*Proof.* Since $x^*$ is a normal efficient solution of $(P)$, by Theorem 3.32.1, there exist $u^*, v^*$, $\lambda^* (= \varphi(x^*))$, $\nu^*$, $\nu_0^*, J_{\nu_0^*}$, $K_{\nu^*\backslash\nu_0^*}, \bar{t}^*$, and $\bar{s}^*$ such that $(x^*, u^*, v^*, \lambda^*, \nu^*, \nu_0^*, J_{\nu_0^*}, K_{\nu^*\backslash\nu_0^*}, \bar{t}^*, \bar{s}^*)$ is a feasible solution of $(DI)$ and $\varphi(x^*) = \lambda^*$. If it were not efficient, then there would exist a feasible solution $(\tilde{x}, \tilde{u}, \tilde{v}, \tilde{\lambda}, \tilde{\nu}, \tilde{\nu}_0, J_{\tilde{\nu}_0}, K_{\tilde{\nu}\backslash\tilde{\nu}_0}, \bar{\tilde{t}}, \bar{\tilde{s}})$ of $(DI)$ such that $\tilde{\lambda} \geqslant \lambda^* = \varphi(x^*)$, contradicting Theorem 3.1. Therefore, $(x^*, u^*, v^*, \lambda^*, \nu^*, \nu_0^*, J_{\nu_0^*}, K_{\nu^*\backslash\nu_0^*}, \bar{t}^*, \bar{s}^*)$ is an efficient solution of $(DI)$. □

We also have the following converse duality result for $(P)$ and $(DI)$.

**Theorem 3.35.** (Strict Converse Duality) Let $x^*$ and $(\tilde{x}, \tilde{u}, \tilde{v}, \tilde{\lambda}, \tilde{\nu}, \tilde{\nu}_0, J_{\tilde{\nu}_0}, K_{\tilde{\nu}\backslash\tilde{\nu}_0}, \bar{\tilde{t}}, \bar{\tilde{s}})$ be arbitrary feasible solutions of (P) and (DI), respectively, such that

$$\sum_{i=1}^{p} \tilde{u}_i \xi_i(x^*, \tilde{x})[f_i(x^*) - \tilde{\lambda}_i g_i(x^*)] = 0, \tag{3.38}$$

where for each $i \in \underline{p}$, $\xi_i$ is a function from $X \times X$ to $\mathbb{R}_+ \backslash \{0\}$. Furthermore, assume that the function $L(\cdot, \tilde{u}, \tilde{v}, \tilde{\lambda}, \bar{\tilde{t}}, \bar{\tilde{s}})$ is strictly $HA(\alpha, \beta, \gamma, \xi, \eta, 0, \theta)$-V-pseudoinvex at $\tilde{x}$ and $\gamma(x^*, \tilde{x}) > 0$. Then $\tilde{x} = x^*$ and $\varphi(x^*) = \tilde{\lambda}$.

*Proof.* (a): Suppose to the contrary that $\tilde{x} \neq x^*$. Now proceeding as in the proof of Theorem 3.33 (with $x$ replaced by $x^*$ and $(y, u, v, \lambda, \nu, \nu_0, J_{\nu_0}, K_{\nu\backslash\nu_0}, \bar{t}, \bar{s})$ by $(\tilde{x}, \tilde{u}, \tilde{v}, \tilde{\lambda}, \tilde{\nu}, \tilde{\nu}_0, J_{\tilde{\nu}_0}, K_{\tilde{\nu}\backslash\tilde{\nu}_0}, \bar{\tilde{t}}, \bar{\tilde{s}})$) and using the conditions set forth above, we arrive at the strict inequality

$$\sum_{i=1}^{p} \tilde{u}_i \xi_i(x^*, \tilde{x})[f_i(x^*) - \tilde{\lambda}_i g_i(x^*)] > 0,$$

which contradicts (38). Therefore, we conclude that $\tilde{x} = x^*$ and $\varphi(x^*) = \tilde{\lambda}$.

□

## 5.1. Duality Model II

In this section, we consider certain variants of $(DI)$ and $(\tilde{D}I)$ that allow for more general types of $HA(\alpha, \beta, \gamma, \xi, \eta, \rho, \theta)$-V-invexity conditions under which duality can be established. These duality models have the following forms:

$$(DII) \qquad \sup_{(y,u,v,\lambda,\nu,\nu_0,J_{\nu_0},K_{\nu\backslash\nu_0},\bar{t},\bar{s}) \in \mathbb{H}} \lambda = (\lambda_1, \cdots, \lambda_p)$$

subject to

$$\sum_{i=1}^{p} u_i[\nabla f_i(y) - \lambda_i \nabla g_i(y)] + \sum_{m=1}^{\nu_0} v_m \nabla G_{j_m}(y, t^m) + \sum_{m=\nu_0+1}^{\nu} v_m \nabla H_{k_m}(y, s^m) = 0, \tag{3.39}$$

$$u_i[f_i(y) - \lambda_i g_i(y)] \geqq 0, \quad i \in \underline{p}, \tag{3.40}$$

$$G_{j_m}(y, t^m) \geqq 0, \quad m \in \underline{\nu_0}, \tag{3.41}$$

$$v_m H_{k_m}(y, s^m) \geqq 0, \quad m \in \underline{v} \backslash \underline{v_0}; \tag{3.42}$$

$(\tilde{D}II)$ $\sup_{(y,u,v,\lambda,\nu,\nu_0,J_{\nu_0},K_{\nu\backslash\nu_0},\bar{t},\bar{s})\in\mathbb{H}} \lambda = (\lambda_1, \cdots, \lambda_p)$

subject to (40) - (42).

The remarks and observations made earlier about the relationships between $(DI)$ and $(\tilde{D}I)$ are, of course, also valid for $(DII)$ and $(\tilde{D}II)$. We shall establish some duality results in which various generalized $HA(\alpha,\beta,\gamma,\xi,\eta,\rho,\theta)$-V-invexity requirements will be placed on certain vector-valued functions, including $(\mathscr{E}_1(\cdot,\lambda,u),\ldots,$
$\mathscr{E}_p(\cdot,\lambda,u))$, where for each $i \in \underline{p}$, the component function $\mathscr{E}_i(\cdot,\lambda,u)$ is defined, for fixed $\lambda$ and $u$, on $X$ by

$$\mathscr{E}_i(z,\lambda,u) = u_i[f_i(z) - \lambda_i g_i(z)].$$

**Theorem 3.36.** Let $x$ and $(y,u,v,\lambda,\nu,\nu_0,J_{\nu_0},K_{\nu\backslash\nu_0},\bar{t},\bar{s})$ be arbitrary feasible solutions of (P) and (DII), respectively, and assume that any one of the following four sets of hypotheses is satisfied:

(a) (i) $(\mathscr{E}_1(\cdot,\lambda,u),\ldots,\mathscr{E}_p(\cdot,\lambda,u))$ is HA($\alpha$, $\beta$, $\bar{\gamma}$, $\xi$, $\eta$, $\bar{\rho}$, $\theta$) - V-pseudoinvex at $y$ and $\bar{\gamma}(x,y) > 0$;

(ii) $(v_1 G_{j_1}(\cdot,t^1),\ldots,v_{\nu_0} G_{j_{\nu_0}}(\cdot,t^{\nu_0}))$ is $HA(\alpha,\beta,\hat{\gamma},\pi,\eta,\hat{\rho},\theta)$-V-quasiinvex at $y$;

(iii) $(v_{\nu_0+1} H_{k_{\nu_0+1}}(\cdot,s^{\nu_0+1}),\ldots,v_\nu H_{k_\nu}(\cdot,s^\nu))$ is HA($\alpha$, $\beta$, $\breve{\gamma}$, $\delta$, $\eta$, $\breve{\rho}$, $\theta$)-V-quasiinvex at $y$;

(iv) $\bar{\rho}(x,y) + \hat{\rho}(x,y) + \breve{\rho}(x,y) \geqq 0$;

(b) (i) $(\mathscr{E}_1(\cdot,\lambda,u),\ldots,\mathscr{E}_p(\cdot,\lambda,u))$ is prestrictly HA($\alpha$, $\beta$, $\bar{\gamma}$, $\xi$, $\eta$, $\bar{\rho}$, $\theta$)-V-quasiinvex at $y$ and $\bar{\gamma}(x,y) > 0$;

(ii) $(v_1 G_{j_1}(\cdot,t^1),\ldots,v_{\nu_0} G_{j_{\nu_0}}(\cdot,t^{\nu_0}))$ is $HA(\alpha,\beta,\hat{\gamma},\pi,\eta,\hat{\rho},\theta)$-V-quasiinvex at $y$;

(iii) $(v_{\nu_0+1} H_{k_{\nu_0+1}}(\cdot,s^{\nu_0+1}),\ldots,v_\nu H_{k_\nu}(\cdot,s^\nu))$ is HA($\alpha$, $\beta$, $\breve{\gamma}$, $\delta$, $\eta$, $\breve{\rho}$, $\theta$)-V-quasiinvex at $y$;

(iv) $\bar{\rho}(x,y) + \hat{\rho}(x,y) + \breve{\rho}(x,y) > 0$;

(c) (i) $(\mathscr{E}_1(\cdot,\lambda,u),\ldots,\mathscr{E}_p(\cdot,\lambda,u))$ is prestrictly HA($\alpha$, $\beta$, $\bar{\gamma}$, $\xi$, $\eta$, $\bar{\rho}$, $\theta$)-V-quasiinvex at $y$ and $\bar{\gamma}(x,y) > 0$;

(ii) $(v_1 G_{j_1}(\cdot,t^1),\ldots,v_{\nu_0} G_{j_{\nu_0}}(\cdot,t^{\nu_0}))$ is strictly HA($\alpha$, $\beta$, $\hat{\gamma}$, $\pi$, $\eta$, $\hat{\rho}$, $\theta$)-V-pseudoinvex at $y$;

(iii) $(v_{\nu_0+1} H_{k_{\nu_0+1}}(\cdot,s^{\nu_0+1}),\ldots,v_\nu H_{k_\nu}(\cdot,s^\nu))$ is HA($\alpha$, $\beta$, $\breve{\gamma}$, $\delta$, $\eta$, $\breve{\rho}$, $\theta$) - V - quasiinvex at $y$;

(iv) $\bar{\rho}(x,y) + \hat{\rho}(x,y) + \breve{\rho}(x,y) \geqq 0$;

(d) (i) $\big(\mathscr{E}_1(\cdot,\lambda,u),\ldots,\mathscr{E}_p(\cdot,\lambda,u)\big)$ is prestrictly HA($\alpha$, $\beta$, $\bar{\gamma}$, $\xi$, $\eta$, $\bar{\rho}$, $\theta$)-V-quasiinvex at $y$ and $\bar{\gamma}(x,y) > 0$;

(ii) $\big(v_1 G_{j_1}(\cdot,t^1),\ldots,v_{\nu_0} G_{j_{\nu_0}}(\cdot,t^{\nu_0})\big)$ is $HA(\alpha,\beta,\hat{\gamma},\pi,\eta,\hat{\rho},\theta)$-V-quasiinvex at $y$;

(iii) $\big(v_{\nu_0+1} H_{k_{\nu_0+1}}(\cdot,s^{\nu_0+1}),\ldots,v_\nu H_{k_\nu}(\cdot,s^\nu)\big)$ is strictly $HA(\alpha,\beta,\breve{\gamma},\delta,\eta,\breve{\rho},\theta)$-V-pseudoinvex at $y$;

(iv) $\bar{\rho}(x,y)+\hat{\rho}(x,y)+\breve{\rho}(x,y) \geqq 0$.

Then $\varphi(x) \nleq \lambda$.

*Proof.* (a): Because of (41) and the primal feasibility of $x$, we have $G_{j_m}(x,t^m) \leqq 0 \leqq G_{j_m}(y,t^m)$.
As $\pi_m(x,y) > 0$ for each $m \in \underline{\nu_0}$, it follows that

$$\sum_{m=1}^{\nu_0} v_m \pi_m(x,y) G_{j_m}(x,t^m) \leqq \sum_{m=1}^{\nu_0} v_m \pi_m(x,x) G_{j_m}(y,t^m),$$

and hence

$$\frac{1}{\alpha(x,y)}\hat{\gamma}(x,y)\Big(e^{\alpha(x,y)\sum_{m=1}^{\nu_0} v_m\pi_m(x,y)[G_{j_m}(x,t^m)-G_{j_m}(y,t^m)]}-1\Big) \leqq 0,$$

which in view of (ii) implies that

$$\frac{1}{\beta(x,y)}\Big\langle \sum_{m=1}^{\nu_0} v_m \nabla G_{j_m}(y,t^m), e^{\beta(x,y)\eta(x,y)}-\mathbf{1}\Big\rangle \leqq -\hat{\rho}(x,y)\|\theta(x,y)\|^2. \tag{3.43}$$

Similarly, we can show that our assumptions in (iii) combined with the primal feasibility of $x$ and (42) lead to the following inequality:

$$\frac{1}{\beta(x,y)}\Big\langle \sum_{m=\nu_0+1}^{\nu} v_m \nabla H_{k_m}(y,s^m), e^{\beta(x,y)\eta(x,y)}-\mathbf{1}\Big\rangle \leqq -\breve{\rho}(x,y)\|\theta(x,y)\|^2. \tag{3.44}$$

Now because of (43), (44), and (iv), (39) yields

$$\frac{1}{\beta(x,y)}\Big\langle \sum_{i=1}^{p} u_i[\nabla f_i(y)-\lambda_i\nabla g_i(y)], e^{\beta(x,y)\eta(x,y)}-\mathbf{1}\Big\rangle \geqq -\bar{\rho}(x,y)\|\theta(x,y)\|^2,$$

which in view of (i) implies that

$$\frac{1}{\alpha(x,y)}\bar{\gamma}(x,y)\Big(e^{\alpha(x,y)\sum_{i=1}^{p} u_i\xi_i(x,y)\{f_i(x)-\lambda_i g_i(x)-[f_i(y)-\lambda_i g_i(y)]\}}-1\Big) \geqq 0.$$

Since $\bar{\gamma}(x,y) > 0$, this inequality implies that

$$\sum_{i=1}^{p} u_i\xi_i(x,y)[f_i(x)-\lambda_i g_i(x)] \geqq \sum_{i=1}^{p} u_i\xi_i(x,y)[f_i(y)-\lambda_i g_i(y)] \geqq 0,$$

where the second inequality follows from (40). Since $u_i\xi_i(x,y) > 0$ for each $i \in \underline{p}$, this inequality implies that

$$\big(f_1(x) - \lambda_1 g_1(x), \dots, f_p(x) - \lambda_p g_p(x)\big) \nleq (0, \dots, 0),$$

which in turn implies that

$$\varphi(x) = \left(\frac{f_1(x)}{g_1(x)}, \dots, \frac{f_p(x)}{g_p(x)}\right) \nleq (\lambda_1, \dots, \lambda_p) = \lambda.$$

(b) : The proof is similar to that of part (a).

(c) : Suppose to the contrary that $\varphi(x) \leqslant \lambda$. This implies that $f_i(x) - \lambda_i g_i(x) \leqq 0$ for each $i \in \underline{p}$ with strict inequality holding for at least one index $\ell \in \underline{p}$. Since $u > 0$, $\xi_i(x,y) > 0$, $i \in \underline{p}$, and (40) holds, it is clear that

$$\sum_{i=1}^{p} \xi_i(x,y)\mathscr{E}_i(x,\lambda,u) < \sum_{i=1}^{p} \xi_i(x,y)\mathscr{E}_i(y,\lambda,u),$$

and therefore it follows that

$$\frac{1}{\alpha(x,y)}\bar{\gamma}(x,y)\left(e^{\alpha(x,y)\sum_{i=1}^{p}\xi_i(x,y)[\mathscr{E}_i(x,\lambda,u)-\mathscr{E}_i(y,\lambda,u)]} - 1\right) < 0,$$

which in view of (i) implies that

$$\frac{1}{\beta(x,y)}\left\langle \sum_{i=1}^{p} u_i[\nabla f_i(y) - \lambda_i \nabla g_i(y)], e^{\beta(x,y)\eta(x,y)} - \mathbf{1}\right\rangle$$
$$\leqq -\bar{\rho}(x,y)\|\theta(x,y)\|^2. \tag{3.45}$$

Proceeding as in the proof of part (a) and using our strict HA($\alpha$, $\beta$, $\hat{\gamma}$, $\pi$, $\eta$, $\hat{\rho}$, $\theta$) - V - pseudoinvexity assumption, we arrive at the strict inequality

$$\frac{1}{\beta(x,y)}\left\langle \sum_{m=1}^{\nu_0} v_m \nabla G_{j_m}(y,t^m), e^{\beta(x,y)\eta(x,y)} - \mathbf{1}\right\rangle < -\hat{\rho}(x,y)\|\theta(x,y)\|^2.$$

Now combining this inequality with (44) (which is valid for the present case because of (iii)) and (39), and using (iv) we obtain

$$\frac{1}{\beta(x,y)}\left\langle \sum_{i=1}^{p} u_i[\nabla f_i(y) - \lambda_i \nabla g_i(y)], e^{\beta(x,y)\eta(x,y)} - \mathbf{1}\right\rangle > -\bar{\rho}(x,y)\|\theta(x,y)\|^2$$

which contradicts (45). Therefore, we conclude that $\varphi(x) \nleq \lambda$.

(d) : The proof is similar to that of part (c).

□

**Theorem 3.37.** (Strong Duality) Let $x^*$ be a normal efficient solution of (P) and assume that for each feasible solution $(y,u,v,\lambda,\nu,\nu_0, J_{\nu_0}, K_{\nu\setminus\nu_0}, \bar{t},\bar{s})$ of (DII), any one of the four sets of conditions specified in Theorem 3.20 is satisfied. Then there exist $u^*, v^*, \lambda^*$, $\nu^*$, $\nu_0^*, J_{\nu_0^*}$, $K_{\nu^*\setminus\nu_0^*}, \bar{t}^*$, and $\bar{s}^*$ such that $(x^*,u^*,v^*,\lambda^*,\nu^*,\nu_0^*, J_{\nu_0^*}, K_{\nu^*\setminus\nu_0^*},\bar{t}^*,\bar{s}^*)$ is an efficient solution of (DII) and $\varphi(x^*) = \lambda^*$.

*Proof.* The proof is similar to that of Theorem 3.18. □

**Theorem 3.38.** (Strict Converse Duality) Let $x^*$ and $\tilde{w} \equiv (\tilde{x}, \tilde{u}, \tilde{v}, \tilde{\lambda}, \tilde{v}, \tilde{v}_0, J_{\tilde{v}_0}, K_{\tilde{v}\backslash\tilde{v}_0}, \bar{\tilde{t}}, \tilde{s})$ be arbitrary feasible solutions of (P) and (DII), respectively, such that

$$\sum_{i=1}^{p} \tilde{u}_i \xi_i(x^*, \tilde{x})[f_i(x^*) - \tilde{\lambda}_i g_i(x^*)] = 0, \tag{3.46}$$

where for each $i \in \underline{p}$, $\xi_i$ is a function from $X \times X$ to $\mathbb{R}_+ \backslash \{0\}$. Furthermore, assume that any one of the following four sets of hypotheses is satisfied:

(a) The assumptions specified in part (a) of Theorem 3.36 are satisfied for the feasible solution $\tilde{w}$ of (DII), and the function $\left(\mathscr{E}_1(\cdot, \tilde{\lambda}, \tilde{u}), \ldots, \mathscr{E}_p(\cdot, \tilde{\lambda}, \tilde{u})\right)$ is strictly HA($\alpha$, $\beta$, $\bar{\gamma}$, $\xi$, $\eta$, $\bar{\rho}$, $\theta$) -V-pseudoinvex at $\tilde{x}$.

(b) The assumptions specified in part (b) of Theorem 3.36 are satisfied for the feasible solution $\tilde{w}$ of (DII), and the function $\left(\mathscr{E}_1(\cdot, \tilde{\lambda}, \tilde{u}), \ldots, \mathscr{E}_p(\cdot, \tilde{\lambda}, \tilde{u})\right)$ is $HA(\alpha, \beta, \bar{\gamma}, \xi, \eta, \bar{\rho}, \theta)$-V-quasiinvex at $\tilde{x}$.

(c) The assumptions specified in part (c) of Theorem 3.36 are satisfied for the feasible solution $\tilde{w}$ of (DII), and the function $\left(\mathscr{E}_1(\cdot, \tilde{\lambda}, \tilde{u}), \ldots, \mathscr{E}_p(\cdot, \tilde{\lambda}, \tilde{u})\right)$ is $HA(\alpha, \beta, \bar{\gamma}, \xi, \eta, \bar{\rho}, \theta)$-V-quasiinvex at $\tilde{x}$.

(d) The assumptions specified in part (d) of Theorem 3.36 are satisfied for the feasible solution $\tilde{w}$ of (DII), and the function $\left(\mathscr{E}_1(\cdot, \tilde{\lambda}, \tilde{u}), \ldots, \mathscr{E}_p(\cdot, \tilde{\lambda}, \tilde{u})\right)$ is $HA(\alpha, \beta, \bar{\gamma}, \xi, \eta, \bar{\rho}, \theta)$-V-quasiinvex at $\tilde{x}$.

Then $\tilde{x} = x^*$ and $\varphi(x^*) = \tilde{\lambda}$.

*Proof.* (a) and (b) : The proofs are similar to that of Theorem 3.36.

(c) : Suppose to the contrary that $\tilde{x} \neq x^*$. Proceeding as in the proof of part (a) of Theorem 3.36, we arrive at the following strict inequality:

$$\frac{1}{\beta(x^*, \tilde{x})}\left\langle \sum_{i=1}^{p} \tilde{u}_i [\nabla f_i(\tilde{x}) - \tilde{\lambda}_i \nabla g_i(\tilde{x})], e^{\beta(x^*, \tilde{x})\eta(x^*, \tilde{x})} - \mathbf{1} \right\rangle > -\bar{\rho}(x^*, \tilde{x}) \|\theta(x^*, \tilde{x})\|^2,$$

which in view of our $HA(\alpha, \beta, \bar{\gamma}, \xi, \eta, \bar{\rho}, \theta)$-V-quasiinvexity hypothesis implies that

$$\frac{1}{\alpha(x^*, \tilde{x})}\bar{\gamma}(x^*, \tilde{x})\left(e^{\alpha(x^*, \tilde{x})\sum_{i=1}^{p} \tilde{u}_i \xi_i(x^*, \tilde{x})\{f_i(x^*) - \tilde{\lambda}_i g_i(x^*) - [f_i(\tilde{x}) - \tilde{\lambda}_i g_i(\tilde{x})]\}} - 1\right) > 0.$$

Since $\bar{\gamma}(x^*, \tilde{x}) > 0$, this inequality implies that

$$\sum_{i=1}^{p} \tilde{u}_i \xi_i(x^*, \tilde{x})[f_i(x^*) - \tilde{\lambda}_i g_i(x^*)] > \sum_{i=1}^{p} \tilde{u}_i \xi_i(x^*, \tilde{x})[f_i(\tilde{x}) - \tilde{\lambda}_i g_i(\tilde{x})] \geqq 0,$$

where the second inequality follows from the dual feasibility of $\tilde{w}$ and (42). This inequality contradicts (42), and, therefore, we conclude that $\tilde{x} = x^*$ and $\varphi(x^*) = \tilde{\lambda}$.

(d) : The proof is similar to that of part (c). □

## 5.2. Duality Model III

In this subsection, we discuss several families of duality results under appropriate generalized $HA(\alpha,\beta,\gamma,\xi,\eta,\rho,\theta)$-V-invexity hypotheses imposed on certain vector functions whose components are comprised of different combinations of the problem functions. This is accomplished by employing a partitioning scheme which was originally proposed in [2] for the purpose of constructing generalized dual problems for nonlinear programming problems. For this we need some additional notations and the following partitioning schemes.

Let $\nu_0$ and $\nu$ be integers, with $1 \leqq \nu_0 \leqq \nu \leqq n+1$, and let $\{J_0,J_1,\ldots,J_M\}$ and $\{K_0,K_1,\ldots,K_M\}$ be partitions of the sets $\underline{\nu_0}$ and $\underline{\nu}\backslash\underline{\nu_0}$, respectively; thus, $J_i \subseteq \underline{\nu_0}$ for each $i \in \underline{M}\cup\{0\}$, $J_i \cap J_j = \emptyset$ for each $i,j \in \underline{M}\cup\{0\}$ with $i \neq j$, and $\cup_{i=0}^{M} J_i = \underline{\nu_0}$. Obviously, similar properties hold for $\{K_0,K_1,\ldots,K_M\}$. Moreover, if $m_1$ and $m_2$ are the numbers of the partitioning sets of $\underline{\nu_0}$ and $\underline{\nu}\backslash\underline{\nu_0}$, respectively, then $M = \max\{m_1,m_2\}$ and $J_i = \emptyset$ or $K_i = \emptyset$ for $i > \min\{m_1,m_2\}$.

In addition, we use the real-valued functions $\Phi_i(\cdot,u,v,\lambda,\bar{t},\bar{s})$, and $\Lambda_\tau(\cdot,v,\bar{t},\bar{s})$ defined, for fixed $u,v,\lambda,\bar{t} \equiv (t^1,t^2,\ldots,t^{\nu_0})$, and $\bar{s} \equiv (s^{\nu_0+1}, s^{\nu_0+2}, \ldots, s^{\nu})$, on $X$ as follows:

$$\Phi_i(z,u,v,\lambda,\bar{t},\bar{s}) = u_i\Big[f_i(z) - \lambda_i g_i(z) + \sum_{m\in J_0} v_m G_{j_m}(z,t^m) + \sum_{m\in K_0} v_m H_{k_m}(z,s^m)\Big],\quad i \in \underline{p},$$

$$\Lambda_\tau(z,v,\bar{t},\bar{s}) = \sum_{m\in J_\tau} v_m G_{j_m}(z,t^m) + \sum_{m\in K_\tau} v_m H_{k_m}(z,s^m), \qquad \tau \in \underline{M}.$$

Making use of the sets and functions defined above, we can state our general duality models as follows:

$(DIII)$ $\qquad \sup\limits_{(y,u,v,\lambda,\nu,\nu_0,J_{\nu_0},K_{\nu\backslash\nu_0},\bar{t},\bar{s})\in\mathbb{H}} \qquad \lambda = (\lambda_1,\cdots,\lambda_p)$

subject to

$$\sum_{i=1}^{p} u_i[\nabla f_i(y) - \lambda_i \nabla g_i(y)] + \sum_{m=1}^{\nu_0} v_m \nabla G_{j_m}(y,t^m) + \sum_{m=\nu_0+1}^{\nu} v_m \nabla H_{k_m}(y,s^m) = 0, \tag{3.47}$$

$$f_i(y) - \lambda_i g_i(y) + \sum_{m\in J_0} v_m G_{j_m}(y,t^m) + \sum_{m\in K_0} v_m H_{k_m}(y,s^m) \geqq 0, \quad i \in \underline{p}, \tag{3.48}$$

$$\sum_{m\in J_\tau} v_m G_{j_m}(y,t^m) + \sum_{m\in K_\tau} v_m H_{k_m}(y,s^m) \geqq 0, \quad \tau \in \underline{M}; \tag{3.49}$$

$(\tilde{D}III)$ $\qquad \sup\limits_{(y,u,v,\lambda,\nu,\nu_0,J_{\nu_0},K_{\nu\backslash\nu_0},\bar{t},\bar{s})\in\mathbb{H}} \qquad \lambda = (\lambda_1,\cdots,\lambda_p)$

subject to (48), and (49).

The remarks and observations made earlier about the relationships between $(DI)$ and $(\tilde{D}I)$ are, of course, also valid for $(DIII)$ and $(\tilde{D}III)$.

The next two theorems show that $(DIII)$ is a dual problem for $(P)$.

**Theorem 3.39.** Let $x$ and $(y,u,v,\lambda,\nu,\nu_0,J_{\nu_0},K_{\nu\backslash\nu_0},\bar{t},\bar{s})$ be arbitrary feasible solutions of (P) and (DIII), respectively, and assume that any one of the following three sets of hypotheses is satisfied:

(a) (i) $\left(\Phi_1(\cdot,u,v,\lambda,\bar{t},\bar{s}),\ldots,\Phi_p(\cdot,u,v,\lambda,\bar{t},\bar{s})\right)$ is HA($\alpha$, $\beta$, $\bar{\gamma}$, $\xi$, $\eta$, $\bar{\rho}$, $\theta$) -V-pseudoinvex at $y$ and $\bar{\gamma}(x,y)>0$;

(ii) $\left(\Lambda_1(\cdot,v,\bar{t},\bar{s}),\ldots,\Lambda_M(\cdot,v,\bar{t},\bar{s})\right)$ is $HA(\alpha,\beta,\tilde{\gamma},\pi,\eta,\tilde{\rho},\theta)$-V-quasiinvex at $y$;

(iii) $\bar{\rho}(x,y)+\tilde{\rho}(x,y)\geqq 0$;

(b) (i) $\left(\Phi_1(\cdot,u,v,\lambda,\bar{t},\bar{s}),\ldots,\Phi_p(\cdot,u,v,\lambda,\bar{t},\bar{s})\right)$ is prestrictly HA($\alpha$, $\beta$, $\bar{\gamma}$, $\xi$, $\eta$, $\bar{\rho}$, $\theta$) - V -
quasiinvex at $y$ and $\bar{\gamma}(x,y)>0$;

(ii) $\left(\Lambda_1(\cdot,v,\bar{t},\bar{s}),\ldots,\Lambda_M(\cdot,v,\bar{t},\bar{s})\right)$ is HA($\alpha$, $\beta$, $\tilde{\gamma}$, $\pi$, $\eta$, $\tilde{\rho}$, $\theta$)-V-quasinvex

(iii) $\bar{\rho}(x,y)+\tilde{\rho}(x,y)>0$;

(c) (i) $\left(\Phi_1(\cdot,u,v,\lambda,\bar{t},\bar{s}),\ldots,\Phi_p(\cdot,u,v,\lambda,\bar{t},\bar{s})\right)$ is prestrictly HA($\alpha$, $\beta$, $\bar{\gamma}$, $\xi$, $\eta$, $\bar{\rho}$, $\theta$) - V - quasiinvex at $y$ and $\bar{\gamma}(x,y)>0$;

(ii) $\left(\Lambda_1(\cdot,v,\bar{t},\bar{s}),\ldots,\Lambda_M(\cdot,v,\bar{t},\bar{s})\right)$ is strictly HA($\alpha$, $\beta$, $\tilde{\gamma}$, $\pi$, $\eta$, $\tilde{\rho}$, $\theta$) -V-pseudoinvex at $y$;

(iii) $\bar{\rho}(x,y)+\tilde{\rho}(x,y)\geqq 0$.

Then $\varphi(x)\nleq\lambda$.

*Proof.* (a): It is clear that (47) can be expressed as follows:

$$\sum_{i=1}^{p}u_i[\nabla f_i(y)-\lambda_i\nabla g_i(y)]+\sum_{m\in J_0}v_m\nabla G_{j_m}(y,t^m)+\sum_{m\in K_0}v_m\nabla H_{k_m}(y,s^m)$$
$$+\sum_{\tau=1}^{M}\Big[\sum_{m\in J_\tau}v_m\nabla G_{j_m}(y,t^m)+\sum_{m\in K_\tau}v_m\nabla H_{k_m}(y,s^m)\Big]=0. \quad (3.50)$$

Since $x\in\mathbb{F}$, $v_m>0$, $m\in\underline{\nu_0}$, it follows that for each $\tau\in\underline{M}$,

$$\begin{aligned}\Lambda_\tau(x,v,\bar{t},\bar{s})&=\sum_{m\in J_\tau}v_mG_{j_m}(x,t^m)+\sum_{m\in K_\tau}v_mH_{k_m}(x,s^m)\\&\leqq 0\\&\leqq\sum_{m\in J_\tau}v_mG_{j_m}(y,t^m)+\sum_{m\in K_\tau}v_mH_{k_m}(y,s^m)\\&=\Lambda_\tau(y,v,\bar{t},\bar{s}).\end{aligned}$$

Because $\pi_\tau(x,y)>0$ for each $\tau\in M$, it is clear that

$$\sum_{\tau=1}^{M}\pi_\tau(x,y)\Lambda_\tau(x,v,\bar{t},\bar{s})\leqq\sum_{\tau=1}^{M}\pi_\tau(x,y)\Lambda_\tau(y,v,\bar{t},\bar{s}),$$

and so

$$\frac{1}{\alpha(x,y)}\tilde{\gamma}(x,y)\left(e^{\alpha(x,y)\sum_{\tau=1}^{M}\pi_\tau(x,y)[\Lambda_\tau(x,v,\bar{t},\bar{s})-\Lambda_\tau(y,v,\bar{t},\bar{s})]}-1\right)\leqq 0,$$

which in view of (ii) implies that

$$\frac{1}{\beta(x,y)}\left\langle\sum_{\tau=1}^{M}\left[\sum_{m\in J_\tau}v_m\nabla G_{j_m}(y,t^m)+\sum_{m\in K_\tau}v_m\nabla H_{k_m}(y,s^m)\right],e^{\beta(x,y)\eta(x,y)}-\mathbf{1}\right\rangle$$

$$\leqq -\tilde{\rho}(x,y)\|\theta(x,y)\|^2. \quad (3.51)$$

Combining (50) and (51), and using (iii) we get

$$\frac{1}{\beta(x,y)}\left\langle\sum_{i=1}^{p}u_i[\nabla f_i(y)-\lambda_i\nabla g_i(y)]+\sum_{m\in J_0}v_m\nabla G_{j_m}(y,t^m)\right.$$
$$\left.+\sum_{m\in K_0}v_m\nabla H_{k_m}(y,s^m),e^{\beta(x,y)\eta(x,y)}-\mathbf{1}\right\rangle$$
$$\geqq \tilde{\rho}(x,y)\|\theta(x,y)\|^2\geqq -\bar{\rho}(x,y)\|\theta(x,y)\|^2, \quad (3.52)$$

which by virtue of (i) implies that

$$\frac{1}{\alpha(x,y)}\bar{\gamma}(x,y)\left(e^{\alpha(x,y)\sum_{i=1}^{p}\xi_i(x,y)[\Phi_i(x,u,v,\lambda,\bar{t},\bar{s})-\Phi_i(y,u,v,\lambda,\bar{t},\bar{s})]}-1\right)\geqq 0. \quad (3.53)$$

Since $\bar{\gamma}(x,y)>0$, we can deduce from this inequality that

$$\sum_{i=1}^{p}\xi_i(x,y)\Phi_i(x,u,v,\lambda,\bar{t},\bar{s})\geqq\sum_{i=1}^{p}\xi_i(x,y)\Phi_i(y,u,v,\lambda,\bar{t},\bar{s})\geqq 0,$$

where the second inequality follows from the dual feasibility of $(y,u,v,\lambda,\nu,\nu_0,J_{\nu_0},K_{\nu\setminus\nu_0},\bar{t},\bar{s})$. Therefore, we have

$$0\leqq\sum_{i=1}^{p}\xi_i(x,y)\Phi_i(x,u,v,\lambda,\bar{t},\bar{s})$$
$$=\sum_{i=1}^{p}u_i\xi_i(x,y)\Big\{f_i(x)-\lambda_i g_i(x)$$
$$+\sum_{m\in J_0}v_m G_{j_m}(x,t^m)+\sum_{m\in K_0}v_m H_{k_m}(x,s^m)\Big\}$$
$$\leqq\sum_{i=1}^{p}u_i\xi_i(x,y)[f_i(x)-\lambda_i g_i(x)] \text{ (by the primal feasibility of } x).$$

As shown in the proof of Theorem 3.17, this inequality leads to the conclusion that $\varphi(x)\nleq\lambda$.

(b): Proceeding in exactly the same manner as in the proof of part (a), we obtain (51) in which the second inequality is strict. By (i), this implies that (53) holds and, therefore, the rest of the proof is identical to that of part (a).

(c): Suppose to the contrary that $\varphi(x) \leqslant \lambda$. This implies that $f_i(x) - \lambda_i g_i(x) \leqq 0$ for each $i \in \underline{p}$, with strict inequality holding for at least one index $\ell \in \underline{p}$. Using these inequalities, we see that

$$\begin{aligned}\Phi_i(x,u,v,\lambda,\bar{t},\bar{s}) &\leqq u_i[f_i(x) - \lambda_i g_i(x)] \text{ (by the primal feasibility of } x\text{)}\\ &\leqq 0\\ &\leqq u_i\Big[f_i(y) - \lambda_i g_i(y) + \sum_{m\in J_0} v_m G_{j_m}(y,t^m)\\ &+ \sum_{m\in K_0} v_m H_{k_m}(y,s^m)\Big] \text{ (by (48))}\\ &= \Phi_i(y,u,v,\lambda,\bar{t},\bar{s}).\end{aligned}$$

Since $u > 0$ and $\xi(x,y) > 0$, the above inequalities yield

$$\sum_{i=1}^{p} \xi_i(x,y)\Phi_i(x,u,v,\lambda,\bar{t},\bar{s}) < \sum_{i=1}^{p} \xi_i(x,y)\Phi_i(y,u,v,\lambda,\bar{t},\bar{s})$$

and so

$$\frac{1}{\alpha(x,y)}\bar{\gamma}(x,y)\left(e^{\alpha(x,y)\sum_{i=1}^{p}\xi_i(x,y)[\Phi_i(x,u,v,\lambda,\bar{t},\bar{s})-\Phi_i(y,u,v,\lambda,\bar{t},\bar{s})]} - 1\right) < 0,$$

which in view of (i) implies that

$$\begin{aligned}\frac{1}{\beta(x,y)}\Big\langle \sum_{i=1}^{p} u_i[\nabla f_i(y) - \lambda_i \nabla g_i(y)] + \sum_{m\in J_0} v_m \nabla G_{j_m}(y,t^m)\\ + \sum_{m\in K_0} v_m \nabla H_{k_m}(y,s^m), e^{\beta(x,y)\eta(x,y)} - \mathbf{1}\Big\rangle\\ \leqq \tilde{\rho}(x,y)\|\theta(x,y)\|^2 \leqq -\bar{\rho}(x,y)\|\theta(x,y)\|^2. \quad (3.54)\end{aligned}$$

Proceeding as in the proof of part (a) and using the strict HA($\alpha$, $\beta$, $\tilde{\gamma}$, $\pi$, $\eta$, $\tilde{\rho}$, $\theta$) -V-pseudoinvexity hypothesis, we arrive at the following inequality:

$$\begin{aligned}\frac{1}{\beta(x,y)}\\ \Big\langle \sum_{\tau=1}^{M}\Big[\sum_{m\in J_\tau} v_m \nabla G_{j_m}(y,t^m) + \sum_{m\in K_\tau} v_m \nabla H_{k_m}(y,s^m)\Big], e^{\beta(x,y)\eta(x,y)} - \mathbf{1}\Big\rangle\\ < -\tilde{\rho}(x,y)\|\theta(x,y)\|^2,\end{aligned}$$

which when combined with (50) yields

$$\begin{aligned}\frac{1}{\beta(x,y)}\Big\langle \sum_{i=1}^{p} u_i[\nabla f_i(y) - \lambda_i \nabla g_i(y)] + \sum_{m\in J_0} v_m \nabla G_{j_m}(y,t^m)\\ + \sum_{m\in K_0} v_m \nabla H_{k_m}(y,s^m), e^{\beta(x,y)\eta(x,y)} - \mathbf{1}\Big\rangle > \tilde{\rho}(x,y)\|\theta(x,y)\|^2\\ \geqq -\bar{\rho}(x,y)\|\theta(x,y)\|^2.\end{aligned}$$

This contradicts (54), and, therefore, we conclude that $\varphi(x) \not\leq \lambda$.

□

**Theorem 3.40.** (Strong Duality) Let $x^*$ be a normal efficient solution of (P) and assume that for each feasible solution $(y, u, v, \lambda, \nu, \nu_0, J_{\nu_0}, K_{\nu\backslash\nu_0}, \bar{t}, \bar{s})$ of (DIII), any one of the three sets of conditions specified in Theorem 3.39 is satisfied. Then there exist $u^*, v^*, \lambda^*, \nu^*, \nu_0^*, J_{\nu_0^*}, K_{\nu^*\backslash\nu_0^*}, \bar{t}^*$, and $\bar{s}^*$ such that $(x^*, u^*, v^*, \lambda^*, \nu^*, \nu_0^*, J_{\nu_0^*}, K_{\nu^*\backslash\nu_0^*}, \bar{t}^*, \bar{s}^*)$ is an efficient solution of (DIII) and $\varphi(x^*) = \lambda^*$.

*Proof.* The proof is similar to that of Theorem 3.18.

□

**Theorem 3.41.** (Strict Converse Duality) Let $x^*$ and $\tilde{w} \equiv (\tilde{x}, \tilde{u}, \tilde{v}, \tilde{\lambda}, \tilde{\nu}, \tilde{\nu}_0, J_{\tilde{\nu}_0}, K_{\tilde{\nu}\backslash\tilde{\nu}_0}, \tilde{\bar{t}}, \tilde{\bar{s}})$ be arbitrary feasible solutions of (P) and (DIII), respectively, such that

$$\sum_{i=1}^{p} \tilde{u}_i \xi_i(x^*, \tilde{x})[f_i(x^*) - \tilde{\lambda}_i g_i(x^*)] = 0,$$

where for each $i \in \underline{p}$, $\xi_i$ is a function from $X \times X$ to $\mathbb{R}_+ \backslash \{0\}$. Furthermore, assume that any one of the following three sets of hypotheses is satisfied:

(a) The assumptions specified in part (a) of Theorem 3.39 are satisfied for the feasible solution $\tilde{w}$ of (DIII), and the function $\big(\Phi_1(\cdot, \tilde{u}, \tilde{v}, \tilde{\lambda}, \tilde{\bar{t}}, \tilde{\bar{s}}), \ldots, \Phi_p(\cdot, \tilde{u}, \tilde{v}, \tilde{\lambda}, \tilde{\bar{t}}, \tilde{\bar{s}})\big)$ is strictly $HA(\alpha, \beta, \bar{\gamma}, \xi, \eta, \bar{\rho}, \theta)$-V-pseudoinvex at $\tilde{x}$.

(b) The assumptions specified in part (b) of Theorem 3.39 are satisfied for the feasible solution $\tilde{w}$ of (DIII), and the function $\big(\Phi_1(\cdot, \tilde{u}, \tilde{v}, \tilde{\lambda}, \tilde{\bar{t}}, \tilde{\bar{s}}), \ldots, \Phi_p(\cdot, \tilde{u}, \tilde{v}, \tilde{\lambda}, \tilde{\bar{t}}, \tilde{\bar{s}})\big)$ is $HA(\alpha, \beta, \bar{\gamma}, \xi, \eta, \bar{\rho}, \theta)$-V-quasiinvex $\tilde{x}$.

(c) The assumptions specified in part (c) of Theorem 3.39 are satisfied for the feasible solution $\tilde{w}$ of (DIII), and the function $\big(\Phi_1(\cdot, \tilde{u}, \tilde{v}, \tilde{\lambda}, \tilde{\bar{t}}, \tilde{\bar{s}}), \ldots, \Phi_p(\cdot, \tilde{u}, \tilde{v}, \tilde{\lambda}, \tilde{\bar{t}}, \tilde{\bar{s}})\big)$ is $HA(\alpha, \beta, \bar{\gamma}, \xi, \eta, \bar{\rho}, \theta)$-V-quasiinvex $\tilde{x}$.

Then $\tilde{x} = x^*$ and $\varphi(x^*) = \tilde{\lambda}$.

*Proof.* The proof is similar to that of Theorem 3.39. □

Each one of the three sets of results given in Theorem 3.39 can be viewed as a family of duality results whose members can easily be identified by appropriate choices of the partitioning sets $J_\mu$ and $K_\mu$, $\mu \in \underline{M} \cup \{0\}$.

## 5.3. Duality Model IV

In this subsection we discuss another collection of duality results for $(P)$ which are different from those stated in Theorems 3.39 - 3.41. In formulating these duality results, we utilize a partition of $\underline{p}$ in addition to those of $\underline{\nu_0}$ and $\underline{\nu}\backslash\underline{\nu_0}$. It appears that this partitioning scheme was first proposed in [3] for a multiobjective fractional programming problem with a finite number of constraints. In our duality theorems, we impose appropriate generalized $HA(\alpha,\beta,\gamma,\xi,\eta,\rho,\theta)$-V-invexity requirements on certain vector functions whose components comprise some combinations of the functions $\mathscr{E}_i(\cdot,\lambda,u)$, $i\in \underline{p}$, $G_j(\cdot,t)$, $j\in \underline{q}$, and $H_k(\cdot,s)$, $k\in \underline{r}$.

Let $\{I_0,I_1,\ldots,I_d\}$, $\{J_0,J_1,\ldots,J_e\}$, and $\{K_0,K_1,\ldots,K_e\}$ be partitions of $\underline{p}$, $\underline{\nu_0}$, and $\underline{\nu}\backslash\underline{\nu_0}$, respectively, such that $D=\{0,1,2,\ldots,d\}\subset E=\{0,1,\ldots,e\}$, and let the function $\Pi_\tau(\cdot,u,v,\lambda,\bar{t},\bar{s}):X\to\mathbb{R}$ be defined, for fixed $u,v,\lambda,\bar{t}$, and $\bar{s}$, by

$$\Pi_\tau(z,u,v,\lambda,\bar{t},\bar{s})=\sum_{i\in I_\tau}u_i[f_i(z)-\lambda_i g_i(z)]+\sum_{m\in J_\tau}v_m G_{j_m}(z,t^m)$$
$$+\sum_{m\in K_\tau}v_m H_{k_m}(z,s^m),\quad \tau\in D.$$

Making use of the sets and functions defined above, we consider the following duality models:

$$(DIV)\qquad \sup_{(y,u,v,\lambda,\nu,\nu_0,J_{\nu_0},K_{\nu\backslash\nu_0},\bar{t},\bar{s})\in\mathbb{H}}\lambda=(\lambda_1,\cdots,\lambda_p)$$

subject to

$$\sum_{i=1}^{p}u_i[\nabla f_i(y)-\lambda_i\nabla g_i(y)]+\sum_{m=1}^{\nu_0}v_m\nabla G_{j_m}(y,t^m)+\sum_{m=\nu_0+1}^{\nu}v_m\nabla H_{k_m}(y,s^m)=0,\tag{3.55}$$

$$\sum_{i\in I_\tau}u_i[f_i(y)-\lambda_i g_i(y)]+\sum_{m\in J_\tau}v_m G_{j_m}(y,t^m)+\sum_{m\in K_\tau}v_m H_{k_m}(y,s^m)\geqq 0,\quad \tau\in D,\tag{3.56}$$

$$\sum_{m\in J_\tau}v_m G_{j_m}(y,t^m)+\sum_{m\in K_\tau}v_m H_{k_m}(y,s^m)\geqq 0,\quad \tau\in E\backslash D;\tag{3.57}$$

$$(\tilde{D}IV)\qquad \sup_{(y,u,v,\lambda,\nu,\nu_0,J_{\nu_0},K_{\nu\backslash\nu_0},\bar{t},\bar{s})\in\mathbb{H}}\lambda=(\lambda_1,\cdots,\lambda_p)$$

subject to (56), and (57).

The remarks and observations made earlier about the relationships between $(DI)$ and $(\tilde{D}I)$ are, of course, also valid for $(DIV)$ and $(\tilde{D}IV)$.

The next two theorems show that $(DIV)$ is a dual problem for $(P)$.

**Theorem 3.42.** Let $x$ and $(y,u,v,\lambda,\nu,\nu_0,J_{\nu_0},K_{\nu\backslash\nu_0},\bar{t},\bar{s})$ be arbitrary feasible solutions of (P) and (DIV), respectively, and assume that any one of the following three sets of hypotheses is satisfied:

(a) (i) $\big(\Pi_0(\cdot,u,v,\lambda,\bar{t},\bar{s}),\ldots,\Pi_d(\cdot,u,v,\lambda,\bar{t},\bar{s})\big)$ is HA($\alpha$, $\beta$, $\bar{\gamma}$, $\xi$, $\eta$, $\bar{\rho}$, $\theta$) -V-pseudoinvex at $y$ and $\bar{\gamma}(x,y)>0$;

(ii) $\big(\Lambda_{d+1}(\cdot,v,\bar{t},\bar{s}),\ldots,\Lambda_e(\cdot,v,\bar{t},\bar{s})\big)$ is $HA(\alpha,\beta,\tilde{\gamma},\pi,\eta,\tilde{\rho},\theta)$ -V-quasiinvex at $y$;

(iii) $\bar{\rho}(x,y)+\tilde{\rho}(x,y)\geqq 0$;

(b) (i) $\left(\Pi_0(\cdot,u,v,\lambda,\bar{t},\bar{s}),\ldots,\Pi_d(\cdot,u,v,\lambda,\bar{t},\bar{s})\right)$ is prestrictly $HA(\alpha,\beta,\bar{\gamma},\xi,\eta,\bar{\rho},\theta)$-V-quasiinvex at $y$ and $\bar{\gamma}(x,y)>0$;

(ii) $\left(\Lambda_{d+1}(\cdot,v,\bar{t},\bar{s}),\ldots,\Lambda_e(\cdot,v,\bar{t},\bar{s})\right)$ is strictly $HA(\alpha,\beta,\tilde{\gamma},\xi,\eta,\tilde{\rho},\theta)$-V-pseudoinvex at $y$;

(iii) $\bar{\rho}(x,y)+\tilde{\rho}(x,y)\geqq 0$;

(c) (i) $\left(\Pi_0(\cdot,u,v,\lambda,\bar{t},\bar{s}),\ldots,\Pi_d(\cdot,u,v,\lambda,\bar{t},\bar{s})\right)$ is prestrictly $HA(\alpha,\beta,\bar{\gamma},\xi,\eta,\bar{\rho},\theta)$-V-quasiinvex at $y$ and $\bar{\gamma}(x,y)>0$;

(ii) $\left(\Lambda_{d+1}(\cdot,v,\bar{t},\bar{s}),\ldots,\Lambda_e(\cdot,v,\bar{t},\bar{s})\right)$ is $HA(\alpha,\beta,\tilde{\gamma},\xi,\eta,\tilde{\rho},\theta)$-V-quasiinvex at $y$;

(iii) $\bar{\rho}(x,y)+\tilde{\rho}(x,y)>0$.

Then $\varphi(x)\not\leqslant\lambda$.

*Proof.* (a): Suppose to the contrary that $\varphi(x)\leqslant\lambda$. Then $f_i(x)-\lambda_i g_i(x)\leqq 0,\quad i\in\underline{p}$, with strict inequality holding for at least one index $\ell\in\underline{p}$. Since $u>0$, we see that for each $\tau\in D$,

$$\sum_{i\in I_\tau}u_i[f_i(x)-\lambda_i g_i(x)]\leqq 0, \tag{3.58}$$

with strict inequality holding for at least one index $\tau\in D$. Now using this inequality, we see that

$$\begin{aligned}
&\Pi_\tau(x,u,v,\lambda,\bar{t},\bar{s})\\
&=\sum_{i\in I_\tau}u_i[f_i(x)-\lambda_i g_i(x)]+\sum_{m\in J_\tau}v_m G_{j_m}(x,t^m)+\sum_{m\in K_\tau}v_m H_{k_m}(x,s^m)\\
&\leqq\sum_{i\in I_\tau}u_i[f_i(x)-\lambda_i g_i(x)]\\
&\quad\text{(by the primal feasibility of } x \text{ and positivity of } v_m,\ m\in\underline{v_0})\\
&\leqq 0\\
&\leqq\sum_{i\in I_\tau}u_i[f_i(y)-\lambda_i g_i(y)]+\sum_{m\in J_\tau}v_m G_{j_m}(y,t^m)\\
&+\sum_{m\in K_\tau}v_m H_{k_m}(y,s^m)\\
&=\Pi_\tau(y,u,v,\lambda,\bar{t},\bar{s}),
\end{aligned}$$

with strict inequality holding for at least one index $\tau\in D$. Since $\xi_\tau(x,y)>0$ for each $\tau\in D$, it is clear that

$$\sum_{\tau\in D}\xi_\tau(x,y)\Pi_\tau(x,u,v,\lambda,\bar{t},\bar{s})<\sum_{\tau\in D}\xi_\tau(x,y)\Pi_\tau(y,u,v,\lambda,\bar{t},\bar{s}),$$

and therefore

$$\frac{1}{\alpha(x,y)}\bar{\gamma}(x,y)\left(e^{\alpha(x,y)\sum_{\tau\in D}\xi_\tau(x,y)[\Pi_\tau(x,u,v,\lambda,\bar{t},\bar{s})-\Pi_\tau(y,u,v,\lambda,\bar{t},\bar{s})]}-1\right)<0,$$

which in view of (i) implies that

$$\frac{1}{\beta(x,y)}\Big\langle \sum_{\tau\in D}\Big[\sum_{m\in J_\tau} v_m\nabla G_{j_m}(y,t^m)+ \sum_{m\in K_\tau} v_m\nabla H_{k_m}(y,s^m)\Big], e^{\beta(x,y)\eta(x,y)}-\mathbf{1}\Big\rangle < -\tilde{\rho}(x,y)\|\theta(x,y)\|^2. \quad (3.59)$$

As shown in the proof of Theorem 3.39, for each $\tau\in E\backslash D$, $\Lambda_\tau(x,v,\bar{t},\bar{s})\leqq\Lambda_\tau(y,v,\bar{t},\bar{s})$, and hence

$$\sum_{\tau\in E\backslash D}\pi_\tau(x,y)\Lambda_\tau(x,v,\bar{t},\bar{s})\leqq\sum_{\tau\in E\backslash D}\pi_\tau(x,y)\Lambda_\tau(y,v,\bar{t},\bar{s}),$$

and

$$\frac{1}{\alpha(x,y)}\tilde{\gamma}(x,y)\left(e^{\alpha(x,y)\sum_{\tau\in E\backslash D}\pi_\tau(x,y)[\Lambda_\tau(x,v,\bar{t},\bar{s})-\Lambda_\tau(y,v,\bar{t},\bar{s})]}-1\right)\leqq 0,$$

which in view of (ii) implies that

$$\frac{1}{\beta(x,y)}\Big\langle \sum_{\tau\in E\backslash D}\Big[\sum_{m\in J_\tau} v_m\nabla G_{j_m}(y,t^m) + \sum_{m\in K_\tau} v_m\nabla H_{k_m}(y,s^m)\Big], e^{\beta(x,y)\eta(x,y)}-\mathbf{1}\Big\rangle \leqq -\tilde{\rho}(x,y)\|\theta(x,y)\|^2. \quad (3.60)$$

Now combining (59) and (60) and using (iii), we see that

$$\frac{1}{\beta(x,y)}\Big\langle \sum_{i=1}^{p}u_i[\nabla f_i(y)-\lambda_i\nabla g_i(y)]+\sum_{m=1}^{\nu_0} v_m\nabla G_{j_m}(y,t^m) + \sum_{m=\nu_0+1}^{\nu} v_m\nabla H_{k_m}(y,s^m), e^{\beta(x,y)\eta(x,y)}-\mathbf{1}\Big\rangle < -[\bar{\rho}(x,y)+\tilde{\rho}(x,y)]\|\theta(x,y)\|^2\leqq 0,$$

which contradicts (55). Therefore, $\varphi(x)\nleq\lambda$.

(b) and (c) : The proofs are similar to that of part (a). □

**Theorem 3.43.** (Strong Duality) Let $x^*$ be a normal efficient solution of (P) and assume that for each feasible solution $(y,u,v,\lambda,\nu,\nu_0,J_{\nu_0},K_{\nu\backslash\nu_0},\bar{t},\bar{s})$ of (DIV), any one of the three sets of conditions specified in Theorem 3.42 is satisfied. Then there exist $u^*,v^*,\lambda^*,\nu^*,\nu_0^*,J_{\nu_0^*}$, $K_{\nu^*\backslash\nu_0^*},\bar{t}^*$, and $\bar{s}^*$ such that $(x^*,u^*,v^*,\lambda^*,\nu^*,\nu_0^*,J_{\nu_0^*},K_{\nu^*\backslash\nu_0^*},\bar{t}^*,\bar{s}^*)$ is an efficient solution of (DIV) and $\varphi(x^*)=\lambda^*$.

*Proof.* The proof is similar to that of Theorem 3.2.

□

**Theorem 3.44.** (Strict Converse Duality) Let $x^*$ and $\tilde{w}\equiv(\tilde{x},\tilde{u},\tilde{v},\tilde{\lambda},\tilde{\nu},\tilde{\nu}_0,J_{\tilde{\nu}_0},K_{\tilde{\nu}\backslash\tilde{\nu}_0},\bar{\tilde{t}},\bar{\tilde{s}})$ be arbitrary feasible solutions of (P) and (DIV), respectively, such that

$$\sum_{i=1}^{p}\tilde{u}_i[f_i(x^*)-\tilde{\lambda}_i g_i(x^*)]=0. \quad (3.61)$$

Furthermore, assume that any one of the following three sets of hypotheses is satisfied:

(a) The assumptions specified in part (a) of Theorem 3.42 are satisfied for the feasible solution $\tilde{w}$ of (DIV), the function $\big(\Pi_0(\cdot,\tilde{u},\tilde{v},\tilde{\lambda},\bar{\tilde{t}},\bar{\tilde{s}}),\ldots,\ \Pi_d(\cdot,\tilde{u},\ \tilde{v},\tilde{\lambda},\bar{\tilde{t}},\bar{\tilde{s}})\big)$ is strictly $HA(\alpha,\beta,\bar{\gamma},\xi,\eta,\bar{\rho},\theta)$-V-pseudoinvex at $\tilde{x}$, and $\xi_0=\xi_1=\ldots=\xi_d=\sigma$.

(b) The assumptions specified in part (b) of Theorem 3.42 are satisfied for the feasible solution $\tilde{w}$ of (DIV), the function $\big(\Pi_0(\cdot,\tilde{u},\tilde{v},\tilde{\lambda},\bar{\tilde{t}},\bar{\tilde{s}}),\ldots,\ \Pi_d(\cdot,\tilde{u},\ \tilde{v},\tilde{\lambda},\bar{\tilde{t}},\bar{\tilde{s}})\big)$ is $HA(\alpha,\beta,\bar{\gamma},\xi,\eta,\bar{\rho},\theta)$-V-quasiinvex at $\tilde{x}$, and $\xi_0=\xi_1=\ldots=\xi_d=\sigma$.

(c) The assumptions specified in part (c) of Theorem 3.42 are satisfied for the feasible solution $\tilde{w}$ of (DIV), the function $\big(\Pi_0(\cdot,\tilde{u},\tilde{v},\tilde{\lambda},\bar{\tilde{t}},\bar{\tilde{s}}),\ldots,\Pi_d(\cdot,\tilde{u},\ \tilde{v},\tilde{\lambda},\bar{\tilde{t}},\bar{\tilde{s}})\big)$ is $HA(\alpha,\beta,\bar{\gamma},\xi,\eta,\bar{\rho},\theta)$-V-quasiinvex at $\tilde{x}$, and $\xi_0=\xi_1=\ldots=\xi_d=\sigma$.

Then $\tilde{x}=x^*$ and $\varphi(x^*)=\tilde{\lambda}$.

*Proof.* (a) : Suppose to the contrary that $\tilde{x}\neq x^*$. By Theorem 3.16, there exist $u^*,v^*,\lambda^*,\nu^*,\nu_0^*,J_{\nu_0^*},K_{\nu^*\backslash\nu_0^*},\bar{t}^*$, and $\bar{s}^*$ such that $(x^*,u^*,v^*,\lambda^*,\nu^*,\nu_0^*,J_{\nu_0^*},K_{\nu^*\backslash\nu_0^*},\bar{t}^*,\bar{s}^*)$ is a feasible solution of $(DIV)$ and $\varphi(x^*)=\lambda^*$.

It is clear that (55) can be expressed as follows:

$$\sum_{\tau\in D}\Big\{\sum_{i\in I_\tau}\tilde{u}_i[\nabla f_i(\tilde{x})-\tilde{\lambda}\nabla g_i(\tilde{x})]+\sum_{m\in J_\tau}\tilde{v}_m\nabla G_{j_m}(\tilde{x},\tilde{t}^m)+\sum_{m\in K_\tau}\tilde{v}_m\nabla H_{k_m}(\tilde{x},\tilde{s}^m)\Big\}+$$
$$\sum_{\tau\in E\backslash D}\Big[\sum_{m\in J_\tau}\tilde{v}_m\nabla G_{j_m}(\tilde{x},\tilde{t}^m)+\sum_{m\in K_\tau}\tilde{v}_m\nabla H_{k_m}(\tilde{x},\tilde{s}^m)\Big]=0. \quad (3.62)$$

Since $x^*\in\mathbb{F}$, $\tilde{v}_m\geqq 0$, $m\in\underline{\tilde{\nu}}$, and (6.3) holds, it follows that for each $\tau\in E\backslash D$,

$$\begin{aligned}\Lambda_\tau(x^*,\tilde{v},\bar{\tilde{t}},\bar{\tilde{s}})&=\sum_{m\in J_\tau}\tilde{v}_mG_{j_m}(x^*,\tilde{t}^m)+\sum_{m\in K_\tau}\tilde{v}_mH_{k_m}(x^*,\tilde{s}^m)\\&\leqq 0\\&\leqq\sum_{m\in J_\tau}\tilde{v}_mG_{j_m}(\tilde{x},\tilde{t}^m)+\sum_{m\in K_\tau}\tilde{v}_mH_{k_m}(\tilde{x},\tilde{s}^m)\\&=\Lambda_\tau(\tilde{x},\tilde{v},\bar{\tilde{t}},\bar{\tilde{s}}),\end{aligned}$$

and hence

$$\sum_{\tau\in E\backslash D}\pi_\tau(x^*,\tilde{x})\Lambda_\tau(x^*,\tilde{v},\bar{\tilde{t}},\bar{\tilde{s}})<\sum_{\tau\in E\backslash D}\pi_\tau(x^*,\tilde{x})\Lambda_\tau(\tilde{x},\tilde{v},\bar{\tilde{t}},\bar{\tilde{s}}).$$

Using this inequality, we get

$$\frac{1}{\alpha(x^*,\tilde{x})}\tilde{\gamma}(x^*,\tilde{x})\Big(e^{\alpha(x^*,\tilde{x})\sum_{\tau\in E\backslash D}\pi_\tau(x^*,\tilde{x})[\Lambda_\tau(x^*,\tilde{v},\bar{\tilde{t}},\bar{\tilde{s}})-\Lambda_\tau(\tilde{x},\tilde{v},\bar{\tilde{t}},\bar{\tilde{s}})]}-1\Big)\leqq 0,$$

which in view of (ii) implies that

$$\frac{1}{\beta(x^*,\tilde{x})}\Big\langle\sum_{\tau\in E\backslash D}\Big[\sum_{m\in J_\tau}\tilde{v}_m\nabla G_{j_m}(\tilde{x},\tilde{t}^m)+$$
$$\sum_{m\in K_\tau}\tilde{v}_m\nabla H_{k_m}(\tilde{x},\tilde{s}^m)\Big],e^{\beta(x^*,\tilde{x})\eta(x^*,\tilde{x})}-\mathbf{1}\Big\rangle$$
$$\leqq-\tilde{\rho}(x^*,\tilde{x})\|\theta(x^*,\tilde{x})\|^2. \quad (3.63)$$

Combining (62) and (63), and using (iii) we get

$$\frac{1}{\beta(x,y)}\Big\langle \sum_{\tau\in D}\Big\{\sum_{i\in I_\tau}\tilde{u}_i[\nabla f_i(\tilde{x})-\tilde{\lambda}\nabla g_i(\tilde{x})]+\sum_{m\in J_\tau}\tilde{v}_m\nabla G_{j_m}(\tilde{x},\tilde{t}^m)$$
$$+\sum_{m\in K_\tau}\tilde{v}_m\nabla H_{k_m}(\tilde{x},\tilde{s}^m)\Big\},e^{\beta(x^*,\tilde{x})\eta(x^*,\tilde{x})}-\mathbf{1}\Big\rangle \geqq \tilde{\rho}(x^*,\tilde{x})\|\theta(x^*,\tilde{x})\|^2$$
$$\geqq -\bar{\rho}(x^*,\tilde{x})\|\theta(x^*,\tilde{x})\|^2, \quad (3.64)$$

which by virtue of our strict $HA(\alpha,\beta,\bar{\gamma},\xi,\eta,\bar{\rho},\theta)$-V-pseudoinvexity hypothesis implies that

$$\frac{1}{\alpha(x^*,\tilde{x})}\bar{\gamma}(x^*,\tilde{x})\Big(e^{\alpha(x^*,\tilde{x})\sigma(x^*,\tilde{x})\sum_{\tau\in D}[\Pi_\tau(x^*,\tilde{u},\tilde{v},\tilde{\lambda},\bar{\tilde{t}},\bar{\tilde{s}})-\Pi_\tau(\tilde{x},\tilde{u},\tilde{v},\tilde{\lambda},\bar{\tilde{t}},\bar{\tilde{s}})]}-1\Big)>0.$$

Since $\bar{\gamma}(x^*,\tilde{x})>0$ and $\sigma(x^*,\tilde{x})>0$, we deduce from this inequality that

$$\sum_{\tau\in D}\Pi_\tau(x^*,\tilde{u},\tilde{v},\tilde{\lambda},\bar{\tilde{t}},\bar{\tilde{s}})>\sum_{\tau\in D}\Pi_\tau(\tilde{x},\tilde{u},\tilde{v},\tilde{\lambda},\bar{\tilde{t}},\bar{\tilde{s}})\geqq 0,$$

where the second inequality follows from the dual feasibility of $\tilde{w}$ and (56). Therefore, we have

$$\begin{aligned}0&<\sum_{\tau\in D}\Pi_\tau(x^*,\tilde{v},\bar{\tilde{t}},\bar{\tilde{s}})\\&=\sum_{\tau\in D}\sum_{i\in I_\tau}\tilde{u}_i[f_i(x^*)-\tilde{\lambda}g_i(x^*)]\\&+\sum_{\tau\in D}\Big[\sum_{m\in J_0}\tilde{v}_mG_{j_m}(x^*,\tilde{t}^m)+\sum_{m\in K_0}\tilde{v}_mH_{k_m}(x^*,\tilde{s}^m)\Big]\\&\leq\sum_{i=1}^{p}\tilde{u}_i[f_i(x^*)-\tilde{\lambda}_ig_i(x^*)]\ \text{(by the primal feasibility of } x^*),\end{aligned}$$

which contradicts (61). Therefore, we conclude that $\tilde{x}=x^*$ and $\varphi(x^*)=\tilde{\lambda}$.

(b) and (c) : The proofs are similar to that of part (a). □

As pointed earlier, one can readily identify numerous special cases and variants of $(DIV)$ and $(\tilde{D}IV)$ by appropriate choices of the partitioning sets $I_\mu$, $\mu\in D$, and $J_\tau$ and $K_\tau$, $\tau\in E$.

We remark that in this subsection we have, based on the work of Zalmai [12-14], established a number of parametric duality results under various generalized $HA(\alpha,\beta,\gamma,\xi,\eta,\rho,\theta)$-V-invexity hypotheses for a semiinfinite multiobjective fractional programming problem. It appears that all these results are new in the area of semiinfinite programming. These results would be very helpful for advancing further quality research even beyond fractional programming problems. All the results obtained here can be considered for collaborative interdisciplinary research applications worldwide. Also, each one of the seven problems designated as $(P1)-(P7)$ in [4], they collectively provide a fairly large number of duality results for various classes of conventional as well as semiinfinite nonlinear programming problems. For example, it seems reasonable to expect that a similar approach can be applied to investigate the optimality and duality aspects of the following semiinfinite minmax fractional programming problem:

$$\underset{x \in \mathbb{F}}{\text{Minimize}} \ \max_{y \in Y} \frac{f(x,y)}{g(x,y)}.$$

## 6. Generalized Duality Results III

This section is primarily based on duality models and results beginning the work in [4] where some information about multiobjective fractional programming is explored, the current status of semiinfinite programming is briefly discussed and numerous key references are cited, further extensions of the variants of the V-invex functions defined recently by Antczak [1] are proposed, and a fairly large number of sets of global parametric sufficient efficiency results under various generalized first-order $HA(\alpha,\beta,\gamma,\xi,\eta,\rho,\theta)$-V-invexity assumptions are established. For the necessary background material and preliminaries, the reader is referred to [4]. Here we shall make use of the sufficient efficiency results developed in [4] and a set of necessary efficiency conditions, which will be recalled in the next section, and construct several first-order parametric duality models for $(P)$ and prove appropriate duality theorems. The second-order counterparts of these duality models and results are discussed in [5].

Although at present time numerous duality models and corresponding duality results exist in the related literature for several classes of multiobjective fractional optimization problems involving certain classes of generalized convex functions, it seems that there is a great research momentum toward establishing results for semiinfinite multiobjective programming problems utilizing any type of generalized $HA(\alpha,\beta,\gamma,\xi,\eta,\rho,\theta)$-V-invexity concepts. As a matter of fact, all the duality models and duality results established in this section are immensely innovative in the area of semiinfinite programming.

This section contains apart from basic definitions and auxiliary results, results related to duality models and duality theorems. This is followed by considering two parametric duality models with somewhat limited constraint structures and proving weak, strong, and strict converse duality theorems under appropriate generalized $HA(\alpha,\beta,\gamma,\xi,\eta,\rho,\theta)$-V-invexity hypotheses. Also, there are two other duality models with different constraint structures which allow for more general types of $HA(\alpha,\beta,\gamma,\xi,\eta,\rho,\theta)$-V-invexity conditions under which duality can be established. Furthermore, this section ends with two partitioning schemes, four generalized parametric duality models, and several duality results using various generalized $HA(\alpha,\beta,\gamma,\xi,\eta,\rho,\theta)$-V-invexity assumptions.

Let us begin with recalling some definitions of certain classes of generalized convex functions which will be needed in the sequel. For a brief discussion of the origins and predecessors of these functions as well as numerous relevant references, the reader is referred to [4].

Let the function $F = (F_1, F_2, \ldots, F_p) : X \to \mathbb{R}^p$ be differentiable at $x^*$. The following notions of V-invexity, V-pseudoinvexity, and V-quasiinvexity for vector-valued functions will be used in this paper. They are substantial generalizations of the class of V-r-invex functions recently introduced by Antczak [1].

**Definition 3.45.** The function $F$ is said to be (strictly) $HA(\alpha,\beta,\gamma,\eta,\rho,\theta)$-V-invex at $x^* \in X$ if there exist functions $\alpha : X\times X \to \mathbb{R}$, $\beta : X\times X \to \mathbb{R}$, $\gamma_i : X\times X \to \mathbb{R}_+ \equiv [0,\infty)$, , $i \in \underline{p}$, $\eta : X\times X \to \mathbb{R}^n$, $\rho_i : X\times X \to \mathbb{R}$, $i \in \underline{p}$, and $\theta : X\times X \to \mathbb{R}^n$ such that for all $x \in X\,(x \neq x^*)$ and $i \in \underline{p}$,

$$\frac{1}{\alpha(x,x^*)}\gamma_i(x,x^*)\left(e^{\alpha(x,x^*)[F_i(x)-F_i(x^*)]}-1\right)$$
$$(>)\geqq \frac{1}{\beta(x,x^*)}\langle \nabla F_i(x^*), e^{\beta(x,x^*)\eta(x,x^*)}-\mathbf{1}\rangle + \rho_i(x,x^*)\|\theta(x,x^*)\|^2$$
$$\text{if } \alpha(x,x^*)\neq 0 \text{ and } \beta(x,x^*)\neq 0 \text{ for all } x\in X,$$

$$\frac{1}{\alpha(x,x^*)}\gamma_i(x,x^*)\left(e^{\alpha(x,x^*)[F_i(x)-F_i(x^*)]}-1\right)(>)\geqq \langle \xi_i(x,x^*)\nabla F_i(x^*), \eta(x,x^*)\rangle$$
$$+\rho_i(x,x^*)\|\theta(x,x^*)\|^2 \quad \text{if } \alpha(x,x^*)\neq 0 \text{ and } \beta(x,x^*)= 0 \text{ for all } x\in X,$$

$$\gamma_i(x,x^*)[F_i(x)-F_i(x^*)](>)\geqq \frac{1}{\beta(x,x^*)}\langle \xi_i(x,x^*)\nabla F_i(x^*), e^{\beta(x,x^*)\eta(x,x^*)}-\mathbf{1}\rangle$$
$$+\rho_i(x,x^*)\|\theta(x,x^*)\|^2 \quad \text{if } \alpha(x,x^*)= 0 \text{ and } \beta(x,x^*)\neq 0 \text{ for all } x\in X,$$

$$\gamma_i(x,x^*)[F_i(x)-F_i(x^*)](>)\geqq \langle \nabla F_i(x^*), \eta(x,x^*)\rangle + \rho_i(x,x^*)\|\theta(x,x^*)\|^2$$
$$\text{if } \alpha(x,x^*)= 0 \text{ and } \beta(x,x^*)= 0 \text{ for all } x\in X,$$

where

$$\left(e^{\beta(x,x^*)\eta(x,x^*)}-\mathbf{1}\right)\equiv\left(e^{\beta(x,x^*)\eta_1(x,x^*)}-1,\ldots,e^{\beta(x,x^*)\eta_n(x,x^*)}-1\right)$$

and $\|\cdot\|$ is a norm on $\mathbb{R}^n$.

The function $F$ is said to be (strictly) $HA(\alpha,\beta,\gamma,\eta,\rho,\theta)$-V-invex on $X$ if it is (strictly) $HA(\alpha,\beta,\gamma,\eta,\rho,\theta)$-V-invex at each $x^*\in X$.

**Definition 3.46.** The function $F$ is said to be (strictly) $HA(\alpha,\beta,\gamma,\eta,\rho,\theta)$-V-pseudoinvex at $x^* \in X$ if there exist functions $\alpha : X\times X \to \mathbb{R}$, $\beta : X\times X \to \mathbb{R}$, $\gamma : X\times X \to \mathbb{R}_+$, $\eta : X\times X \to \mathbb{R}^n$, $\rho : X\times X \to \mathbb{R}$, and $\theta : X\times X \to \mathbb{R}^n$ such that for all $x \in X\,(x\neq x^*)$,

$$\frac{1}{\beta(x,x^*)}\left\langle \sum_{i=1}^{p}\nabla F_i(x^*), e^{\beta(x,x^*)\eta(x,x^*)}-\mathbf{1}\right\rangle \geqq -\rho(x,x^*)\|\theta(x,x^*)\|^2$$
$$\Rightarrow\ \frac{1}{\alpha(x,x^*)}\gamma(x,x^*)\left(e^{\alpha(x,x^*)\sum_{i=1}^{p}\xi_i(x,x^*)[F_i(x)-F_i(x^*)]}-1\right)(>)\geqq 0$$
$$\text{if } \alpha(x,x^*)\neq 0 \text{ and } \beta(x,x^*)\neq 0 \text{ for all } x\in X,$$

$$\left\langle \sum_{i=1}^{p}\nabla F_i(x^*), \eta(x,x^*)\right\rangle \geqq -\rho(x,x^*)\|\theta(x,x^*)\|^2$$
$$\Rightarrow\ \frac{1}{\alpha(x,x^*)}\gamma(x,x^*)\left(e^{\alpha(x,x^*)\sum_{i=1}^{p}\xi_i(x,x^*)[F_i(x)-F_i(x^*)]}-1\right)(>)\geqq 0$$
$$\text{if } \alpha(x,x^*)\neq 0 \text{ and } \beta(x,x^*)= 0 \text{ for all } x\in X,$$

$$\frac{1}{\beta(x,x^*)}\left\langle \sum_{i=1}^{p} \nabla F_i(x^*), e^{\beta(x,x^*)\eta(x,x^*)} - \mathbf{1} \right\rangle \geqq -\rho(x,x^*)\|\theta(x,x^*)\|^2$$
$$\Rightarrow \gamma(x,x^*)\sum_{i=1}^{p}[F_i(x) - F_i(x^*)](>) \geqq 0$$
$$\text{if } \alpha(x,x^*) = 0 \text{ and } \beta(x,x^*) \neq 0 \text{ for all } x \in X,$$

$$\left\langle \sum_{i=1}^{p} \nabla F_i(x^*), \eta(x,x^*) \right\rangle \geqq -\rho(x,x^*)\|\theta(x,x^*)\|^2$$
$$\Rightarrow \gamma(x,x^*)\sum_{i=1}^{p}[F_i(x) - F_i(x^*)](>) \geqq 0$$
$$\text{if } \alpha(x,x^*) = 0 \text{ and } \beta(x,x^*) = 0 \text{ for all } x \in X.$$

The function $F$ is said to be (strictly) $HA(\alpha,\beta,\gamma,\eta,\rho,\theta)$-V-pseudoinvex on $X$ if it is (strictly) $HA(\alpha,\beta,\gamma,\eta,\rho,\theta)$-V-pseudoinvex at each $x^* \in X$.

**Definition 3.47.** The function $F$ is said to be (prestrictly) HA($\alpha$, $\beta$, $\gamma$, $\xi$, $\eta$, $\rho$, $\theta$) -V-quasiinvex at $x^* \in X$ if there exist functions $\alpha : X \times X \to \mathbb{R}$, $\beta : X \times X \to \mathbb{R}$, $\gamma : X \times X \to \mathbb{R}_+$, $i \in \underline{p}$, $\eta : X \times X \to \mathbb{R}^n$, $\rho : X \times X \to \mathbb{R}$, and $\theta : X \times X \to \mathbb{R}^n$ such that for all $x \in X$,

$$\frac{1}{\alpha(x,x^*)}\gamma(x,x^*)\left(e^{\alpha(x,x^*)\sum_{i=1}^{p}[F_i(x)-F_i(x^*)]} - 1\right)(<) \leqq 0$$
$$\Rightarrow \frac{1}{\beta(x,x^*)}\left\langle \sum_{i=1}^{p} \nabla F_i(x^*), e^{\beta(x,x^*)\eta(x,x^*)} - \mathbf{1} \right\rangle \leqq -\rho(x,x^*)\|\theta(x,x^*)\|^2$$
$$\text{if } \alpha(x,x^*) \neq 0 \text{ and } \beta(x,x^*) \neq 0 \text{ for all } x \in X,$$

$$\frac{1}{\alpha(x,x^*)}\gamma(x,x^*)\left(e^{\alpha(x,x^*)\sum_{i=1}^{p}\xi_i(x,x^*)[F_i(x)-F_i(x^*)]} - 1\right)(<) \leqq 0$$
$$\Rightarrow \left\langle \sum_{i=1}^{p} \nabla F_i(x^*), \eta(x,x^*) \right\rangle \leqq -\rho(x,x^*)\|\theta(x,x^*)\|^2$$
$$\text{if } \alpha(x,x^*) \neq 0 \text{ and } \beta(x,x^*) = 0 \text{ for all } x \in X,$$

$$\gamma(x,x^*)\sum_{i=1}^{p}[F_i(x) - F_i(x^*)](<) \leqq 0$$
$$\Rightarrow \frac{1}{\beta(x,x^*)}\left\langle \sum_{i=1}^{p} \nabla F_i(x^*), e^{\beta(x,x^*)\eta(x,x^*)} - \mathbf{1} \right\rangle$$
$$\leqq -\rho(x,x^*)\|\theta(x,x^*)\|^2$$
$$\text{if } \alpha(x,x^*) = 0 \text{ and } \beta(x,x^*) \neq 0 \text{ for all } x \in X,$$

$$\gamma(x,x^*)\sum_{i=1}^{p}[F_i(x)-F_i(x^*)](<)$$
$$\leqq 0 \;\Rightarrow\; \langle \nabla F_i(x^*), \eta(x,x^*)\rangle \leqq -\rho(x,x^*)\|\theta(x,x^*)\|^2$$
$$\text{if } \alpha(x,x^*)=0 \text{ and } \beta(x,x^*)=0 \text{ for all } x\in X.$$

The function $F$ is said to be (prestrictly) $HA(\alpha,\beta,\gamma,\eta,\rho,\theta)$-V-quasiinvex on $X$ if it is (prestrictly) $HA(\alpha,\beta,\gamma,\eta,\rho,\theta)$-V-quasiinvex at each $x^*\in X$.

In the proofs of the duality theorems (specially designed), sometimes it may be more convenient to use certain alternative but equivalent forms of the above definitions. These are obtained by considering the contrapositive statements. For example, $HA(\alpha,\beta,\gamma,\eta,\rho,\theta)$-V-pseudoinvexity (when $\alpha(x,x^*)\neq 0$ and $\beta(x,x^*)\neq 0$ for all $x\in X$) can be defined in the following equivalent way:

The function $f$ is said to be $HA(\alpha,\beta,\gamma,\eta,\rho,\theta)$-V-pseudoinvex at $x^*\in X$ if there exist functions $\alpha: X\times X\to\mathbb{R}$, $\beta: X\times X\to\mathbb{R}$, $\gamma: X\times X\to\mathbb{R}_+$, $\eta: X\times X\to\mathbb{R}^n$, $\rho: X\times X\to\mathbb{R}$, and $\theta: X\times X\to\mathbb{R}^n$ such that for all $x\in X$,

$$\frac{1}{\alpha(x,x^*)}\gamma(x,x^*)\left(e^{\alpha(x,x^*)\sum_{i=1}^{p}\xi_i(x,x^*)[F_i(x)-F_i(x^*)]}-1\right)<0$$
$$\Rightarrow \frac{1}{\beta(x,x^*)}\langle\sum_{i=1}^{p}\nabla F_i(x^*), e^{\beta(x,x^*)\eta(x,x^*)}-\mathbf{1}\rangle < -\rho(x,x^*)\|\theta(x,x^*)\|^2.$$

In the sequel, we shall also need a consistent notation for vector inequalities. For $a,b\in\mathbb{R}^m$, the following order notation will be used: $a\geqq b$ if and only if $a_i\geqq b_i$ for all $i\in\underline{m}$; $a\geqslant b$ if and only if $a_i\geqq b_i$ for all $i\in\underline{m}$, but $a\neq b$; $a>b$ if and only if $a_i>b_i$ for all $i\in\underline{m}$; and $a\ngeqslant b$ is the negation of $a\geqslant b$.

Consider the multiobjective problem

$$(P^*)\qquad \underset{x\in\mathbb{F}}{\text{Minimize}}\; F(x)=(F_1(x),\ldots,F_p(x)),$$

where $F_i$, $i\in\underline{p}$, are real-valued functions defined on $X$.

An element $x^\circ\in\mathbb{F}$ is said to be an *efficient (Pareto optimal, nondominated, noninferior)* solution of $(P^*)$ if there exists no $x\in\mathbb{F}$ such that $F(x)\leqslant F(x^\circ)$. There are several versions of the concept of efficiency that are currently used in the area of multiobjective programming. However, throughout this section we shall deal exclusively with the efficient solutions of $(P)$ in the sense defined above.

We may need the following result on a set of parametric necessary efficiency conditions for $(P)$ which will be needed for proving strong duality theorems.

**Theorem 3.48.** [6] Let $x^*\in\mathbb{F}$, $\lambda^*=\varphi(x^*)$, for each $i\in\underline{p}$, let $f_i$ and $g_i$ be continuously differentiable at $x^*$, for each $j\in\underline{q}$, let the function $G_j(\cdot,t)$ be continuously differentiable at $x^*$ for all $t\in T_j$, and for each $k\in\underline{r}$, let the function $H_k(\cdot,s)$ be continuously differentiable at $x^*$ for all $s\in S_k$. If $x^*$ is an efficient solution of (P), if the generalized Guignard constraint qualification holds at $x^*$, and if for each $i_0\in\underline{p}$, the set $cone(\{\nabla G_j(x^*,t): t\in\hat{T}_j(x^*), j\in\underline{q}\}\cup\{\nabla f_i(x^*)-\lambda_i^*\nabla g_i(x^*): i\in\underline{p}, i\neq i_0\})+span(\{\nabla H_k(x^*,s): s\in S_k, k\in\underline{r}\})$ is closed, then there exist $u^*\in U$ and integers $\nu_0$ and $\nu$, with $0\leqq\nu_0\leqq\nu\leqq n+1$, such that there exist

$\nu_0$ indices $j_m$, with $1 \leqq j_m \leqq q$, together with $\nu_0$ points $t^m \in \hat{T}_{j_m}(x^*)$, $m \in \underline{\nu_0}$, $\nu - \nu_0$ indices $k_m$, with $1 \leqq k_m \leqq r$, together with $\nu - \nu_0$ points $s^m \in S_{k_m}$ for $m \in \underline{\nu} \backslash \underline{\nu_0}$, and $\nu$ real numbers $v_m^*$, with $v_m^* > 0$ for $m \in \underline{\nu_0}$, with the property that

$$\sum_{i=1}^{p} u_i^*[\nabla f_i(x^*) - \lambda_i^* \nabla g_i(x^*)] + \sum_{m=1}^{\nu_0} v_m^* \nabla G_{j_m}(x^*, t^m)$$
$$+ \sum_{m=\nu_0+1}^{\nu} v_m^* \nabla H_{k_m}(x^*, s^m) = 0,$$

where $cone(V)$ is the conic hull of the set $V \subset \mathbb{R}^n$ (i.e., the smallest convex cone containing $V$), $span(V)$ is the linear hull of $V$ (i.e., the smallest subspace containing $V$), $\hat{T}_j(x^*) = \{t \in T_j : G_j(x^*, t) = 0\}$, $U = \{u \in \mathbb{R}^p : u > 0, \sum_{i=1}^{p} u_i = 1\}$, and $\underline{\nu^*} \backslash \underline{\nu_0^*}$ is the complement of the set $\underline{\nu_0^*}$ relative to the set $\underline{\nu^*}$.

For brevity, we shall henceforth refer to an efficient solution $x^* \in \mathbb{F}$ as a *normal* efficient solution of $(P)$ if the generalized Guignard constraint qualification is satisfied at $x^*$ and for each $i_0 \in \underline{p}$, the set $cone(\{\nabla G_j(x^*, t) : t \in \hat{T}_j(x^*), j \in \underline{q}\} \cup \{\nabla f_i(x^*) - \lambda_i^* \nabla g_i(x^*) : i \in \underline{p}, i \neq i_0\}) + span(\{\nabla H_k(x^*, s) : s \in S_k, k \in \underline{r}\})$ is closed, where $\lambda_i^* = f_i(x^*)/g_i(x^*)$, $i \in \underline{p}$.

We further assume that the functions $f_i$, $g_i$, $i \in \underline{p}$, $G_j(\cdot, t)$, and $H_k(\cdot, s)$, are continuously differentiable on $X$ for all $t \in T_j$, $j \in \underline{q}$, and $s \in S_k$, $k \in \underline{r}$.

With regard to the choice of the type of generalized $HA(\alpha, \beta, \gamma, \xi, \eta, \rho, \theta)$-V-invex functions, specified in Definitions 3.1 - 3.3, to be used in the statements and proofs of our duality theorems, we shall consistently use the cases in which the functions $\alpha$ and $\beta$ are nonzero for all $(x, y) \in X \times X$.

### 6.1. Duality Model I

In this section, we consider two dual problems with relatively simple constraint structures and prove weak, strong, and strict converse duality theorems under appropriate $HA(\alpha, \beta, \gamma, \xi, \eta, \rho, \theta)$-V-invexity conditions. More general duality models and results for $(P)$ will be discussed in the subsequent sections.

Let

$$\mathbb{H} = \Big\{ (y, u, v, \lambda, \nu, \nu_0, J_{\nu_0}, K_{\nu \backslash \nu_0}, \bar{t}, \bar{s}) : y \in X;\ u \in U;\ \lambda \in \mathbb{R}^p;\ 0 \leqq \nu_0 \leqq \nu \leqq n+1;$$
$$v \in \mathbb{R}^{\nu}, v_i > 0, 1 \leqq i \leqq \nu_0;\ J_{\nu_0} = (j_1, j_2, \ldots, j_{\nu_0}), 1 \leqq j_i \leqq q;\ K_{\nu \backslash \nu_0} = (k_{\nu_0+1}, \ldots, k_{\nu}),$$
$$1 \leqq k_i \leqq r;\ \bar{t} = (t^1, t^2, \ldots, t^{\nu_0}), t^i \in T_{j_i};\ \bar{s} = (s^{\nu_0+1}, \ldots, s^{\nu}), s^i \in S_{k_i} \Big\}.$$

Consider the following two problems:

$$(DI) \quad \sup_{(y,u,v,\lambda,\nu,\nu_0,J_{\nu_0},K_{\nu \backslash \nu_0},\bar{t},\bar{s}) \in \mathbb{H}} \lambda = (\lambda_1, \cdots, \lambda_p)$$

subject to

$$\sum_{i=1}^{p} u_i[\nabla f_i(y) - \lambda_i \nabla g_i(y)] + \sum_{m=1}^{\nu_0} v_m \nabla G_{j_m}(y,t^m)$$

$$+ \sum_{m=\nu_0+1}^{\nu} v_m \nabla H_{k_m}(y,s^m) = 0, \tag{3.65}$$

$$u_i[f_i(y) - \lambda_i g_i(y)] + \sum_{m=1}^{\nu_0} v_m G_{j_m}(y,t^m)$$

$$+ \sum_{m=\nu_0+1}^{\nu} v_m H_{k_m}(y,s^m) \geqq 0, \ \ i \in \underline{p}; \tag{3.66}$$

$$(\tilde{D}I) \qquad \sup_{(y,u,v,\lambda,\nu,\nu_0,J_{\nu_0},K_{\nu\backslash\nu_0},\bar{t},\bar{s})\in\mathbb{H}} \qquad \lambda = (\lambda_1,\cdots,\lambda_p)$$

subject to (66) and

$$\frac{1}{\beta(x,y)}\Big\langle \sum_{i=1}^{p} u_i[\nabla f_i(y) - \lambda_i \nabla g_i(y)] + \sum_{m=1}^{\nu_0} v_m \nabla G_{j_m}(y,t^m)$$

$$+ \sum_{m=\nu_0+1}^{\nu} v_m \nabla H_{k_m}(y,s^m), e^{\beta(x,y)\eta(x,y)} - \mathbf{1}\Big\rangle \geqq 0 \text{ for all } x \in \mathbb{F}, \tag{3.67}$$

where $\beta : X \times X \to \mathbb{R}$ and $\eta : X \times X \to \mathbb{R}^n$ are given functions.

The structures of the two problems designated above as $(DI)$ and $(\tilde{D}I)$, which can be proved under appropriate $HA(\alpha,\beta,\gamma,\xi,\eta,\rho,\theta)$-V-invexity hypotheses to be dual problems for $(P)$, are based directly on the form and contents of the necessary efficiency conditions of Theorem 3.48. This is, of course, the standard method for constructing Wolfe-type dual problems. Comparing $(DI)$ and $(\tilde{D}I)$, we see that $(\tilde{D}I)$ is relatively more general than $(DI)$ in the sense that any feasible solution of $(DI)$ is also feasible for $(\tilde{D}I)$, but the converse is not necessarily true. Furthermore, we observe that (65) is a system of $n$ equations, whereas (67) is a single inequality. Clearly, from a computational point of view, $(DI)$ is preferable to $(\tilde{D}I)$ because of the dependence of (67) on the feasible set of $(P)$. Despite these apparent differences, it turns out that the statements and proofs of all the duality theorems for $(P)-(DI)$ and $(P)-(\tilde{D}I)$ are almost identical and, therefore, we shall consider only the pair $(P)-(DI)$.

The next two theorems show that $(DI)$ is a dual problem for $(P)$.

**Theorem 3.49.** Let $x$ and $(y,u,v,\lambda,\nu,\nu_0,J_{\nu_0},K_{\nu\backslash\nu_0},\bar{t},\bar{s})$ be arbitrary feasible solutions of (P) and (DI), respectively, and assume that the Lagrangian-type function $\big(L_1(\cdot,u,v,\lambda,\bar{t},\bar{s}),\ldots,L_p(\cdot,u,v,\lambda,\bar{t},\bar{s})\big) : X \to \mathbb{R}^p$ is $HA(\alpha,\beta,\gamma,\xi,\eta,0,\theta)$-V-pseudoinvex and $\gamma(x,y) > 0$,

where

$$L_i(z,u,v,\lambda,\bar{t},\bar{s}) = u_i\Big[f_i(z) - \lambda_i g_i(z) + \sum_{m=1}^{\nu_0} v_m G_{j_m}(z,t^m)$$

$$+ \sum_{m=\nu_0+1}^{\nu} v_m H_{k_m}(z,s^m)\Big], \quad i \in \underline{p}.$$

Then $\varphi(x) \nleq \lambda$.

*Proof.* From (65) we observe that

$$\frac{1}{\beta(x,y)}\Big\langle \sum_{i=1}^{p} u_i[\nabla f_i(y) - \lambda_i \nabla g_i(y)] + \sum_{m=1}^{\nu_0} v_m \nabla G_{j_m}(y,t^m)$$

$$+ \sum_{m=\nu_0+1}^{\nu} v_m \nabla H_{k_m}(y,s^m), e^{\beta(x,y)\eta(x,y)} - \mathbf{1}\Big\rangle = 0,$$

which in view of our $HA(\alpha,\beta,\gamma,\xi,\eta,0,\theta)$-V-pseudoinvexity assumption implies that

$$\frac{1}{\alpha(x,y)}\gamma(x,y)\Big(e^{\alpha(x,y)\sum_{i=1}^{p}\xi_i(x,y)[L_i(x,u,v,\lambda,\bar{t},\bar{s}) - L_i(y,u,v,\lambda,\bar{t},\bar{s})]} - 1\Big) \geqq 0.$$

We need to consider two cases: $\alpha(x,y) > 0$ and $\alpha(x,y) < 0$. If we assume that $\alpha(x,y) > 0$ and recall that $\gamma(x,y) > 0$, then the above inequality becomes

$$e^{\alpha(x,y)\sum_{i=1}^{p}\xi_i(x,y)[L_i(x,u,v,\lambda,\bar{t},\bar{s}) - L_i(y,u,v,\lambda,\bar{t},\bar{s})]} \geqq 1,$$

which implies that

$$\sum_{i=1}^{p}\xi_i(x,y)L_i(x,u,v,\lambda,\bar{t},\bar{s}) \geqq \sum_{i=1}^{p}\xi_i(x,y)L_i(y,u,v,\lambda,\bar{t},\bar{s}).$$

Because of the dual feasibility of $(y,u,v,\lambda,\nu,\nu_0,J_{\nu_0},K_{\nu\setminus\nu_0},\bar{t},\bar{s})$ and (66), the right-hand side of the above inequality is greater than or equal to zero, and hence we have $L(x,u,v,\lambda,\bar{t},\bar{s}) \geqq 0$. Inasmuch as $x \in \mathbb{F}$, and $v_m > 0$, $m \in \underline{\nu_0}$, this inequality simplifies to

$$\sum_{i=1}^{q} u_i\xi_i(x,y)[f_i(x) - \lambda_i g_i(x)] \geqq 0. \tag{3.68}$$

Since $u > 0$ and $\xi_i(x,y) > 0$, $i \in \underline{p}$, the above inequality implies that

$$\big(f_1(x) - \lambda_1 g_1(x), \ldots, f_p(x) - \lambda_p g_p(x)\big) \nleq (0,\ldots,0),$$

which in turn implies that

$$\varphi(x) = \Big(\frac{f_1(x)}{g_1(x)}, \ldots, \frac{f_p(x)}{g_p(x)}\Big) \nleq (\lambda_1,\ldots,\lambda_p) = \lambda.$$

If we assume that $\alpha(x,x^*) < 0$, we arrive at the same conclusion.

□

**Theorem 3.50.** (Strong Duality) Let $x^*$ be a normal efficient solution of (P) and assume that for each feasible solution $(y,u,v,\lambda,\nu,\nu_0,J_{\nu_0},K_{\nu\backslash\nu_0},\bar{t},\bar{s})$ of (DI), the conditions specified in Theorem 3.49 are satisfied. Then there exist $u^*,v^*,\lambda^*,\nu^*,\nu_0^*,J_{\nu_0^*},K_{\nu^*\backslash\nu_0^*},\bar{t}^*$, and $\bar{s}^*$ such that $(x^*,u^*,v^*,\lambda^*,\nu^*,\nu_0^*,J_{\nu_0^*},K_{\nu^*\backslash\nu_0^*},\bar{t}^*,\bar{s}^*)$ is an efficient solution of (DI) and $\varphi(x^*)=\lambda^*$.

*Proof.* Since $x^*$ is a normal efficient solution of $(P)$, by Theorem 3.16, there exist $u^*,v^*$, $\lambda^*(=\varphi(x^*))$, $\nu^*$, $\nu_0^*,J_{\nu_0^*},K_{\nu^*\backslash\nu_0^*},\bar{t}^*$, and $\bar{s}^*$ such that $(x^*,u^*,v^*,\lambda^*,\nu^*,\nu_0^*,J_{\nu_0^*},K_{\nu^*\backslash\nu_0^*},\bar{t}^*,\bar{s}^*)$ is a feasible solution of $(DI)$ and $\varphi(x^*)=\lambda^*$. If it were not efficient, then there would exist a feasible solution $(\tilde{x},\tilde{u},\tilde{v},\tilde{\lambda},\tilde{\nu},\tilde{\nu}_0,J_{\tilde{\nu}_0},K_{\tilde{\nu}\backslash\tilde{\nu}_0},\bar{\tilde{t}},\bar{\tilde{s}})$ of $(DI)$ such that $\tilde{\lambda}\geqslant\lambda^*=\varphi(x^*)$, contradicting Theorem 3.1. Therefore, $(x^*,u^*,v^*,\lambda^*,\nu^*,\nu_0^*,J_{\nu_0^*},K_{\nu^*\backslash\nu_0^*},\bar{t}^*,\bar{s}^*)$ is an efficient solution of $(DI)$. □

We also have the following converse duality result for $(P)$ and $(DI)$.

**Theorem 3.51.** (Strict Converse Duality) Let $x^*$ and $(\tilde{x},\tilde{u},\tilde{v},\tilde{\lambda},\tilde{\nu},\tilde{\nu}_0,J_{\tilde{\nu}_0},K_{\tilde{\nu}\backslash\tilde{\nu}_0},\bar{\tilde{t}},\bar{\tilde{s}})$ be arbitrary feasible solutions of (P) and (DI), respectively, such that

$$\sum_{i=1}^{p}\tilde{u}_i\xi_i(x^*,\tilde{x})[f_i(x^*)-\tilde{\lambda}_i g_i(x^*)]=0, \tag{3.69}$$

where for each $i\in\underline{p}$, $\xi_i$ is a function from $X\times X$ to $\mathbb{R}_+\backslash\{0\}$. Furthermore, assume that the function $L(\cdot,\tilde{u},\tilde{v},\tilde{\lambda},\bar{\tilde{t}},\bar{\tilde{s}})$ is strictly $HA(\alpha,\beta,\gamma,\xi,\eta,0,\theta)$-V-pseudoinvex at $\tilde{x}$ and $\gamma(x^*,\tilde{x})>0$. Then $\tilde{x}=x^*$ and $\varphi(x^*)=\tilde{\lambda}$.

*Proof.* (a): Suppose to the contrary that $\tilde{x}\neq x^*$. Now proceeding as in the proof of Theorem 3.49 (with $x$ replaced by $x^*$ and $(y,u,v,\lambda,\nu,\nu_0,J_{\nu_0},K_{\nu\backslash\nu_0},\bar{t},\bar{s})$ by $(\tilde{x},\tilde{u},\tilde{v},\tilde{\lambda},\tilde{\nu},\tilde{\nu}_0,J_{\tilde{\nu}_0},K_{\tilde{\nu}\backslash\tilde{\nu}_0},\bar{\tilde{t}},\bar{\tilde{s}})$) and using the conditions set forth above, we arrive at the strict inequality

$$\sum_{i=1}^{p}\tilde{u}_i\xi_i(x^*,\tilde{x})[f_i(x^*)-\tilde{\lambda}_i g_i(x^*)]>0,$$

which contradicts (69). Therefore, we conclude that $\tilde{x}=x^*$ and $\varphi(x^*)=\tilde{\lambda}$.

□

## 6.2. Duality Model II

In this subsection, we consider certain variants of $(DI)$ and $(\tilde{D}I)$ that allow for more general types of $HA(\alpha,\beta,\gamma,\xi,\eta,\rho,\theta)$-V-invexity conditions under which duality can be established. These duality models may have the following forms:

$$(DII)\qquad \sup_{(y,u,v,\lambda,\nu,\nu_0,J_{\nu_0},K_{\nu\backslash\nu_0},\bar{t},\bar{s})\in\mathbb{H}}\lambda=(\lambda_1,\cdots,\lambda_p)$$

subject to

$$\sum_{i=1}^{p}u_i[\nabla f_i(y)-\lambda_i\nabla g_i(y)]+\sum_{m=1}^{\nu_0}v_m\nabla G_{j_m}(y,t^m)+\sum_{m=\nu_0+1}^{\nu}v_m\nabla H_{k_m}(y,s^m)=0, \tag{3.70}$$

$$u_i[f_i(y) - \lambda_i g_i(y)] \geqq 0, \quad i \in \underline{p}, \tag{3.71}$$

$$G_{j_m}(y,t^m) \geqq 0, \quad m \in \underline{\nu_0}, \tag{3.72}$$

$$v_m H_{k_m}(y,s^m) \geqq 0, \quad m \in \underline{\nu} \backslash \underline{\nu_0}; \tag{3.73}$$

$$(\tilde{D}II) \qquad \sup_{(y,u,v,\lambda,\nu,\nu_0,J_{\nu_0},K_{\nu\backslash\nu_0},\bar{t},\bar{s})\in\mathbb{III}} \lambda = (\lambda_1,\cdots,\lambda_p)$$

subject to and (71) - (73).

The remarks and observations made earlier about the relationships between $(DI)$ and $(\tilde{D}I)$ are, of course, also valid for $(DII)$ and $(\tilde{D}II)$. Since the constraint inequalities of $(DII)$ are formed by splitting the inequality (66) into three inequalities (71), (72), and (73), it is clear that Theorems 3.52 - 3.54 are valid for the pair $(P) - (DII)$. Below, we shall establish some duality results in which various generalized $HA(\alpha,\beta,\gamma,\xi,\eta,\rho,\theta)$-V-invexity requirements will be placed on certain vector-valued functions, including $(\mathscr{E}_1(\cdot,\lambda,u),\ldots,\mathscr{E}_p(\cdot,\lambda,u))$, where for each $i \in \underline{p}$, the component function $\mathscr{E}_i(\cdot,\lambda,u)$ is defined, for fixed $\lambda$ and $u$, on $X$ by

$$\mathscr{E}_i(z,\lambda,u) = u_i[f_i(z) - \lambda_i g_i(z)].$$

**Theorem 3.52.** Let $x$ and $(y,u,v,\lambda,\nu,\nu_0,J_{\nu_0},K_{\nu\backslash\nu_0},\bar{t},\bar{s})$ be arbitrary feasible solutions of (P) and (DII), respectively, and assume that any one of the following four sets of hypotheses is satisfied:

(a) (i) $(\mathscr{E}_1(\cdot,\lambda,u),\ldots,\mathscr{E}_p(\cdot,\lambda,u))$ is $HA(\alpha,\beta,\bar{\gamma},\xi,\eta,\bar{\rho},\theta)$-V-pseudoinvex at $y$ and $\bar{\gamma}(x,y) > 0$;

(ii) $(v_1 G_{j_1}(\cdot,t^1),\ldots,v_{\nu_0} G_{j_{\nu_0}}(\cdot,t^{\nu_0}))$ is $HA(\alpha,\beta,\hat{\gamma},\pi,\eta,\hat{\rho},\theta)$-V-quasiinvex at $y$;

(iii) $(v_{\nu_0+1} H_{k_{\nu_0+1}}(\cdot,s^{\nu_0+1}),\ldots,v_\nu H_{k_\nu}(\cdot,s^\nu))$ is HA($\alpha$, $\beta$, $\breve{\gamma}$, $\delta$, $\eta$, $\breve{\rho}$, $\theta$) -V-quasiinvex at $y$;

(iv) $\bar{\rho}(x,y) + \hat{\rho}(x,y) + \breve{\rho}(x,y) \geqq 0$;

(b) (i) $(\mathscr{E}_1(\cdot,\lambda,u),\ldots,\mathscr{E}_p(\cdot,\lambda,u))$ is prestrictly HA($\alpha$, $\beta$, $\bar{\gamma}$, $\xi$, $\eta$, $\bar{\rho}$, $\theta$) -V-quasiinvex at $y$ and $\bar{\gamma}(x,y) > 0$;

(ii) $(v_1 G_{j_1}(\cdot,t^1),\ldots,v_{\nu_0} G_{j_{\nu_0}}(\cdot,t^{\nu_0}))$ is $HA(\alpha,\beta,\hat{\gamma},\pi,\eta,\hat{\rho},\theta)$-V-quasiinvex at $y$;

(iii) $(v_{\nu_0+1} H_{k_{\nu_0+1}}(\cdot,s^{\nu_0+1}),\ldots,v_\nu H_{k_\nu}(\cdot,s^\nu))$ is HA($\alpha$, $\beta$, $\breve{\gamma}$, $\delta$, $\eta$, $\breve{\rho}$, $\theta$) -V-quasiinvex at $y$;

(iv) $\bar{\rho}(x,y) + \hat{\rho}(x,y) + \breve{\rho}(x,y) > 0$;

(c) (i) $(\mathscr{E}_1(\cdot,\lambda,u),\ldots,\mathscr{E}_p(\cdot,\lambda,u))$ is prestrictly HA($\alpha$, $\beta$, $\bar{\gamma}$, $\xi$, $\eta$, $\bar{\rho}$, $\theta$) -V-quasiinvex at $y$ and $\bar{\gamma}(x,y) > 0$;

(ii) $\left(v_1 G_{j_1}(\cdot,t^1),\ldots,v_{v_0} G_{j_{v_0}}(\cdot,t^{v_0})\right)$ is strictly $HA(\alpha,\beta,\hat{\gamma},\pi,\eta,\hat{\rho},\theta)$-V-pseudoinvex at $y$;

(iii) $\left(v_{v_0+1} H_{k_{v_0+1}}(\cdot,s^{v_0+1}),\ldots,v_v H_{k_v}(\cdot,s^v)\right)$ is HA($\alpha$, $\beta$, $\breve{\gamma}$, $\delta$, $\eta$, $\breve{\rho}$, $\theta$) -V-quasiinvex at $y$;

(iv) $\bar{\rho}(x,y)+\hat{\rho}(x,y)+\breve{\rho}(x,y) \geqq 0$;

(d) (i) $\left(\mathscr{E}_1(\cdot,\lambda,u),\ldots,\mathscr{E}_p(\cdot,\lambda,u)\right)$ is prestrictly HA($\alpha$, $\beta$, $\bar{\gamma}$, $\xi$, $\eta$, $\bar{\rho}$, $\theta$) -V-quasiinvex at $y$ and $\bar{\gamma}(x,y) > 0$;

(ii) $\left(v_1 G_{j_1}(\cdot,t^1),\ldots,v_{v_0} G_{j_{v_0}}(\cdot,t^{v_0})\right)$ is $HA(\alpha,\beta,\hat{\gamma},\pi,\eta,\hat{\rho},\theta)$-V-quasiinvex at $y$;

(iii) $\left(v_{v_0+1} H_{k_{v_0+1}}(\cdot,s^{v_0+1}),\ldots,v_v H_{k_v}(\cdot,s^v)\right)$ is strictly $HA(\alpha,\beta,\breve{\gamma},\delta,\eta,\breve{\rho},\theta)$-V-pseudoinvex at $y$;

(iv) $\bar{\rho}(x,y)+\hat{\rho}(x,y)+\breve{\rho}(x,y) \geqq 0$.

Then $\varphi(x) \nleq \lambda$.

*Proof.* (a): Because of (72) and the primal feasibility of $x$, we have $G_{j_m}(x,t^m) \leqq 0 \leqq G_{j_m}(y,t^m)$. As $\pi_m(x,y) > 0$ for each $m \in \underline{v_0}$, it follows that

$$\sum_{m=1}^{v_0} v_m \pi_m(x,y) G_{j_m}(x,t^m) \leqq \sum_{m=1}^{v_0} v_m \pi_m(x,x) G_{j_m}(y,t^m),$$

and hence

$$\frac{1}{\alpha(x,y)}\hat{\gamma}(x,y)\left(e^{\alpha(x,y)\sum_{m=1}^{v_0} v_m \pi_m(x,y)[G_{j_m}(x,t^m)-G_{j_m}(y,t^m)]}-1\right) \leqq 0,$$

which in view of (ii) implies that

$$\frac{1}{\beta(x,y)}\left\langle \sum_{m=1}^{v_0} v_m \nabla G_{j_m}(y,t^m), e^{\beta(x,y)\eta(x,y)}-\mathbf{1}\right\rangle \leqq -\hat{\rho}(x,y)\|\theta(x,y)\|^2. \tag{3.74}$$

Similarly, we can show that our assumptions in (iii) combined with the primal feasibility of $x$ and (73) lead to the following inequality:

$$\frac{1}{\beta(x,y)}\left\langle \sum_{m=v_0+1}^{v} v_m \nabla H_{k_m}(y,s^m), e^{\beta(x,y)\eta(x,y)}-\mathbf{1}\right\rangle \leqq -\breve{\rho}(x,y)\|\theta(x,y)\|^2. \tag{3.75}$$

Now because of (74), (75), and (iv), (70) yields

$$\frac{1}{\beta(x,y)}\left\langle \sum_{i=1}^{p} u_i[\nabla f_i(y)-\lambda_i \nabla g_i(y)], e^{\beta(x,y)\eta(x,y)}-\mathbf{1}\right\rangle \geqq -\bar{\rho}(x,y)\|\theta(x,y)\|^2,$$

which in view of (i) implies that

$$\frac{1}{\alpha(x,y)}\bar{\gamma}(x,y)\left(e^{\alpha(x,y)\sum_{i=1}^{p} u_i \xi_i(x,y)\{f_i(x)-\lambda_i g_i(x)-[f_i(y)-\lambda_i g_i(y)]\}}-1\right) \geqq 0.$$

Since $\bar{\gamma}(x,y) > 0$, this inequality implies that

$$\sum_{i=1}^{p} u_i\xi_i(x,y)[f_i(x) - \lambda_i g_i(x)] \geqq \sum_{i=1}^{p} u_i\xi_i(x,y)[f_i(y) - \lambda_i g_i(y)] \geqq 0,$$

where the second inequality follows from (71). Since $u_i\xi_i(x,y) > 0$ for each $i \in \underline{p}$, this inequality implies that

$$\left(f_1(x) - \lambda_1 g_1(x), \ldots, f_p(x) - \lambda_p g_p(x)\right) \nleq (0,\ldots,0),$$

which in turn implies that

$$\varphi(x) = \left(\frac{f_1(x)}{g_1(x)}, \ldots, \frac{f_p(x)}{g_p(x)}\right) \nleq (\lambda_1,\ldots,\lambda_p) = \lambda.$$

(b) : The proof is similar to that of part (a).

(c) : Suppose to the contrary that $\varphi(x) \leqslant \lambda$. This implies that $f_i(x) - \lambda_i g_i(x) \leqq 0$ for each $i \in \underline{p}$ with strict inequality holding for at least one index $\ell \in \underline{p}$. Since $u > 0$, $\xi_i(x,y) > 0$, $i \in \underline{p}$, and (71) holds, it is clear that

$$\sum_{i=1}^{p} \xi_i(x,y)\mathscr{E}_i(x,\lambda,u) < \sum_{i=1}^{p} \xi_i(x,y)\mathscr{E}_i(y,\lambda,u),$$

and therefore it follows that

$$\frac{1}{\alpha(x,y)}\bar{\gamma}(x,y)\left(e^{\alpha(x,y)\sum_{i=1}^{p}\xi_i(x,y)[\mathscr{E}_i(x,\lambda,u)-\mathscr{E}_i(y,\lambda,u)]} - 1\right) < 0,$$

which in view of (i) implies that

$$\frac{1}{\beta(x,y)}\left\langle \sum_{i=1}^{p} u_i[\nabla f_i(y) - \lambda_i \nabla g_i(y)], e^{\beta(x,y)\eta(x,y)} - \mathbf{1}\right\rangle \leqq -\bar{\rho}(x,y)\|\theta(x,y)\|^2. \tag{3.76}$$

Proceeding as in the proof of part (a) and using our strict $HA(\alpha,\beta,\hat{\gamma},\pi,\eta,\hat{\rho},\theta)$-V-pseudoinvexity assumption, we arrive at the strict inequality

$$\frac{1}{\beta(x,y)}\left\langle \sum_{m=1}^{v_0} v_m \nabla G_{j_m}(y,t^m), e^{\beta(x,y)\eta(x,y)} - \mathbf{1}\right\rangle < -\hat{\rho}(x,y)\|\theta(x,y)\|^2.$$

Now combining this inequality with (75) (which is valid for the present case because of (iii)) and (70), and using (iv) we obtain

$$\frac{1}{\beta(x,y)}\left\langle \sum_{i=1}^{p} u_i[\nabla f_i(y) - \lambda_i \nabla g_i(y)], e^{\beta(x,y)\eta(x,y)} - \mathbf{1}\right\rangle > -\bar{\rho}(x,y)\|\theta(x,y)\|^2$$

which contradicts (76). Therefore, we conclude that $\varphi(x) \nleq \lambda$.

(d) : The proof is similar to that of part (c).

□

**Theorem 3.53.** (Strong Duality) Let $x^*$ be a normal efficient solution of (P) and assume that for each feasible solution $(y,u,v,\lambda,\nu,\nu_0,J_{\nu_0},K_{\nu\setminus\nu_0},\bar{t},\bar{s})$ of (DII), any one of the four sets of conditions specified in Theorem 3.52 is satisfied. Then there exist $u^*,v^*,\lambda^*,\ \nu^*,\nu_0^*,J_{\nu_0^*}$, $K_{\nu^*\setminus\nu_0^*},\bar{t}^*$, and $\bar{s}^*$ such that $(x^*,u^*,v^*,\lambda^*,\nu^*,\nu_0^*,\ J_{\nu_0^*},\ K_{\nu^*\setminus\nu_0^*},\bar{t}^*,\bar{s}^*)$ is an efficient solution of (DII) and $\varphi(x^*)=\lambda^*$.

*Proof.* The proof is similar to that of Theorem 3.52.

□

**Theorem 3.54.** (Strict Converse Duality) Let $x^*$ and $\tilde{w}\equiv(\tilde{x},\tilde{u},\tilde{v},\tilde{\lambda},\tilde{\nu},\tilde{\nu}_0,\ J_{\tilde{\nu}_0},\ K_{\tilde{\nu}\setminus\tilde{\nu}_0},\ \bar{\tilde{t}},\tilde{\bar{s}})$ be arbitrary feasible solutions of (P) and (DII), respectively, such that

$$\sum_{i=1}^{p}\tilde{u}_i\xi_i(x^*,\tilde{x})[f_i(x^*)-\tilde{\lambda}_i g_i(x^*)]=0, \tag{3.77}$$

where for each $i\in\underline{p}$, $\xi_i$ is a function from $X\times X$ to $\mathbb{R}_+\setminus\{0\}$. Furthermore, assume that any one of the following four sets of hypotheses is satisfied:

(a) The assumptions specified in part (a) of Theorem 3.52 are satisfied for the feasible solution $\tilde{w}$ of (DII), and the function $\big(\mathscr{E}_1(\cdot,\tilde{\lambda},\tilde{u}),\ldots,\mathscr{E}_p(\cdot,\tilde{\lambda},\tilde{u})\big)$ is strictly HA($\alpha$, $\beta$, $\bar{\gamma}$, $\xi$, $\eta$, $\bar{\rho}$, $\theta$) - V-pseudoinvex at $\tilde{x}$.

(b) The assumptions specified in part (b) of Theorem 3.52 are satisfied for the feasible solution $\tilde{w}$ of (DII), and the function $\big(\mathscr{E}_1(\cdot,\tilde{\lambda},\tilde{u}),\ldots,\mathscr{E}_p(\cdot,\tilde{\lambda},\tilde{u})\big)$ is $HA(\alpha,\beta,\bar{\gamma},\xi,\eta,\bar{\rho},\theta)$-V-quasiinvex at $\tilde{x}$.

(c) The assumptions specified in part (c) of Theorem 3.52 are satisfied for the feasible solution $\tilde{w}$ of (DII), and the function $\big(\mathscr{E}_1(\cdot,\tilde{\lambda},\tilde{u}),\ldots,\mathscr{E}_p(\cdot,\tilde{\lambda},\tilde{u})\big)$ is $HA(\alpha,\beta,\bar{\gamma},\xi,\eta,\bar{\rho},\theta)$-V-quasiinvex at $\tilde{x}$.

(d) The assumptions specified in part (d) of Theorem 3.52 are satisfied for the feasible solution $\tilde{w}$ of (DII), and the function $\big(\mathscr{E}_1(\cdot,\tilde{\lambda},\tilde{u}),\ldots,\mathscr{E}_p(\cdot,\tilde{\lambda},\tilde{u})\big)$ is $HA(\alpha,\beta,\bar{\gamma},\xi,\eta,\bar{\rho},\theta)$-V-quasiinvex at $\tilde{x}$.

Then $\tilde{x}=x^*$ and $\varphi(x^*)=\tilde{\lambda}$.

*Proof.* (a) and (b) : The proofs are similar to that of Theorem 3.51.

(c) : Suppose to the contrary that $\tilde{x}\neq x^*$. Proceeding as in the proof of part (a) of Theorem 3.52, we arrive at the following strict inequality:

$$\frac{1}{\beta(x^*,\tilde{x})}\Big\langle\sum_{i=1}^{p}\tilde{u}_i[\nabla f_i(\tilde{x})-\tilde{\lambda}_i\nabla g_i(\tilde{x})],e^{\beta(x^*,\tilde{x})\eta(x^*,\tilde{x})}-\mathbf{1}\Big\rangle>-\bar{\rho}(x^*,\tilde{x})\|\theta(x^*,\tilde{x})\|^2,$$

which in view of our $HA(\alpha,\beta,\bar{\gamma},\xi,\eta,\bar{\rho},\theta)$-V-quasiinvexity hypothesis implies that

$$\frac{1}{\alpha(x^*,\tilde{x})}\bar{\gamma}(x^*,\tilde{x})\Big(e^{\alpha(x^*,\tilde{x})\sum_{i=1}^{p}\tilde{u}_i\xi_i(x^*,\tilde{x})\{f_i(x^*)-\tilde{\lambda}_i g_i(x^*)-[f_i(\tilde{x})-\tilde{\lambda}_i g_i(\tilde{x})]\}}-1\Big)>0.$$

Since $\bar{\gamma}(x^*,\tilde{x}) > 0$, this inequality implies that

$$\sum_{i=1}^{p} \tilde{u}_i\xi_i(x^*,\tilde{x})[f_i(x^*) - \tilde{\lambda}_i g_i(x^*)] > \sum_{i=1}^{p} \tilde{u}_i\xi_i(x^*,\tilde{x})[f_i(\tilde{x}) - \tilde{\lambda}_i g_i(\tilde{x})] \geqq 0,$$

where the second inequality follows from the dual feasibility of $\tilde{w}$ and (71). This inequality contradicts (77), and, therefore, we conclude that $\tilde{x} = x^*$ and $\varphi(x^*) = \tilde{\lambda}$.

(d) : The proof is similar to that of part (c).

□

### 6.3. Duality Model III

In this subsection, we discuss several families of duality results under appropriate generalized $HA(\alpha,\beta,\gamma,\xi,\eta,\rho,\theta)$-V-invexity hypotheses imposed on certain vector functions whose components are comprised of different combinations of the problem functions. This is accomplished by employing a partitioning scheme which was originally proposed in [2] for the purpose of constructing generalized dual problems for nonlinear programming problems. Now we introduce, based on partitioning scheme, additional notations.

Let $\nu_0$ and $\nu$ be integers, with $1 \leqq \nu_0 \leqq \nu \leqq n+1$, and let $\{J_0, J_1, \ldots, J_M\}$ and $\{K_0, K_1, \ldots, K_M\}$ be partitions of the sets $\underline{\nu_0}$ and $\underline{\nu}\backslash\underline{\nu_0}$, respectively; thus, $J_i \subseteq \underline{\nu_0}$ for each $i \in \underline{M} \cup \{0\}$, $J_i \cap J_j = \emptyset$ for each $i, j \in \underline{M} \cup \{0\}$ with $i \neq j$, and $\cup_{i=0}^{M} J_i = \underline{\nu_0}$. Obviously, similar properties hold for $\{K_0, K_1, \ldots, K_M\}$. Moreover, if $m_1$ and $m_2$ are the numbers of the partitioning sets of $\underline{\nu_0}$ and $\underline{\nu}\backslash\underline{\nu_0}$, respectively, then $M = \max\{m_1, m_2\}$ and $J_i = \emptyset$ or $K_i = \emptyset$ for $i > \min\{m_1, m_2\}$.

In addition, we use the real-valued functions $\Phi_i(\cdot, u, v, \lambda, \bar{t}, \bar{s})$, and $\Lambda_\tau(\cdot, v, \bar{t}, \bar{s})$ defined, for fixed $u, v, \lambda, \bar{t} \equiv (t^1, t^2, \ldots, t^{\nu_0})$, and $\bar{s} \equiv (s^{\nu_0+1}, s^{\nu_0+2}, \ldots, s^{\nu})$, on $X$ as follows:

$$\Phi_i(z,u,v,\lambda,\bar{t},\bar{s}) = u_i\Big[f_i(z) - \lambda_i g_i(z) + \sum_{m\in J_0} v_m G_{j_m}(z,t^m)$$
$$+ \sum_{m\in K_0} v_m H_{k_m}(z,s^m)\Big], \quad i \in \underline{p},$$

$$\Lambda_\tau(z,v,\bar{t},\bar{s}) = \sum_{m\in J_\tau} v_m G_{j_m}(z,t^m) + \sum_{m\in K_\tau} v_m H_{k_m}(z,s^m), \quad \tau \in \underline{M}.$$

Making use of the sets and functions defined above, we can state our general duality models as follows:

$$(DIII) \qquad \sup_{(y,u,v,\lambda,\nu,\nu_0,J_{\nu_0},K_{\nu\backslash\nu_0},\bar{t},\bar{s})\in\mathbb{H}} \qquad \lambda = (\lambda_1, \cdots, \lambda_p)$$

subject to

$$\sum_{i=1}^{p} u_i[\nabla f_i(y) - \lambda_i \nabla g_i(y)] + \sum_{m=1}^{\nu_0} v_m \nabla G_{j_m}(y,t^m)$$
$$+ \sum_{m=\nu_0+1}^{\nu} v_m \nabla H_{k_m}(y,s^m) = 0, \tag{3.78}$$

$$f_i(y) - \lambda_i g_i(y) + \sum_{m \in J_0} v_m G_{j_m}(y, t^m)$$

$$+ \sum_{m \in K_0} v_m H_{k_m}(y, s^m) \geqq 0, \quad i \in \underline{p}, \tag{3.79}$$

$$\sum_{m \in J_\tau} v_m G_{j_m}(y, t^m)$$

$$+ \sum_{m \in K_\tau} v_m H_{k_m}(y, s^m) \geqq 0, \quad \tau \in \underline{M}; \tag{3.80}$$

$(\tilde{D}III) \qquad \sup_{(y,u,v,\lambda,\nu,\nu_0,J_{\nu_0},K_{\nu\backslash\nu_0},\bar{t},\bar{s}) \in \mathbb{H}} \qquad \lambda = (\lambda_1, \cdots, \lambda_p)$

subject to (79), and (80).

The remarks and observations made earlier about the relationships between $(DI)$ and $(\tilde{D}I)$ are, of course, also valid for $(DIII)$ and $(\tilde{D}III)$.

The next two theorems show that $(DIII)$ is a dual problem for $(P)$.

**Theorem 3.55.** Let $x$ and $(y, u, v, \lambda, \nu, \nu_0, J_{\nu_0}, K_{\nu\backslash\nu_0}, \bar{t}, \bar{s})$ be arbitrary feasible solutions of (P) and (DIII), respectively, and assume that any one of the following three sets of hypotheses is satisfied:

(a) (i) $\big(\Phi_1(\cdot, u, v, \lambda, \bar{t}, \bar{s}), \ldots, \Phi_p(\cdot, u, v, \lambda, \bar{t}, \bar{s})\big)$ is HA($\alpha$, $\beta$, $\bar{\gamma}$, $\xi$, $\eta$, $\bar{\rho}$, $\theta$) - V-pseudoinvex at $y$ and $\bar{\gamma}(x, y) > 0$;

(ii) $\big(\Lambda_1(\cdot, v, \bar{t}, \bar{s}), \ldots, \Lambda_M(\cdot, v, \bar{t}, \bar{s})\big)$ is $HA(\alpha, \beta, \tilde{\gamma}, \pi, \eta, \tilde{\rho}, \theta)$-V-quasiinvex at $y$;

(iii) $\bar{\rho}(x, y) + \tilde{\rho}(x, y) \geqq 0$;

(b) (i) $\big(\Phi_1(\cdot, u, v, \lambda, \bar{t}, \bar{s}), \ldots, \Phi_p(\cdot, u, v, \lambda, \bar{t}, \bar{s})\big)$ is prestrictly HA($\alpha$, $\beta$, $\bar{\gamma}$, $\xi$, $\eta$, $\bar{\rho}$, $\theta$) - V-
quasiinvex at $y$ and $\bar{\gamma}(x, y) > 0$;

(ii) $\big(\Lambda_1(\cdot, v, \bar{t}, \bar{s}), \ldots, \Lambda_M(\cdot, v, \bar{t}, \bar{s})\big)$ is $HA(\alpha, \beta, \gamma, \pi, \eta, \tilde{\rho}, \theta)$-V-quasiinvex

(iii) $\bar{\rho}(x, y) + \tilde{\rho}(x, y) > 0$;

(c) (i) $\big(\Phi_1(\cdot, u, v, \lambda, \bar{t}, \bar{s}), \ldots, \Phi_p(\cdot, u, v, \lambda, \bar{t}, \bar{s})\big)$ is prestrictly HA($\alpha$, $\beta$, $\bar{\gamma}$, $\xi$, $\eta$, $\bar{\rho}$, $\theta$) - V-
quasiinvex at $y$ and $\bar{\gamma}(x, y) > 0$;

(ii) $\big(\Lambda_1(\cdot, v, \bar{t}, \bar{s}), \ldots, \Lambda_M(\cdot, v, \bar{t}, \bar{s})\big)$ is strictly HA($\alpha$, $\beta$, $\tilde{\gamma}$, $\pi$, $\eta$, $\tilde{\rho}$, $\theta$) - V-pseudoinvex at $y$;

(iii) $\bar{\rho}(x, y) + \tilde{\rho}(x, y) \geqq 0$.

Then $\varphi(x) \nleq \lambda$.

*Proof.* (a): It is clear that (78) can be expressed as follows:

$$\sum_{i=1}^{p} u_i[\nabla f_i(y) - \lambda_i \nabla g_i(y)] + \sum_{m\in J_0} v_m \nabla G_{j_m}(y,t^m) + \sum_{m\in K_0} v_m \nabla H_{k_m}(y,s^m) + \sum_{\tau=1}^{M}\Big[\sum_{m\in J_\tau} v_m \nabla G_{j_m}(y,t^m) + \sum_{m\in K_\tau} v_m \nabla H_{k_m}(y,s^m)\Big] = 0. \quad (3.81)$$

Since $x \in \mathbb{F}$, $v_m > 0$, $m \in \underline{v_0}$, and (79) holds, it follows that for each $\tau \in \underline{M}$,

$$\begin{aligned}\Lambda_\tau(x,v,\bar{t},\bar{s}) &= \sum_{m\in J_\tau} v_m G_{j_m}(x,t^m) + \sum_{m\in K_\tau} v_m H_{k_m}(x,s^m)\\ &\leqq 0\\ &\leqq \sum_{m\in J_\tau} v_m G_{j_m}(y,t^m) + \sum_{m\in K_\tau} v_m H_{k_m}(y,s^m)\\ &= \Lambda_\tau(y,v,\bar{t},\bar{s}).\end{aligned}$$

Because $\pi_\tau(x,y) > 0$ for each $\tau \in M$, it is clear that

$$\sum_{\tau=1}^{M} \pi_\tau(x,y)\Lambda_\tau(x,v,\bar{t},\bar{s}) \leqq \sum_{\tau=1}^{M} \pi_\tau(x,y)\Lambda_\tau(y,v,\bar{t},\bar{s}),$$

and so

$$\frac{1}{\alpha(x,y)}\tilde{\gamma}(x,y)\Big(e^{\alpha(x,y)\sum_{\tau=1}^{M}\pi_\tau(x,y)[\Lambda_\tau(x,v,\bar{t},\bar{s})-\Lambda_\tau(y,v,\bar{t},\bar{s})]} - 1\Big) \leqq 0,$$

which in view of (ii) implies that

$$\frac{1}{\beta(x,y)}\Big\langle \sum_{\tau=1}^{M}\Big[\sum_{m\in J_\tau} v_m \nabla G_{j_m}(y,t^m) + \sum_{m\in K_\tau} v_m \nabla H_{k_m}(y,s^m)\Big], e^{\beta(x,y)\eta(x,y)} - \mathbf{1}\Big\rangle \leqq -\tilde{\rho}(x,y)\|\theta(x,y)\|^2. \quad (3.82)$$

Combining (81) and (82), and using (iii) we get

$$\frac{1}{\beta(x,y)}\Big\langle \sum_{i=1}^{p} u_i[\nabla f_i(y) - \lambda_i \nabla g_i(y)] + \sum_{m\in J_0} v_m \nabla G_{j_m}(y,t^m) + \sum_{m\in K_0} v_m \nabla H_{k_m}(y,s^m), e^{\beta(x,y)\eta(x,y)} - \mathbf{1}\Big\rangle \geqq \tilde{\rho}(x,y)\|\theta(x,y)\|^2 \geqq -\bar{\rho}(x,y)\|\theta(x,y)\|^2, \quad (3.83)$$

which by virtue of (i) implies that

$$\frac{1}{\alpha(x,y)}\bar{\gamma}(x,y)\Big(e^{\alpha(x,y)\sum_{i=1}^{p}\xi_i(x,y)[\Phi_i(x,u,v,\lambda,\bar{t},\bar{s})-\Phi_i(y,u,v,\lambda,\bar{t},\bar{s})]} - 1\Big) \geqq 0. \quad (3.84)$$

Since $\bar{\gamma}(x,y) > 0$, we can deduce from this inequality that

$$\sum_{i=1}^{p} \xi_i(x,y)\Phi_i(x,u,v,\lambda,\bar{t},\bar{s}) \geqq \sum_{i=1}^{p} \xi_i(x,y)\Phi_i(y,u,v,\lambda,\bar{t},\bar{s}) \geqq 0,$$

where the second inequality follows from the dual feasibility of $(y, u, v, \lambda, \nu, \nu_0, J_{\nu_0}, K_{\nu\setminus\nu_0}, \bar{t}, \bar{s})$ and (79). Therefore, we have

$$\begin{aligned}0 &\leqq \sum_{i=1}^{p} \xi_i(x,y)\Phi_i(x,u,v,\lambda,\bar{t},\bar{s}) \\ &= \sum_{i=1}^{p} u_i\xi_i(x,y)\Big\{f_i(x)-\lambda_i g_i(x)+\sum_{m\in J_0} v_m G_{j_m}(x,t^m)+\sum_{m\in K_0} v_m H_{k_m}(x,s^m)\Big\} \\ &\leqq \sum_{i=1}^{p} u_i\xi_i(x,y)[f_i(x)-\lambda_i g_i(x)] \ \text{(by the primal feasibility of } x).\end{aligned}$$

As shown in the proof of Theorem 3.55, this inequality leads to the conclusion that $\varphi(x) \nleq \lambda$.

(b): Proceeding in exactly the same manner as in the proof of part (a), we obtain (83) in which the second inequality is strict. By (i), this implies that (84) holds and, therefore, the rest of the proof is identical to that of part (a).

(c): Suppose to the contrary that $\varphi(x) \leqslant \lambda$. This implies that $f_i(x)-\lambda_i g_i(x) \leqq 0$ for each $i \in \underline{p}$, with strict inequality holding for at least one index $\ell \in \underline{p}$. Using these inequalities, we see that

$$\begin{aligned}\Phi_i(x,u,v,\lambda,\bar{t},\bar{s}) &\leqq u_i[f_i(x)-\lambda_i g_i(x)] \ \text{(by the primal feasibility of } x) \\ &\leqq 0 \\ &\leqq u_i\Big[f_i(y)-\lambda_i g_i(y)+\sum_{m\in J_0} v_m G_{j_m}(y,t^m) \\ &\quad+\sum_{m\in K_0} v_m H_{k_m}(y,s^m)\Big] \ \text{(by (79))} \\ &= \Phi_i(y,u,v,\lambda,\bar{t},\bar{s}).\end{aligned}$$

Since $u > 0$ and $\xi(x,y) > 0$, the above inequalities yield

$$\sum_{i=1}^{p} \xi_i(x,y)\Phi_i(x,u,v,\lambda,\bar{t},\bar{s}) < \sum_{i=1}^{p} \xi_i(x,y)\Phi_i(y,u,v,\lambda,\bar{t},\bar{s})$$

and so

$$\frac{1}{\alpha(x,y)}\bar{\gamma}(x,y)\Big(e^{\alpha(x,y)\sum_{i=1}^{p}\xi_i(x,y)[\Phi_i(x,u,v,\lambda,\bar{t},\bar{s})-\Phi_i(y,u,v,\lambda,\bar{t},\bar{s})]}-1\Big) < 0,$$

which in view of (i) implies that

$$\begin{aligned}\frac{1}{\beta(x,y)}\Big\langle \sum_{i=1}^{p} u_i[\nabla f_i(y)-\lambda_i\nabla g_i(y)]+\sum_{m\in J_0} v_m\nabla G_{j_m}(y,t^m) \\ +\sum_{m\in K_0} v_m\nabla H_{k_m}(y,s^m), e^{\beta(x,y)\eta(x,y)}-\mathbf{1}\Big\rangle \leqq \tilde{\rho}(x,y)\|\theta(x,y)\|^2 \\ \leqq -\bar{\rho}(x,y)\|\theta(x,y)\|^2. \quad (3.85)\end{aligned}$$

Proceeding as in the proof of part (a) and using the strict $HA(\alpha,\beta,\tilde{\gamma},\pi,\eta,\tilde{\rho},\theta)$-V-pseudoinvexity hypothesis, we arrive at the following inequality:

$$\frac{1}{\beta(x,y)}\Big\langle\sum_{\tau=1}^{M}\Big[\sum_{m\in J_\tau}v_m\nabla G_{j_m}(y,t^m)+\sum_{m\in K_\tau}v_m\nabla H_{k_m}(y,s^m)\Big],e^{\beta(x,y)\eta(x,y)}-\mathbf{1}\Big\rangle$$
$$<-\tilde{\rho}(x,y)\|\theta(x,y)\|^2,$$

which when combined with (81) yields

$$\frac{1}{\beta(x,y)}\Big\langle\sum_{i=1}^{p}u_i[\nabla f_i(y)-\lambda_i\nabla g_i(y)]+\sum_{m\in J_0}v_m\nabla G_{j_m}(y,t^m)$$
$$+\sum_{m\in K_0}v_m\nabla H_{k_m}(y,s^m),e^{\beta(x,y)\eta(x,y)}-\mathbf{1}\Big\rangle>\tilde{\rho}(x,y)\|\theta(x,y)\|^2\geqq-\bar{\rho}(x,y)\|\theta(x,y)\|^2.$$

This contradicts (85), and, therefore, we conclude that $\varphi(x)\nleq\lambda$. □

**Theorem 3.56.** (Strong Duality) Let $x^*$ be a normal efficient solution of (P) and assume that for each feasible solution $(y,u,v,\lambda,\nu,\nu_0,J_{\nu_0},K_{\nu\backslash\nu_0},\bar{t},\bar{s})$ of (DIII), any one of the three sets of conditions specified in Theorem 3.55 is satisfied. Then there exist $u^*,v^*,\lambda^*$, $\nu^*,\nu_0^*,J_{\nu_0^*}$, $K_{\nu^*\backslash\nu_0^*},\bar{t}^*$, and $\bar{s}^*$ such that $(x^*,u^*,v^*,\lambda^*,\nu^*,\nu_0^*,J_{\nu_0^*},K_{\nu^*\backslash\nu_0^*},\bar{t}^*,\bar{s}^*)$ is an efficient solution of (DIII) and $\varphi(x^*)=\lambda^*$.

*Proof.* The proof is similar to that of Theorem 3.55. □

**Theorem 3.57.** (Strict Converse Duality) Let $x^*$ and $\tilde{w}\equiv(\tilde{x},\tilde{u},\tilde{v},\tilde{\lambda},\tilde{\nu},\tilde{\nu}_0,J_{\tilde{\nu}_0},K_{\tilde{\nu}\backslash\tilde{\nu}_0},\bar{\tilde{t}},\bar{\tilde{s}})$ be arbitrary feasible solutions of (P) and (DIII), respectively, such that

$$\sum_{i=1}^{p}\tilde{u}_i\xi_i(x^*,\tilde{x})[f_i(x^*)-\tilde{\lambda}_i g_i(x^*)]=0,$$

where for each $i\in\underline{p}$, $\xi_i$ is a function from $X\times X$ to $\mathbb{R}_+\backslash\{0\}$. Furthermore, assume that any one of the following three sets of hypotheses is satisfied:

(a) The assumptions specified in part (a) of Theorem 5.1 are satisfied for the feasible solution $\tilde{w}$ of (DIII), and the function $\big(\Phi_1(\cdot,\tilde{u},\tilde{v},\tilde{\lambda},\bar{\tilde{t}},\bar{\tilde{s}}),\ldots,\Phi_p(\cdot,\tilde{u},\tilde{v},\tilde{\lambda},\bar{\tilde{t}},\bar{\tilde{s}})\big)$ is strictly $HA(\alpha,\beta,\bar{\gamma},\xi,\eta,\bar{\rho},\theta)$-V-pseudoinvex at $\tilde{x}$.

(b) The assumptions specified in part (b) of Theorem 3.55 are satisfied for the feasible solution $\tilde{w}$ of (DIII), and the function $\big(\Phi_1(\cdot,\tilde{u},\tilde{v},\tilde{\lambda},\bar{\tilde{t}},\bar{\tilde{s}}),\ldots,\Phi_p(\cdot,\tilde{u},\tilde{v},\tilde{\lambda},\bar{\tilde{t}},\bar{\tilde{s}})\big)$ is $HA(\alpha,\beta,\bar{\gamma},\xi,\eta,\bar{\rho},\theta)$-V-quasiinvex $\tilde{x}$.

(c) The assumptions specified in part (c) of Theorem 3.55 are satisfied for the feasible solution $\tilde{w}$ of (DIII), and the function $\big(\Phi_1(\cdot,\tilde{u},\tilde{v},\tilde{\lambda},\bar{\tilde{t}},\bar{\tilde{s}}),\ldots,\Phi_p(\cdot,\tilde{u},\tilde{v},\tilde{\lambda},\bar{\tilde{t}},\bar{\tilde{s}})\big)$ is $HA(\alpha,\beta,\bar{\gamma},\xi,\eta,\bar{\rho},\theta)$-V-quasiinvex $\tilde{x}$.

Then $\tilde{x} = x^*$ and $\varphi(x^*) = \tilde{\lambda}$.

*Proof.* The proof is similar to that of Theorem 3.55. □

Each one of the three sets of results given in Theorem 3.55 can be viewed as a family of duality results whose members can easily be identified by appropriate choices of the partitioning sets $J_\mu$ and $K_\mu$, $\mu \in \underline{M} \cup \{0\}$.

## 6.4. Duality Model IV

In this subsection we discuss another collection of duality results for $(P)$ which are different from those stated in Theorems 3.55 - 3.57. In formulating these duality results, we utilize a partition of $\underline{p}$ in addition to those of $\underline{\nu_0}$ and $\underline{\nu}\backslash\underline{\nu_0}$. It appears that this partitioning scheme was first proposed in [3] for a multiobjective fractional programming problem with a finite number of constraints. In our duality theorems, we impose appropriate generalized $HA(\alpha,\beta,\gamma,\xi,\eta,\rho,\theta)$-V-invexity requirements on certain vector functions whose components comprise some combinations of the functions $\mathscr{E}_i(\cdot,\lambda,u)$, $i \in \underline{p}$, $G_j(\cdot,t)$, $j \in \underline{q}$, and $H_k(\cdot,s)$, $k \in \underline{r}$.

Let $\{I_0, I_1, \ldots, I_d\}$, $\{J_0, J_1, \ldots, J_e\}$, and $\{K_0, K_1, \ldots, K_e\}$ be partitions of $\underline{p}$, $\underline{\nu_0}$, and $\underline{\nu}\backslash\underline{\nu_0}$, respectively, such that $D = \{0,1,2,\ldots,d\} \subset E = \{0,1,\ldots,e\}$, and let the function $\Pi_\tau(\cdot,u,v,\lambda,\bar{t},\bar{s}) : X \to \mathbb{R}$ be defined, for fixed $u, v, \lambda, \bar{t}$, and $\bar{s}$, by

$$\Pi_\tau(z,u,v,\lambda,\bar{t},\bar{s}) = \sum_{i \in I_\tau} u_i[f_i(z) - \lambda_i g_i(z)] + \sum_{m \in J_\tau} v_m G_{j_m}(z,t^m) + \sum_{m \in K_\tau} v_m H_{k_m}(z,s^m), \quad \tau \in D.$$

Making use of the sets and functions defined above, we consider the following duality models:

$$(DIV) \qquad \sup_{(y,u,v,\lambda,\nu,\nu_0,J_{\nu_0},K_{\nu\backslash\nu_0},\bar{t},\bar{s}) \in \mathbb{H}} \lambda = (\lambda_1,\cdots,\lambda_p)$$

subject to

$$\sum_{i=1}^{p} u_i[\nabla f_i(y) - \lambda_i \nabla g_i(y)] + \sum_{m=1}^{\nu_0} v_m \nabla G_{j_m}(y,t^m) + \sum_{m=\nu_0+1}^{\nu} v_m \nabla H_{k_m}(y,s^m) = 0, \tag{3.86}$$

$$\sum_{i \in I_\tau} u_i[f_i(y) - \lambda_i g_i(y)] + \sum_{m \in J_\tau} v_m G_{j_m}(y,t^m) + \sum_{m \in K_\tau} v_m H_{k_m}(y,s^m) \geqq 0, \quad \tau \in D, \tag{3.87}$$

$$\sum_{m \in J_\tau} v_m G_{j_m}(y,t^m) + \sum_{m \in K_\tau} v_m H_{k_m}(y,s^m) \geqq 0, \quad \tau \in E\backslash D; \tag{3.88}$$

$$(\tilde{D}IV) \qquad \sup_{(y,u,v,\lambda,\nu,\nu_0,J_{\nu_0},K_{\nu\backslash\nu_0},\bar{t},\bar{s}) \in \mathbb{H}} \lambda = (\lambda_1,\cdots,\lambda_p)$$

subject to (87) and (88).

The remarks and observations made earlier about the relationships between $(DI)$ and $(\tilde{D}I)$ are, of course, also valid for $(DIV)$ and $(\tilde{D}IV)$.

The next two theorems show that $(DIV)$ is a dual problem for $(P)$.

**Theorem 3.58.** Let $x$ and $(y,u,v,\lambda,\nu,\nu_0,J_{\nu_0},K_{\nu\setminus\nu_0},\bar{t},\bar{s})$ be arbitrary feasible solutions of (P) and (DIV), respectively, and assume that any one of the following three sets of hypotheses is satisfied:

(a) (i) $\big(\Pi_0(\cdot,u,v,\lambda,\bar{t},\bar{s}),\ldots,\Pi_d(\cdot,u,v,\lambda,\bar{t},\bar{s})\big)$ is HA($\alpha$, $\beta$, $\bar{\gamma}$, $\xi$, $\eta$, $\bar{\rho}$, $\theta$) -V-pseudoinvex at $y$ and $\gamma(x,y)>0$;

(ii) $\big(\Lambda_{d+1}(\cdot,v,\bar{t},\bar{s}),\ldots,\Lambda_e(\cdot,v,\bar{t},\bar{s})\big)$ is $HA(\alpha,\beta,\tilde{\gamma},\pi,\eta,\tilde{\rho},\theta)$-V-quasiinvex at $y$;

(iii) $\bar{\rho}(x,y)+\tilde{\rho}(x,y)\geqq 0$;

(b) (i) $\big(\Pi_0(\cdot,u,v,\lambda,\bar{t},\bar{s}),\ldots,\Pi_d(\cdot,u,v,\lambda,\bar{t},\bar{s})\big)$ is prestrictly HA($\alpha$, $\beta$, $\bar{\gamma}$, $\xi$, $\eta$, $\bar{\rho}$, $\theta$) -V-
quasiinvex at $y$ and $\bar{\gamma}(x,y)>0$;

(ii) $\big(\Lambda_{d+1}(\cdot,v,\bar{t},\bar{s}),\ldots,\Lambda_e(\cdot,v,\bar{t},\bar{s})\big)$ is strictly $HA(\alpha,\beta,\tilde{\gamma},\xi,\eta,\tilde{\rho},\theta)$-V-pseudoinvex at $y$;

(iii) $\bar{\rho}(x,y)+\tilde{\rho}(x,y)\geqq 0$;

(c) (i) $\big(\Pi_0(\cdot,u,v,\lambda,\bar{t},\bar{s}),\ldots,\Pi_d(\cdot,u,v,\lambda,\bar{t},\bar{s})\big)$ is prestrictly HA($\alpha$, $\beta$, $\bar{\gamma}$, $\xi$, $\eta$, $\bar{\rho}$, $\theta$) -V-
quasiinvex at $y$ and $\bar{\gamma}(x,y)>0$;

(ii) $\big(\Lambda_{d+1}(\cdot,v,\bar{t},\bar{s}),\ldots,\Lambda_e(\cdot,v,\bar{t},\bar{s})\big)$ is $HA(\alpha,\beta,\tilde{\gamma},\xi,\eta,\tilde{\rho},\theta)$-V-quasiinvex at $y$;

(iii) $\bar{\rho}(x,y)+\tilde{\rho}(x,y)>0$.

Then $\varphi(x)\nleq\lambda$.

*Proof.* (a): Suppose to the contrary that $\varphi(x)\leqslant\lambda$. Then $f_i(x)-\lambda_i g_i(x)\leqq 0,\quad i\in\underline{p}$, with strict inequality holding for at least one index $\ell\in\underline{p}$. Since $u>0$, we see that for each $\tau\in D$,

$$\sum_{i\in I_\tau} u_i[f_i(x)-\lambda_i g_i(x)]\leqq 0, \tag{3.89}$$

with strict inequality holding for at least one index $\tau\in D$. Now using this inequality, we see that

$$\begin{aligned}
&\Pi_\tau(x,u,v,\lambda,\bar{t},\bar{s})\\
&= \sum_{i\in I_\tau} u_i[f_i(x)-\lambda_i g_i(x)] + \sum_{m\in J_\tau} v_m G_{j_m}(x,t^m) + \sum_{m\in K_\tau} v_m H_{k_m}(x,s^m)\\
&\leqq \sum_{i\in I_\tau} u_i[f_i(x)-\lambda_i g_i(x)]\\
&\quad \text{(by the primal feasibility of } x \text{ and positivity of } v_m,\ m\in \underline{v_0})\\
&\leqq 0 \quad \text{(by (89))}\\
&\leqq \sum_{i\in I_\tau} u_i[f_i(y)-\lambda_i g_i(y)] + \sum_{m\in J_\tau} v_m G_{j_m}(y,t^m) + \sum_{m\in K_\tau} v_m H_{k_m}(y,s^m) \quad \text{(by (87))}\\
&= \Pi_\tau(y,u,v,\lambda,\bar{t},\bar{s}),
\end{aligned}$$

with strict inequality holding for at least one index $\tau \in D$. Since $\xi_\tau(x,y) > 0$ for each $\tau \in D$, it is clear that

$$\sum_{\tau\in D} \xi_\tau(x,y)\Pi_\tau(x,u,v,\lambda,\bar{t},\bar{s}) < \sum_{\tau\in D} \xi_\tau(x,y)\Pi_\tau(y,u,v,\lambda,\bar{t},\bar{s}),$$

and therefore

$$\frac{1}{\alpha(x,y)}\bar{\gamma}(x,y)\left(e^{\alpha(x,y)\sum_{\tau\in D}\xi_\tau(x,y)[\Pi_\tau(x,u,v,\lambda,\bar{t},\bar{s})-\Pi_\tau(y,u,v,\lambda,\bar{t},\bar{s})]}-1\right) < 0,$$

which in view of (i) implies that

$$\frac{1}{\beta(x,y)}\Big\langle \sum_{\tau\in D}\Big[\sum_{m\in J_\tau} v_m\nabla G_{j_m}(y,t^m) + \sum_{m\in K_\tau} v_m\nabla H_{k_m}(y,s^m)\Big], e^{\beta(x,y)\eta(x,y)}-\mathbf{1}\Big\rangle < -\tilde{\rho}(x,y)\|\theta(x,y)\|^2. \quad (3.90)$$

As shown in the proof of Theorem 3.55, for each $\tau \in E\backslash D$, $\Lambda_\tau(x,v,\bar{t},\bar{s}) \leqq \Lambda_\tau(y,v,\bar{t},\bar{s})$, and hence

$$\sum_{\tau\in E\backslash D} \pi_\tau(x,y)\Lambda_\tau(x,v,\bar{t},\bar{s}) \leqq \sum_{\tau\in E\backslash D} \pi_\tau(x,y)\Lambda_\tau(y,v,\bar{t},\bar{s}),$$

and

$$\frac{1}{\alpha(x,y)}\tilde{\gamma}(x,y)\left(e^{\alpha(x,y)\sum_{\tau\in E\backslash D}\pi_\tau(x,y)[\Lambda_\tau(x,v,\bar{t},\bar{s})-\Lambda_\tau(y,v,\bar{t},\bar{s})]}-1\right) \leqq 0,$$

which in view of (ii) implies that

$$\frac{1}{\beta(x,y)}\Big\langle \sum_{\tau\in E\backslash D}\Big[\sum_{m\in J_\tau} v_m\nabla G_{j_m}(y,t^m) + \sum_{m\in K_\tau} v_m\nabla H_{k_m}(y,s^m)\Big], e^{\beta(x,y)\eta(x,y)}-\mathbf{1}\Big\rangle \leqq -\tilde{\rho}(x,y)\|\theta(x,y)\|^2. \quad (3.91)$$

Now combining (90) and (91) and using (iii), we see that

$$\frac{1}{\beta(x,y)}\Big\langle \sum_{i=1}^{p} u_i[\nabla f_i(y)-\lambda_i\nabla g_i(y)] + \sum_{m=1}^{v_0} v_m\nabla G_{j_m}(y,t^m) + \sum_{m=v_0+1}^{v} v_m\nabla H_{k_m}(y,s^m), e^{\beta(x,y)\eta(x,y)}-\mathbf{1}\Big\rangle < -[\bar{\rho}(x,y)+\tilde{\rho}(x,y)]\|\theta(x,y)\|^2 \leqq 0,$$

which contradicts (86). Therefore, $\varphi(x) \nleq \lambda$.

(b) and (c) : The proofs are similar to that of part (a).

□

**Theorem 3.59.** (Strong Duality) Let $x^*$ be a normal efficient solution of (P) and assume that for each feasible solution $(y,u,v,\lambda,\nu,\nu_0,J_{\nu_0},K_{\nu\backslash\nu_0},\bar{t},\bar{s})$ of (DIV), any one of the three sets of conditions specified in Theorem 3.58 is satisfied. Then there exist $u^*,v^*,\lambda^*,\ \nu^*,\nu_0^*,J_{\nu_0^*}$, $K_{\nu^*\backslash\nu_0^*},\bar{t}^*$, and $\bar{s}^*$ such that $(x^*,u^*,v^*,\lambda^*,\nu^*,\nu_0^*,\ J_{\nu_0^*},\ K_{\nu^*\backslash\nu_0^*},\bar{t}^*,\bar{s}^*)$ is an efficient solution of (DIV) and $\varphi(x^*)=\lambda^*$.

*Proof.* The proof is similar to that of Theorem 3.58. □

**Theorem 3.60.** (Strict Converse Duality) Let $x^*$ and $\tilde{w}\equiv(\tilde{x},\tilde{u},\tilde{v},\tilde{\lambda},\tilde{\nu},\tilde{\nu}_0,\ J_{\tilde{\nu}_0},\ K_{\tilde{\nu}\backslash\tilde{\nu}_0},\ \bar{\tilde{t}},\bar{\tilde{s}})$ be arbitrary feasible solutions of (P) and (DIV), respectively, such that

$$\sum_{i=1}^{p}\tilde{u}_i[f_i(x^*)-\tilde{\lambda}_i g_i(x^*)]=0. \tag{3.92}$$

Furthermore, assume that any one of the following three sets of hypotheses is satisfied:

(a) The assumptions specified in part (a) of Theorem 3.58 are satisfied for the feasible solution $\tilde{w}$ of (DIV), the function $\big(\Pi_0(\cdot,\tilde{u},\tilde{v},\tilde{\lambda},\bar{\tilde{t}},\bar{\tilde{s}}),\ldots,\ \Pi_d(\cdot,\tilde{u},\ \tilde{v},\tilde{\lambda},\bar{\tilde{t}},\bar{\tilde{s}})\big)$ is strictly $HA(\alpha,\beta,\bar{\gamma},\xi,\eta,\bar{\rho},\theta)$-V-pseudoinvex at $\tilde{x}$, and $\xi_0=\xi_1=\ldots=\xi_d=\sigma$.

(b) The assumptions specified in part (b) of Theorem 3.58 are satisfied for the feasible solution $\tilde{w}$ of (DIV), the function $\big(\Pi_0(\cdot,\tilde{u},\tilde{v},\tilde{\lambda},\bar{\tilde{t}},\bar{\tilde{s}}),\ldots,\ \Pi_d(\cdot,\tilde{u},\ \tilde{v},\tilde{\lambda},\bar{\tilde{t}},\bar{\tilde{s}})\big)$ is $HA(\alpha,\beta,\bar{\gamma},\xi,\eta,\bar{\rho},\theta)$-V-quasiinvex at $\tilde{x}$, and $\xi_0=\xi_1=\ldots=\xi_d=\sigma$.

(c) The assumptions specified in part (c) of Theorem 3.58 are satisfied for the feasible solution $\tilde{w}$ of (DIV), the function $\big(\Pi_0(\cdot,\tilde{u},\tilde{v},\tilde{\lambda},\bar{\tilde{t}},\bar{\tilde{s}}),\ldots,\Pi_d(\cdot,\tilde{u},\ \tilde{v},\tilde{\lambda},\bar{\tilde{t}},\bar{\tilde{s}})\big)$ is $HA(\alpha,\beta,\bar{\gamma},\xi,\eta,\bar{\rho},\theta)$-V-quasiinvex at $\tilde{x}$, and $\xi_0=\xi_1=\ldots=\xi_d=\sigma$.

Then $\tilde{x}=x^*$ and $\varphi(x^*)=\tilde{\lambda}$.

*Proof.* (a) : Suppose to the contrary that $\tilde{x}\neq x^*$. By Theorem 3.58, there exist $u^*,v^*,\lambda^*,\nu^*,\nu_0^*,\ J_{\nu_0^*},\ K_{\nu^*\backslash\nu_0^*},\bar{t}^*$, and $\bar{s}^*$ such that $(x^*,u^*,v^*,\lambda^*,\nu^*,\nu_0^*,J_{\nu_0^*},K_{\nu^*\backslash\nu_0^*},\bar{t}^*,\bar{s}^*)$ is a feasible solution of $(DIV)$ and $\varphi(x^*)=\lambda^*$.

It is clear that (86) can be expressed as follows:

$$\sum_{\tau\in D}\Big\{\sum_{i\in I_\tau}\tilde{u}_i[\nabla f_i(\tilde{x})-\tilde{\lambda}\nabla g_i(\tilde{x})]+\sum_{m\in J_\tau}\tilde{\nu}_m\nabla G_{j_m}(\tilde{x},\tilde{t}^m)+\sum_{m\in K_\tau}\tilde{\nu}_m\nabla H_{k_m}(\tilde{x},\tilde{s}^m)\Big\}+$$
$$\sum_{\tau\in E\backslash D}\Big[\sum_{m\in J_\tau}\tilde{\nu}_m\nabla G_{j_m}(\tilde{x},\tilde{t}^m)+\sum_{m\in K_\tau}\tilde{\nu}_m\nabla H_{k_m}(\tilde{x},\tilde{s}^m)\Big]=0. \tag{3.93}$$

Since $x^* \in \mathbb{F}$, $\tilde{v}_m \geqq 0$, $m \in \underline{\tilde{\nu}}$, and (88) holds, it follows that for each $\tau \in E \backslash D$,

$$\begin{aligned}\Lambda_\tau(x^*,\tilde{v},\bar{\bar{t}},\bar{\bar{s}}) &= \sum_{m\in J_\tau} \tilde{v}_m G_{j_m}(x^*,\tilde{t}^m) + \sum_{m\in K_\tau} \tilde{v}_m H_{k_m}(x^*,\tilde{s}^m)\\ &\leqq 0\\ &\leqq \sum_{m\in J_\tau} \tilde{v}_m G_{j_m}(\tilde{x},\tilde{t}^m) + \sum_{m\in K_\tau} \tilde{v}_m H_{k_m}(\tilde{x},\tilde{s}^m)\\ &= \Lambda_\tau(\tilde{x},\tilde{v},\bar{\bar{t}},\bar{\bar{s}}),\end{aligned}$$

and hence

$$\sum_{\tau\in E\backslash D} \pi_\tau(x^*,\tilde{x})\Lambda_\tau(x^*,\tilde{v},\bar{\bar{t}},\bar{\bar{s}}) \leq \sum_{\tau\in E\backslash D} \pi_\tau(x^*,\tilde{x})\Lambda_\tau(\tilde{x},\tilde{v},\bar{\bar{t}},\bar{\bar{s}}).$$

Using this inequality, we get

$$\frac{1}{\alpha(x^*,\tilde{x})}\tilde{\gamma}(x^*,\tilde{x})\left(e^{\alpha(x^*,\tilde{x})\sum_{\tau\in E\backslash D}\pi_\tau(x^*,\tilde{x})[\Lambda_\tau(x^*,\tilde{v},\bar{\bar{t}},\bar{\bar{s}})-\Lambda_\tau(\tilde{x},\tilde{v},\bar{\bar{t}},\bar{\bar{s}})]}-1\right)\leqq 0,$$

which in view of (ii) implies that

$$\begin{aligned}&\frac{1}{\beta(x^*,\tilde{x})}\\ &\qquad\Big\langle \sum_{\tau\in E\backslash D}\Big[\sum_{m\in J_\tau}\tilde{v}_m\nabla G_{j_m}(\tilde{x},\tilde{t}^m)+\sum_{m\in K_\tau}\tilde{v}_m\nabla H_{k_m}(\tilde{x},\tilde{s}^m)\Big], e^{\beta(x^*,\tilde{x})\eta(x^*,\tilde{x})}\\ &\qquad\qquad -\mathbf{1}\Big\rangle\\ &\qquad\qquad\qquad\leqq -\tilde{\rho}(x^*,\tilde{x})\|\theta(x^*,\tilde{x})\|^2. \qquad (3.94)\end{aligned}$$

Combining (93) and (94), and using (iii) we get

$$\begin{aligned}&\frac{1}{\beta(x,y)}\Big\langle\sum_{\tau\in D}\Big\{\sum_{i\in I_\tau}\tilde{u}_i[\nabla f_i(\tilde{x})-\tilde{\lambda}\nabla g_i(\tilde{x})]+\sum_{m\in J_\tau}\tilde{v}_m\nabla G_{j_m}(\tilde{x},\tilde{t}^m)\\ &\quad+\sum_{m\in K_\tau}\tilde{v}_m\nabla H_{k_m}(\tilde{x},\tilde{s}^m)\Big\}, e^{\beta(x^*,x)\eta(x^*,\tilde{x})}-\mathbf{1}\Big\rangle \geqq \rho(x^*,x)\|\theta(x^*,\tilde{x})\|^2\\ &\qquad\qquad\geqq -\bar{\rho}(x^*,\tilde{x})\|\theta(x^*,\tilde{x})\|^2, \qquad (3.95)\end{aligned}$$

which by virtue of our strict $HA(\alpha,\beta,\bar{\gamma},\xi,\eta,\bar{\rho},\theta)$-V-pseudoinvexity hypothesis implies that

$$\frac{1}{\alpha(x^*,\tilde{x})}\bar{\gamma}(x^*,\tilde{x})\left(e^{\alpha(x^*,\tilde{x})\sigma(x^*,\tilde{x})\sum_{\tau\in D}[\Pi_\tau(x^*,\tilde{u},\tilde{v},\tilde{\lambda},\bar{\bar{t}},\bar{\bar{s}})-\Pi_\tau(\tilde{x},\tilde{u},\tilde{v},\tilde{\lambda},\bar{\bar{t}},\bar{\bar{s}})]}-1\right)>0.$$

Since $\bar{\gamma}(x^*,\tilde{x})>0$ and $\sigma(x^*,\tilde{x})>0$, we deduce from this inequality that

$$\sum_{\tau\in D}\Pi_\tau(x^*,\tilde{u},\tilde{v},\tilde{\lambda},\bar{\bar{t}},\bar{\bar{s}}) > \sum_{\tau\in D}\Pi_\tau(\tilde{x},\tilde{u},\tilde{v},\tilde{\lambda},\bar{\bar{t}},\bar{\bar{s}})\geqq 0,$$

where the second inequality follows from the dual feasibility of $\tilde{w}$ and (97). Therefore, we have

$$\begin{aligned}
0 &< \sum_{\tau\in D} \Pi_\tau(x^*, \tilde{v}, \bar{\bar{t}}, \bar{\bar{s}}) \\
&= \sum_{\tau\in D}\sum_{i\in I_\tau} \tilde{u}_i[f_i(x^*) - \tilde{\lambda} g_i(x^*)] \\
&+ \sum_{\tau\in D}\Big[\sum_{m\in J_0} \tilde{v}_m G_{j_m}(x^*, \tilde{t}^m) + \sum_{m\in K_0} \tilde{v}_m H_{k_m}(x^*, \tilde{s}^m)\Big] \\
&< \sum_{i=1}^{p} \tilde{u}_i[f_i(x^*) - \tilde{\lambda}_i g_i(x^*)] \text{ (by the primal feasibility of } x^*),
\end{aligned}$$

which contradicts (92). Therefore, we conclude that $\tilde{x} = x^*$ and $\varphi(x^*) = \tilde{\lambda}$.

(b) and (c) : The proofs are similar to that of part (a). □

As pointed earlier, one can readily identify numerous special cases and variants of $(DIV)$ and $(\tilde{D}IV)$ by appropriate choices of the partitioning sets $I_\mu$, $\mu \in D$, and $J_\tau$ and $K_\tau$, $\tau \in E$.

We have established a number of parametric duality results under various generalized $HA(\alpha,\beta,\gamma,\xi,\eta,\rho,\theta)$-V-invexity hypotheses for a semiinfinite multiobjective fractional programming problem. It appears that all these results are new in the area of semiinfinite programming. Since all the results obtained here can be modified and restated in a straightforward manner for each one of the seven problems designated as $(P1)-(P7)$ in [4], they collectively provide a fairly large number of duality results for various classes of conventional as well as semiinfinite nonlinear programming problems. Plus, the obtained results can be applied to establish similar results for some other related classes of optimization problems. For example, it is reasonable to expect that a similar approach can be applied to investigate the optimality and duality aspects of the following semiinfinite minmax fractional programming problem:

$$\underset{x\in\mathbb{F}}{\text{Minimize}}\ \max_{y\in Y} \frac{f(x,y)}{g(x,y)}.$$

## 6.5. Special Duality Model

In this section we discuss another collection of duality results for $(P)$ which are different from those stated in Theorems 3.23 - 3.25. In formulating these duality results, we utilize a partition of $\underline{p}$ in addition to those of $\underline{\nu_0}$ and $\underline{\nu}\backslash\underline{\nu_0}$. It appears that this partitioning scheme was first proposed in [10] for a multiobjective fractional programming problem with a finite number of constraints. In our theorems, we impose appropriate generalized $HA(\alpha,\beta,\eta,\omega,\rho,\theta)$-V-sonvexity requirements on certain vector functions whose components comprise some combinations of the functions $f_i$, $g_i$, $i \in \underline{p}$, $G_j(\cdot,t)$, $j \in \underline{q}$, and $H_k(\cdot,s)$, $k \in \underline{r}$.

Let $\{I_0, I_1, \ldots, I_d\}$, $\{J_0, J_1, \ldots, J_e\}$, and $\{K_0, K_1, \ldots, K_e\}$ be partitions of $\underline{p}$, $\underline{\nu_0}$, and $\underline{\nu}\backslash\underline{\nu_0}$, respectively, such that $D = \{0,1,2,\ldots,d\} \subset E = \{0,1,\ldots,e\}$, and let the function

$\Pi_\tau(\cdot,u,v,\lambda,\bar{t},\bar{s}):X\to\mathbb{R}$ be defined, for fixed $u,v,\lambda,\bar{t}$, and $\bar{s}$ by

$$\Pi_\tau(\zeta,u,v,\lambda,\bar{t},\bar{s})=\sum_{i\in I_\tau}u_i[f_i(\zeta)-\lambda_i g_i(\zeta)]+\sum_{m\in J_\tau}v_m G_{j_m}(\zeta,t^m)+\sum_{m\in K_\tau}v_m H_{k_m}(\zeta,s^m),\ \ \tau\in D.$$

Making use of the sets and functions defined above, we consider the following second-order duality models:

$(DIV)$ $\sup_{(y,z,u,v,\lambda,\nu,\nu_0,J_{\nu_0},K_{\nu\backslash\nu_0},\bar{t},\bar{s})\in\mathbb{H}}\ \lambda=(\lambda_1,\cdots,\lambda_p)$

subject to

$$\sum_{i=1}^{p}u_i[\nabla f_i(y)-\lambda_i\nabla g_i(y)]+\sum_{m=1}^{\nu_0}v_m\nabla G_{j_m}(y,t^m)+\sum_{m=\nu_0+1}^{\nu}v_m\nabla H_{k_m}(y,s^m)+\Big\{\sum_{i=1}^{p}u_i[\nabla^2 f_i(y)-\lambda_i\nabla^2 g_i(y)]+\sum_{m=1}^{\nu_0}v_m\nabla^2 G_{j_m}(y,t^m)+\sum_{m=\nu_0+1}^{\nu}v_m\nabla^2 H_{k_m}(y,s^m)\Big\}z=0,\quad (3.96)$$

$$\sum_{i\in I_\tau}u_i[f_i(y)-\lambda_i g_i(y)]+\sum_{m\in J_\tau}v_m G_{j_m}(y,t^m)+\sum_{m\in K_\tau}v_m H_{k_m}(y,s^m)-\frac{1}{2}\Big\langle\omega(x,y),\Big\{\sum_{i\in I_\tau}u_i[\nabla^2 f_i(y)-\lambda_i\nabla^2 g_i(y)]+\sum_{m\in J_\tau}v_m\nabla^2 G_{j_m}(y,t^m)+\sum_{m\in K_\tau}v_m\nabla^2 H_{k_m}(y,s^m)\Big\}z\Big\rangle\geqq 0,\ \ \tau\in D, and\ x\in\mathbb{F},\quad (3.97)$$

$$\sum_{m\in J_\tau}v_m G_{j_m}(y,t^m)+\sum_{m\in K_\tau}v_m H_{k_m}(y,s^m)-\frac{1}{2}\Big\langle\omega(x,y),\Big[\sum_{m\in J_\tau}v_m\nabla^2 G_{j_m}(y,t^m)+\sum_{m\in K_\tau}v_m\nabla^2 H_{k_m}(y,s^m)\Big]z\Big\rangle\geqq 0,\ \ \tau\in E\backslash D, x\in\mathbb{F};\quad (3.98)$$

$(\tilde{D}IV)$ $\sup_{(y,z,u,v,\lambda,\nu,\nu_0,J_{\nu_0},K_{\nu\backslash\nu_0},\bar{t},\bar{s})\in\mathbb{H}}\ \lambda=(\lambda_1,\ldots,\lambda_p)$

subject to (3), (97), and (98).

The remarks and observations made earlier about the relationships between $(DI)$ and $(\tilde{D}I)$ are, of course, also valid for $(DIV)$ and $(\tilde{D}IV)$.

The next two theorems show that $(DIV)$ is a dual problem for $(P)$.

**Theorem 3.61.** (Weak Duality) Let $x$ and $(y,z,u,v,\lambda,\nu,\nu_0,J_{\nu_0},K_{\nu\backslash\nu_0},\bar{t},\bar{s})$ be arbitrary feasible solutions of (P) and (DIV), respectively, and assume that any one of the following three sets of hypotheses is satisfied:

(a) (i) $\left(\Pi_0(\cdot,u,v,\lambda,\bar{t},\bar{s}),\ldots,\Pi_d(\cdot,u,v,\lambda,\bar{t},\bar{s})\right)$ is HA($\alpha$, $\beta$, $\eta$, $\omega$, $\bar{\rho}$, $\theta$)-V-pseudosonvex at $y$ and $\bar{\gamma}(x,y) > 0$;

(ii) $\left(\Lambda_{d+1}(\cdot,v,\bar{t},\bar{s}),\ldots,\Lambda_e(\cdot,v,\bar{t},\bar{s})\right)$ is HA($\alpha$, $\beta$, $\pi$, $\eta$, $\omega$, $\tilde{\rho}$, $\theta$)- V-quasisonvex at $y$;

(iii) $\bar{\rho}(x,y) + \tilde{\rho}(x,y) \geqq 0$;

(b) (i) $\left(\Pi_0(\cdot,u,v,\lambda,\bar{t},\bar{s}),\ldots,\Pi_d(\cdot,u,v,\lambda,\bar{t},\bar{s})\right)$ is prestrictly HA($\alpha$, $\beta$, $\eta$, $\omega$, $\bar{\rho}$, $\theta$)-V-quasisonvex at $y$ and $\bar{\gamma}(x,y) > 0$;

(ii) $\left(\Lambda_{d+1}(\cdot,v,\bar{t},\bar{s}),\ldots,\Lambda_e(\cdot,v,\bar{t},\bar{s})\right)$ is strictly HA($\alpha$, $\beta$, $\eta$, $\omega$, $\tilde{\rho}$, $\theta$)-V-pseudosonvex at $y$;

(iii) $\bar{\rho}(x,y) + \tilde{\rho}(x,y) \geqq 0$;

(c) (i) $\left(\Pi_0(\cdot,u,v,\lambda,\bar{t},\bar{s}),\ldots,\Pi_d(\cdot,u,v,\lambda,\bar{t},\bar{s})\right)$ is prestrictly HA($\alpha$, $\beta$, $\eta$, $\omega$, $\bar{\rho}$, $\theta$)-V-quasisonvex at $y$ and $\bar{\gamma}(x,y) > 0$;

(ii) $\left(\Lambda_{d+1}(\cdot,v,\bar{t},\bar{s}),\ldots,\Lambda_e(\cdot,v,\bar{t},\bar{s})\right)$ is HA($\alpha$, $\beta$, $\eta$, $\omega$, $\tilde{\rho}$, $\theta$)-V-quasisonvex at $y$;

(iii) $\bar{\rho}(x,y) + \tilde{\rho}(x,y) > 0$.

Then $\varphi(x) \nleq \lambda$.

*Proof.* (a): Suppose to the contrary that $\varphi(x) \leqslant \lambda$. Then $f_i(x) - \lambda_i g_i(x) \leqq 0$, $i \in \underline{p}$, with strict inequality holding for at least one index $\ell \in \underline{p}$. Since $u > 0$, we see that for each $\tau \in D$,

$$\sum_{i\in I_\tau} u_i[f_i(x) - \lambda_i g_i(x)] \leqq 0, \tag{3.99}$$

with strict inequality holding for at least one index $\tau \in D$.

Now using this inequality, we see that

$$\begin{aligned}
&\Pi_\tau(x,u,v,\lambda,\bar{t},\bar{s})\\
&= \sum_{i\in I_\tau} u_i[f_i(x) - \lambda_i g_i(x)] + \sum_{m\in J_\tau} v_m G_{j_m}(x,t^m) + \sum_{m\in K_\tau} v_m H_{k_m}(x,s^m)\\
&\leqq \sum_{i\in I_\tau} u_i[f_i(x) - \lambda_i g_i(x)]\\
&;\text{(by the primal feasibility of } x \text{ and positivity of } v_m,\ m \in \underline{v_0})\\
&\leqq 0 \quad \text{(by (99))}\\
&\leqq \Pi_\tau(y,u,v,\lambda,\bar{t},\bar{s}) - \frac{1}{2}\langle \omega(x,y), \nabla^2\Pi_\tau(y,u,v,\lambda,\bar{t},\bar{s})z\rangle \quad \text{(by (97))},
\end{aligned}$$

with strict inequality holding for at least one index $\tau \in D$. Since $\xi_\tau(x,y) > 0$ for each $\tau \in D$, the above inequalities yield

$$\begin{aligned}
&\sum_{\tau\in D} \xi_\tau(x,y)\Pi_\tau(x,u,v,\lambda,\bar{t},\bar{s})\\
&< \sum_{\tau\in D} \xi_\tau(x,y)\left[\Pi_\tau(y,u,v,\lambda,\bar{t},\bar{s}) - \frac{1}{2}\langle \omega(x,y), \nabla^2\Pi_\tau(y,u,v,\lambda,\bar{t},\bar{s})z\rangle\right],
\end{aligned}$$

and hence

$$\frac{1}{\alpha(x,y)}\left(e^{\alpha(x,y)\sum_{\tau\in D}\xi_\tau(x,y)[\Pi_\tau(x,u,v,\lambda,\bar{t},\bar{s})-\Pi_\tau(y,u,v,\lambda,\bar{t},\bar{s})+\frac{1}{2}\langle\omega(x,y),\nabla^2\Pi_\tau(y,u,v,\lambda,\bar{t},\bar{s})z\rangle]}-1\right)<0,$$

which in view of (i) implies that

$$\begin{aligned}\frac{1}{\beta(x,y)}\Big\langle \sum_{i=1}^{p}u_i[\nabla f_i(y)-\lambda_i\nabla g_i(y)]+\sum_{\tau\in D}\Big[\sum_{m\in J_\tau}v_m\nabla G_{j_m}(y,t^m)\\ +\sum_{m\in K_\tau}v_m\nabla H_{k_m}(y,s^m)\Big]+\Big\{\sum_{i=1}^{p}u_i[\nabla^2 f_i(y)-\lambda_i\nabla^2 g_i(y)]\\ +\sum_{\tau\in D}\Big[\sum_{m\in J_\tau}v_m\nabla^2 G_{j_m}(y,t^m)\\ +\sum_{m\in K_\tau}v_m\nabla^2 H_{k_m}(y,s^m)\Big]\Big\}z,e^{\beta(x,y)\eta(x,y)}-\mathbf{1}\Big\rangle\\ <-\bar{\rho}(x,y)\|\theta(x,y)\|^2.\quad(3.100)\end{aligned}$$

As shown in the proof of Theorem 3.20, for each $\tau\in E\backslash D$,

$$\Lambda_\tau(x,v,\bar{t},\bar{s})\leqq\Lambda_\tau(y,v,\bar{t},\bar{s})-\frac{1}{2}\langle\omega(x,y),\nabla^2\Lambda_\tau(y,v,\bar{t},\bar{s})z\rangle,$$

and so

$$\begin{aligned}&\sum_{\tau\in E\backslash D}\pi_\tau(x,y)\Lambda_\tau(x,v,\bar{t},\bar{s})\\ &\leqq\sum_{\tau\in E\backslash D}\pi_\tau(x,y)\big[\Lambda_\tau(y,v,\bar{t},\bar{s})\\ &-\frac{1}{2}\langle\omega(x,y),\nabla^2\Lambda_\tau(y,v,\bar{t},\bar{s})z\rangle\big],\end{aligned}$$

and

$$\frac{1}{\alpha(x,y)}\left(e^{\alpha(x,y)\sum_{\tau\in E\backslash D}\pi_\tau(x,y)[\Lambda_\tau(x,v,\bar{t},\bar{s})-\Lambda_\tau(y,v,\bar{t},\bar{s})+\frac{1}{2}\langle\omega(x,y),\nabla^2\Lambda_\tau(y,v,\bar{t},\bar{s})z\rangle]}-1\right)\leqq 0,$$

which in view of (ii) implies that

$$\begin{aligned}\frac{1}{\beta(x,y)}\Big\langle\sum_{\tau\in E\backslash D}\Big\{\sum_{m\in J_\tau}v_m\nabla G_{j_m}(y,t^m)+\sum_{m\in K_\tau}v_m\nabla H_{k_m}(y,s^m)+\\ \Big[\sum_{m\in J_\tau}v_m\nabla^2 G_{j_m}(y,t^m)+\sum_{m\in K_\tau}v_m\nabla^2 H_{k_m}(y,s^m)\Big]z\Big\},e^{\beta(x,y)\eta(x,y)}-\mathbf{1}\Big\rangle\\ \leqq-\tilde{\rho}(x,y)\|\theta(x,y)\|^2.\quad(3.101)\end{aligned}$$

Now combining (100) and (101) and using (iii), we see that

$$\frac{1}{\beta(x,y)}\Big\langle \sum_{i=1}^{p} u_i[\nabla f_i(y)-\lambda_i\nabla g_i(y)]+\sum_{m=1}^{\nu_0} v_m\nabla G_{j_m}(y,t^m)$$
$$+\sum_{m=\nu_0+1}^{\nu} v_m\nabla H_{k_m}(y,s^m)+\Big\{\sum_{i=1}^{p} u_i[\nabla^2 f_i(y)-\lambda_i\nabla^2 g_i(y)]$$
$$+\sum_{m=1}^{\nu_0} v_m\nabla^2 G_{j_m}(y,t^m)$$
$$+\sum_{m=\nu_0+1}^{\nu} v_m\nabla^2 H_{k_m}(y,s^m)\Big\}z, e^{\beta(x,y)\eta(x,y)}-\mathbf{1}\Big\rangle < -[\rho(x,y)+\bar{\rho}(x,y)]\|\theta(x,y)\|^2 \leqq 0,$$

which contradicts (96). Therefore, $\varphi(x) \nleq \lambda$.

(b) and (c) : The proofs are similar to that of part (a).

□

**Theorem 3.62.** (Strong Duality) Let $x^*$ be a normal efficient solution of (P) and assume that for each feasible solution $(y,z,u,v,\lambda,\nu,\nu_0, J_{\nu_0},K_{\nu\backslash\nu_0},\bar{t},\bar{s})$ of (DIV), any one of the three sets of conditions specified in Theorem 3.26 is satisfied. Then there exist $u^*,v^*,\lambda^*$, $\nu^*,\nu_0^*,J_{\nu_0^*}, K_{\nu^*\backslash\nu_0^*},\bar{t}^*$, and $\bar{s}^*$ such that $(x^*,z^*=0,u^*,v^*,\lambda^*,\nu^*,\nu_0^*, J_{\nu_0^*}, K_{\nu^*\backslash\nu_0^*},\bar{t}^*,\bar{s}^*)$ is an efficient solution of (DIV) and $\varphi(x^*)=\lambda^*$.

*Proof.* The proof is similar to that of Theorem 3.18.

□

**Theorem 3.63.** (Strict Converse Duality) Let $x^*$ and $\tilde{w}\equiv(\tilde{x},\tilde{z},\tilde{u},\tilde{v},\tilde{\lambda},\tilde{\nu},\tilde{\nu}_0, J_{\tilde{\nu}_0}, K_{\tilde{\nu}\backslash\tilde{\nu}_0},\bar{\tilde{t}},\bar{\tilde{s}})$ be arbitrary feasible solutions of (P) and (DIV), respectively, such that

$$\sum_{i=1}^{p}\tilde{u}_i[f_i(x^*)-\tilde{\lambda}_i g_i(x^*)]=0. \tag{3.102}$$

Furthermore, assume that any one of the following three sets of hypotheses is satisfied:

(a) The assumptions specified in part (a) of Theorem 3.60 are satisfied for the feasible solution $\tilde{w}$ of (DIV), the function $\big(\Pi_0(\cdot,\tilde{u},\tilde{v},\tilde{\lambda},\bar{\tilde{t}},\bar{\tilde{s}}),\ldots,\Pi_d(\cdot,\tilde{u},\tilde{v},\tilde{\lambda},\bar{\tilde{t}},\bar{\tilde{s}})\big)$ is strictly $HA(\alpha,\beta,\eta,\omega,\bar{\rho},\theta)$-V-pseudosonvex at $\tilde{x}$, and $\xi_0=\xi_1=\ldots=\xi_d=\sigma$.

(b) The assumptions specified in part (b) of Theorem 3.60 are satisfied for the feasible solution $\tilde{w}$ of (DIV), the function $\big(\Pi_0(\cdot,\tilde{u},\tilde{v},\tilde{\lambda},\bar{\tilde{t}},\bar{\tilde{s}}),\ldots,\Pi_d(\cdot,\tilde{u},\tilde{v},\tilde{\lambda},\bar{\tilde{t}},\bar{\tilde{s}})\big)$ is $HA(\alpha,\beta,\eta,\omega,\bar{\rho},\theta)$-V-quasisonvex at $\tilde{x}$, and $\xi_0=\xi_1=\ldots=\xi_d=\sigma$.

(c) The assumptions specified in part (c) of Theorem 3.60 are satisfied for the feasible solution $\tilde{w}$ of (DIV), the function $\big(\Pi_0(\cdot,\tilde{u},\tilde{v},\tilde{\lambda},\bar{\tilde{t}},\bar{\tilde{s}}),\ldots,\Pi_d(\cdot,\tilde{u},\tilde{v},\tilde{\lambda},\bar{\tilde{t}},\bar{\tilde{s}})\big)$ is $HA(\alpha,\beta,eta,\omega,\bar{\rho},\theta)$-V-quasisonvex at $\tilde{x}$, and $\xi_0=\xi_1=\ldots=\xi_d=\sigma$.

Then $\tilde{x}=x^*$ and $\varphi(x^*)=\tilde{\lambda}$.

*Proof.* (a) : Suppose to the contrary that $\tilde{x} \neq x^*$. It is clear that (96) can be expressed as follows:

$$\sum_{\tau\in D}\Big\{\sum_{i\in I_\tau}\tilde{u}_i[\nabla f_i(\tilde{x})-\tilde{\lambda}_i\nabla g_i(\tilde{x})]+\sum_{m\in J_\tau}\tilde{v}_m\nabla G_{j_m}(\tilde{x},\tilde{t}^m)+\sum_{m\in K_\tau}\tilde{v}_m\nabla H_{k_m}(\tilde{x},\tilde{s}^m)+$$
$$\Big\{\sum_{i\in I_\tau}\tilde{u}_i[\nabla^2 f_i(\tilde{x})-\tilde{\lambda}_i\nabla^2 g_i(\tilde{x})]+\sum_{m\in J_\tau}\tilde{v}_m\nabla^2 G_{j_m}(\tilde{x},\tilde{t}^m)+\sum_{m\in K_\tau}\tilde{v}_m\nabla^2 H_{k_m}(\tilde{x},\tilde{s}^m)\Big\}\tilde{z}\Big\}+$$
$$\sum_{\tau\in E\backslash D}\Big\{\sum_{m\in J_\tau}\tilde{v}_m\nabla G_{j_m}(\tilde{x},\tilde{t}^m)+\sum_{m\in K_\tau}\tilde{v}_m\nabla H_{k_m}(\tilde{x},\tilde{s}^m)+\Big[\sum_{m\in J_\tau}\tilde{v}_m\nabla^2 G_{j_m}(\tilde{x},\tilde{t}^m)$$
$$+\sum_{m\in K_\tau}\tilde{v}_m\nabla^2 H_{k_m}(\tilde{x},\tilde{s}^m)\Big]\tilde{z}\Big\}=0. \quad (3.103)$$

Since $x^* \in \mathbb{F}$, $\tilde{v}_m \geqq 0$, $m \in \underline{\nu}$, and (98) holds, it follows that for each $\tau \in E\backslash D$,

$$\begin{aligned}\Lambda_\tau(x^*,\tilde{v},\bar{\tilde{t}},\bar{\tilde{s}}) &= \sum_{m\in J_\tau}\tilde{v}_m G_{j_m}(x^*,\tilde{t}^m)+\sum_{m\in K_\tau}\tilde{v}_m H_{k_m}(x^*,\tilde{s}^m)\\ &\leqq 0\\ &\leqq \Lambda_\tau(\tilde{x},\tilde{v},\bar{\tilde{t}},\bar{\tilde{s}})-\frac{1}{2}\langle\omega(x^*,\tilde{x}),\nabla^2\Lambda_\tau(\tilde{x},\tilde{v},\bar{\tilde{t}},\bar{\tilde{s}})\tilde{z}\rangle \text{ (by (97))},\end{aligned}$$

and hence

$$\begin{aligned}&\sum_{\tau\in E\backslash D}\pi_\tau(x^*,\tilde{x})\Lambda_\tau(x^*,\tilde{v},\bar{\tilde{t}},\bar{\tilde{s}})\\ &\leqq \sum_{\tau\in E\backslash D}\pi_\tau(x^*,\tilde{x})\big[\Lambda_\tau(\tilde{x},\tilde{v},\bar{\tilde{t}},\bar{\tilde{s}})-\frac{1}{2}\big\langle\omega(x^*,\tilde{x}),\nabla^2\Lambda_\tau(\tilde{x},\tilde{v},\bar{\tilde{t}},\bar{\tilde{s}})\tilde{z}\big\rangle\big].\end{aligned}$$

Using this inequality in conjunction with the properties of the functions $\alpha$, $\tilde{\gamma}$, and $\zeta \to e^\zeta$, we get

$$\begin{aligned}&\frac{1}{\alpha(x^*,\tilde{x})}\tilde{\gamma}(x^*,\tilde{x})\\ &\Big(e^{\alpha(x^*,\tilde{x})\sum_{\tau\in E\backslash D}\pi_\tau(x^*,\tilde{x})[\Lambda_\tau(x^*,\tilde{v},\bar{\tilde{t}},\bar{\tilde{s}})-\Lambda_\tau(\tilde{x},\tilde{v},\bar{\tilde{t}},\bar{\tilde{s}})+\frac{1}{2}\langle\omega(x^*,\tilde{x}),\nabla^2\Lambda_\tau(\tilde{x},\tilde{v},\bar{\tilde{t}},\bar{\tilde{s}})\tilde{z}\rangle]}-1\Big)\\ &\leqq 0,\end{aligned}$$

which in view of (ii) implies that

$$\frac{1}{\beta(x^*,\tilde{x})}\Big\langle\sum_{\tau\in E\backslash D}\Big\{\sum_{m\in J_\tau}\tilde{v}_m\nabla G_{j_m}(\tilde{x},\tilde{t}^m)$$
$$+\sum_{m\in K_\tau}\tilde{v}_m\nabla H_{k_m}(\tilde{x},\tilde{s}^m)+\Big[\sum_{m\in J_\tau}\tilde{v}_m\nabla^2 G_{j_m}(\tilde{x},\tilde{t}^m)$$
$$+\sum_{m\in K_\tau}\tilde{v}_m\nabla^2 H_{k_m}(\tilde{x},\tilde{s}^m)\Big]\tilde{z}\Big\},e^{\beta(x^*,\tilde{x})\eta(x^*,\tilde{x})}-\mathbf{1}\Big\rangle\leqq-\tilde{\rho}(x^*,\tilde{x})\|\theta(x^*,\tilde{x})\|^2. \quad (3.104)$$

Combining (103) and (104), and using (iii) we get

$$\frac{1}{\beta(x,y)}\Big\langle \sum_{\tau\in D}\Big\{\sum_{i\in I_\tau}\tilde{u}_i[\nabla f_i(\tilde{x})-\tilde{\lambda}_i\nabla g_i(\tilde{x})]+\sum_{m\in J_\tau}\tilde{v}_m\nabla G_{j_m}(\tilde{x},\tilde{t}^m)$$

$$+\sum_{m\in K_\tau}\tilde{v}_m\nabla H_{k_m}(\tilde{x},\tilde{s}^m)+\Big\{\sum_{i\in I_\tau}\tilde{u}_i[\nabla^2 f_i(\tilde{x})-\tilde{\lambda}\nabla^2 g_i(\tilde{x})]$$

$$+\sum_{m\in J_\tau}\tilde{v}_m\nabla^2 G_{j_m}(\tilde{x},\tilde{t}^m)$$

$$+\sum_{m\in K_\tau}\tilde{v}_m\nabla^2 H_{k_m}(\tilde{x},\tilde{s}^m)\Big\}\tilde{z}\Big\}, e^{\beta(x^*,\tilde{x})\eta(x^*,\tilde{x})}-\mathbf{1}\Big\rangle$$

$$\geqq \tilde{\rho}(x^*,\tilde{x})\|\theta(x^*,\tilde{x})\|^2\geqq -\bar{\rho}(x^*,\tilde{x})\|\theta(x^*,\tilde{x})\|^2, \quad (3.105)$$

which by virtue of our strict $HA(\alpha,\beta,\eta,\omega,\bar{\rho},\theta)$-V-pseudosonvexity hypothesis implies that

$$\frac{1}{\alpha(x^*,\tilde{x})}\Big(e^{\alpha(x^*,\tilde{x})\sigma(x^*,\tilde{x})\sum_{\tau\in D}[\Pi_\tau(x^*,\tilde{u},\tilde{v},\tilde{\lambda},\bar{\tilde{t}},\bar{\tilde{s}})-\Pi_\tau(\tilde{x},\tilde{u},\tilde{v},\tilde{\lambda},\bar{\tilde{t}},\bar{\tilde{s}})+\frac{1}{2}\langle\omega(x^*,\tilde{x}),\nabla^2\Pi_\tau(\tilde{x},\tilde{u},\tilde{v},\tilde{\lambda},\bar{\tilde{t}},\bar{\tilde{s}})\tilde{z}\rangle]}-1\Big)>0.$$

We deduce from this inequality that

$$\sum_{\tau\in D}\Pi_\tau(x^*,\tilde{u},\tilde{v},\tilde{\lambda},\bar{\tilde{t}},\bar{\tilde{s}})$$
$$>\sum_{\tau\in D}\Big[\Pi_\tau(\tilde{x},\tilde{u},\tilde{v},\tilde{\lambda},\bar{\tilde{t}},\bar{\tilde{s}})-\frac{1}{2}\langle\omega(x^*,\tilde{x}),\nabla^2\Pi_\tau(\tilde{x},\tilde{u},\tilde{v},\tilde{\lambda},\bar{\tilde{t}},\bar{\tilde{s}})\tilde{z}\rangle\Big]\geqq 0,$$

where the second inequality follows from the dual feasibility of $\tilde{w}$ and (97). Therefore, we have

$$0 < \sum_{\tau \in D} \Pi_\tau(x^*, \tilde{u}, \tilde{v}, \tilde{\lambda}, \bar{\tilde{t}}, \bar{\tilde{s}})$$

$$= \sum_{\tau \in D} \sum_{i \in I_\tau} \tilde{u}_i [f_i(x^*) - \tilde{\lambda}_i g_i(x^*)] + \sum_{\tau \in D} \Big[ \sum_{m \in J_\tau} \tilde{v}_m G_{j_m}(x^*, \tilde{t}^m)$$

$$+ \sum_{m \in K_\tau} \tilde{v}_m H_{k_m}(x^*, \tilde{s}^m) \Big]$$

$$\leqq \sum_{i=1}^{p} \tilde{u}_i [f_i(x^*) - \tilde{\lambda}_i g_i(x^*)] \text{ (by the primal feasibility of } x^*),$$

which contradicts (102). Therefore, we conclude that $\tilde{x} = x^*$ and $\varphi(x^*) = \tilde{\lambda}$.
(b) and (c) : The proofs are similar to that of part (a). □

# 7. General Remarks

The duality models $(DIV)$ and $(\tilde{D}IV)$ are two families of dual problems for $(P)$ whose members can easily be identified by appropriate choices of the partitioning sets $I_\mu$, $\mu \in D$, and $J_\tau$ and $K_\tau$, $\tau \in E$. These duality models and their special cases contain the semiinfinite analogues of various dual problems proposed and investigated previously for many types of nonlinear programming problems, as well as numerous other duality formulations that appear here for the first time.

Based on our observation, semiinfinite multiobjective programming is one of the most rapidly fast-expanding fields of research in mathematical fractional programming, while several second-order duality results exist in the related literature for several classes of mathematical programming problems with a finite number of constraints.

It is worth mentioning that all the Definitions 3.4 - 3.16 can be mildly generalized to the case of HA($\alpha$, $\beta$, $\gamma$, $\xi$, $\eta$, $\omega$, $\rho$, $\theta$, $m$)-V-invexity, where $m$ is a positive integer. For example, we consider the following:

**Definition 3.64.** The function $F$ is said to be (strictly) HA($\alpha$, $\beta$, $\gamma$, $\xi$, $\eta$, $\omega$, $\rho$, $\theta$, $m$) - V-sonvex at $x^* \in X$ if there exist functions $\alpha : X \times X \to \mathbb{R}$, $\beta : X \times X \to \mathbb{R}$, $\gamma_i : X \times X \to \mathbb{R}_+$, $\xi_i : X \times X \to \mathbb{R}_+$, $i \in \underline{p}$, $\eta, \omega : X \times X \to \mathbb{R}^n$, $\rho_i : X \times X \to \mathbb{R}$, $i \in \underline{p}$, and $\theta : X \times X \to \mathbb{R}^n$ such that for all $x \in X (x \neq x^*)$, $z \in \mathbb{R}^n$, and $i \in \underline{p}$,

$$\frac{1}{\alpha(x,x^*)} \gamma_i(x,x^*) \left( e^{\alpha(x,x^*)[F_i(x) - F_i(x^*) + \frac{1}{2}\langle \omega(x,x^*), \nabla^2 F_i(x^*) z \rangle]} - 1 \right) (>) \geqq$$
$$\frac{1}{\beta(x,x^*)} \langle \xi_i(x,x^*)[\nabla F_i(x^*) + \nabla^2 F_i(x^*) z], e^{\beta(x,x^*)\eta(x,x^*)} - \mathbf{1} \rangle + \rho_i(x,x^*) \|\theta(x,x^*)\|^m$$
$$\text{if } \alpha(x,x^*) \neq 0 \text{ and } \beta(x,x^*) \neq 0 \text{ for all } x \in X,$$

$$\frac{1}{\alpha(x,x^*)}\gamma_i(x,x^*)\left(e^{\alpha(x,x^*)[F_i(x)-F_i(x^*)+\frac{1}{2}\langle\omega(x,x^*),\nabla^2 F_i(x^*)z\rangle]}-1\right)(>)\geqq$$
$$\langle\xi_i(x,x^*)[\nabla F_i(x^*)+\nabla^2 F_i(x^*)z],\eta(x,x^*)\rangle+\rho_i(x,x^*)\|\theta(x,x^*)\|^m$$
$$\text{if } \alpha(x,x^*)\neq 0 \text{ and } \beta(x,x^*)=0 \text{ for all } x\in X,$$

$$\gamma_i(x,x^*)[F_i(x)-F_i(x^*)+\frac{1}{2}\langle\omega(x,x^*),\nabla^2 F_i(x^*)z\rangle](>)\geqq$$
$$\frac{1}{\beta(x,x^*)}\langle\xi_i(x,x^*)[\nabla F_i(x^*)+\nabla^2 F_i(x^*)z],e^{\beta(x,x^*)\eta(x,x^*)}-\mathbf{1}\rangle+\rho_i(x,x^*)\|\theta(x,x^*)\|^m$$
$$\text{if } \alpha(x,x^*)=0 \text{ and } \beta(x,x^*)\neq 0 \text{ for all } x\in X,$$

$$\gamma_i(x,x^*)[F_i(x)-F_i(x^*)+\frac{1}{2}\langle\omega(x,x^*),\nabla^2 F_i(x^*)z\rangle](>)\geqq$$
$$\langle\xi_i(x,x^*)[\nabla F_i(x^*)+\nabla^2 F_i(x^*)z],\eta(x,x^*)\rangle+\rho_i(x,x^*)\|\theta(x,x^*)\|^m$$
$$\text{if } \alpha(x,x^*)=0 \text{ and } \beta(x,x^*)=0 \text{ for all } x\in X,$$

where
$$\left(e^{\beta(x,x^*)\eta(x,x^*)}-\mathbf{1}\right)\equiv\left(e^{\beta(x,x^*)\eta_1(x,x^*)}-1,\ldots,e^{\beta(x,x^*)\eta_n(x,x^*)}-1\right)$$
and $\|\cdot\|$ is a norm on $\mathbb{R}^n$.

The function $F$ is said to be (strictly) HA($\alpha$, $\beta$, $\gamma$, $\xi$, $\eta$, $\omega$, $\rho$, $\theta$, $m$) - V-sonvex on $X$ if it is (strictly) HA($\alpha$, $\beta$, $\gamma$, $\xi$, $\eta$, $\omega$, $\rho$, $\theta$. $m$)-V-sonvex at each point $x^*\in X$.

**Definition 3.65.** The function $F$ is said to be (strictly) HA($\alpha$, $\beta$, $\gamma$, $\xi$, $\eta$, $\omega$, $\rho$, $\theta$, $m$) - V-pseudosonvex at $x^*\in X$ if there exist functions $\alpha: X\times X\to\mathbb{R}$, $\beta: X\times X\to\mathbb{R}$, $\gamma: X\times X\to\mathbb{R}_+$, $\xi_i: X\times X\to\mathbb{R}_+$, $i\in\underline{p}$, $\eta,\omega: X\times X\to\mathbb{R}^n$, $\rho: X\times X\to\mathbb{R}$, and $\theta: X\times X\to\mathbb{R}^n$ such that for all $x\in X(x\neq x^*)$ and $z\in\mathbb{R}^n$,

$$\frac{1}{\beta(x,x^*)}\left\langle\sum_{i=1}^{p}\xi_i(x,x^*)[\nabla F_i(x^*)+\nabla^2 F_i(x^*)z],e^{\beta(x,x^*)\eta(x,x^*)}-\mathbf{1}\right\rangle$$
$$\geqq-\rho(x,x^*)\|\theta(x,x^*)\|^m$$
$$\Rightarrow\quad\frac{1}{\alpha(x,x^*)}\gamma(x,x^*)\left(e^{\alpha(x,x^*)\sum_{i=1}^{p}[F_i(x)-F_i(x^*)+\frac{1}{2}\langle\omega(x,x^*),\nabla^2 F_i(x^*)z\rangle]}-1\right)(>)\geqq 0$$
$$\text{if } \alpha(x,x^*)\neq 0 \text{ and } \beta(x,x^*)\neq 0 \text{ for all } x\in X,$$

$$\left\langle\sum_{i=1}^{p}\xi_i(x,x^*)[\nabla F_i(x^*)+\nabla^2 F_i(x^*)z],\eta(x,x^*)\right\rangle\geqq-\rho(x,x^*)\|\theta(x,x^*)\|^m$$
$$\Rightarrow\quad\frac{1}{\alpha(x,x^*)}\gamma(x,x^*)\left(e^{\alpha(x,x^*)\sum_{i=1}^{p}[F_i(x)-F_i(x^*)+\frac{1}{2}\langle\omega(x,x^*),\nabla^2 F_i(x^*)z\rangle]}-1\right)(>)\geqq 0$$
$$\text{if } \alpha(x,x^*)\neq 0 \text{ and } \beta(x,x^*)=0 \text{ for all } x\in X,$$

$$\frac{1}{\beta(x,x^*)}\Big\langle \sum_{i=1}^{p} \xi_i(x,x^*)[\nabla F_i(x^*)+\nabla^2 F_i(x^*)z], e^{\beta(x,x^*)\eta(x,x^*)}-\mathbf{1}\Big\rangle$$
$$\geqq -\rho(x,x^*)\|\theta(x,x^*)\|^m$$
$$\Rightarrow \quad \gamma(x,x^*)\sum_{i=1}^{p}[F_i(x)-F_i(x^*)+\frac{1}{2}\langle\omega(x,x^*),\nabla^2 F_i(x^*)z\rangle](>)\geqq 0$$
$$\text{if } \alpha(x,x^*)=0 \text{ and } \beta(x,x^*)\neq 0 \text{ for all } x\in X,$$

$$\Big\langle \sum_{i=1}^{p} \xi_i(x,x^*)[\nabla F_i(x^*)+\nabla^2 F_i(x^*)z], \eta(x,x^*)\Big\rangle \geqq -\rho(x,x^*)\|\theta(x,x^*)\|^m$$
$$\Rightarrow \quad \gamma(x,x^*)\sum_{i=1}^{p}[F_i(x)-F_i(x^*)+\frac{1}{2}\langle\omega(x,x^*),\nabla^2 F_i(x^*)z\rangle](>)\geqq 0$$
$$\text{if } \alpha(x,x^*)=0 \text{ and } \beta(x,x^*)=0 \text{ for all } x\in X.$$

The function $F$ is said to be (strictly) $HA(\alpha,\beta,\gamma,\xi,\eta,\omega,\rho,\theta,m)$-V-pseudosonvex on $X$ if it is (strictly) $HA(\alpha,\beta,\gamma,\xi,\eta,\omega,\rho,\theta,m)$-V-pseudosonvex at each point $x^*\in X$.

**Definition 3.66.** The function $F$ is said to be (prestrictly) $HA(\alpha,\beta,\gamma,\xi,\eta,\omega,\rho,\theta,m)$-V-quasisonvex at $x^*\in X$ if there exist functions $\alpha: X\times X\to\mathbb{R}$, $\beta: X\times X\to\mathbb{R}$, $\gamma: X\times X\to\mathbb{R}_+$, $\xi_i: X\times X\to\mathbb{R}_+$, $i\in\underline{p}$, $\eta,\omega: X\times X\to\mathbb{R}^n$, $\rho: X\times X\to\mathbb{R}$, and $\theta: X\times X\to\mathbb{R}^n$ such that for all $x\in X$ and $z\in\mathbb{R}^n$,

$$\frac{1}{\alpha(x,x^*)}\gamma(x,x^*)\Big(e^{\alpha(x,x^*)\sum_{i=1}^{p}[F_i(x)-F_i(x^*)+\frac{1}{2}\langle\omega(x,x^*),\nabla^2 F_i(x^*)z\rangle]}-1\Big)(<)\leqq 0$$
$$\Rightarrow \quad \frac{1}{\beta(x,x^*)}\Big\langle \sum_{i=1}^{p} \xi_i(x,x^*)[\nabla F_i(x^*)+\nabla^2 F_i(x^*)z], e^{\beta(x,x^*)\eta(x,x^*)}-\mathbf{1}\Big\rangle$$
$$\leqq -\rho(x,x^*)\|\theta(x,x^*)\|^m$$
$$\text{if } \alpha(x,x^*)\neq 0 \text{ and } \beta(x,x^*)\neq 0 \text{ for all } x\in X,$$

$$\frac{1}{\alpha(x,x^*)}\gamma(x,x^*)\Big(e^{\alpha(x,x^*)\sum_{i=1}^{p}[F_i(x)-F_i(x^*)+\frac{1}{2}\langle\omega(x,x^*),\nabla^2 F_i(x^*)z\rangle]}-1\Big)(<)\leqq 0$$
$$\Rightarrow \quad \Big\langle \sum_{i=1}^{p} \xi_i(x,x^*)[\nabla F_i(x^*)+\nabla^2 F_i(x^*)z], \eta(x,x^*)\Big\rangle \leqq -\rho(x,x^*)\|\theta(x,x^*)\|^m$$
$$\text{if } \alpha(x,x^*)\neq 0 \text{ and } \beta(x,x^*)=0 \text{ for all } x\in X,$$

$$\gamma(x,x^*)\sum_{i=1}^{p}[F_i(x)-F_i(x^*)+\frac{1}{2}\langle\omega(x,x^*),\nabla^2F_i(x^*)z\rangle](<)\leqq 0$$

$$\Rightarrow \frac{1}{\beta(x,x^*)}\Big\langle\sum_{i=1}^{p}\xi_i(x,x^*)[\nabla F_i(x^*)+\nabla^2F_i(x^*)z],e^{\beta(x,x^*)\eta(x,x^*)}-\mathbf{1}\Big\rangle$$
$$\leqq -\rho(x,x^*)\|\theta(x,x^*)\|^m$$
$$\text{if } \alpha(x,x^*)=0 \text{ and } \beta(x,x^*)\neq 0 \text{ for all } x\in X,$$

$$\gamma(x,x^*)\sum_{i=1}^{p}[F_i(x)-F_i(x^*)+\frac{1}{2}\langle\omega(x,x^*),\nabla^2F_i(x^*)z\rangle](<)\leqq 0$$

$$\Rightarrow \Big\langle\sum_{i=1}^{p}\xi_i(x,x^*)[\nabla F_i(x^*)+\nabla^2F_i(x^*)z],\eta(x,x^*)\Big\rangle\leqq -\rho(x,x^*)\|\theta(x,x^*)\|^m$$
$$\text{if } \alpha(x,x^*)=0 \text{ and } \beta(x,x^*)=0 \text{ for all } x\in X.$$

The function $F$ is said to be (prestrictly) $HA(\alpha,\beta,\gamma,\xi,\eta,\omega,\rho,\theta,m)$-V-quasisonvex on $X$ if it is (prestrictly) $HA(\alpha,\beta,\gamma,\xi,\eta,\omega,\rho,\theta,m)$-V-quasisonvex at each point $x^*\in X$.

**Definition 3.67.** The function $F$ is said to be (strictly) $HA(\alpha,\beta,\gamma,\xi,\eta,\rho,\theta,m)$-V-sonvex at $x^*\in X$ if there exist functions $\alpha: X\times X\to\mathbb{R}$, $\beta: X\times X\to\mathbb{R}$, $\gamma_i: X\times X\to\mathbb{R}_+$, $\xi_i: X\times X\to\mathbb{R}_+$, $i\in \underline{p}$, $\eta: X\times X\to\mathbb{R}^n$, $\rho_i: X\times X\to\mathbb{R}$, $i\in\underline{p}$, and $\theta: X\times X\to\mathbb{R}^n$ such that for all $x\in X(x\neq x^*)$, $z\in\mathbb{R}^n$, and $i\in\underline{p}$,

$$\frac{1}{\alpha(x,x^*)}\gamma_i(x,x^*)\Big(e^{\alpha(x,x^*)[F_i(x)-F_i(x^*)+\frac{1}{2}\langle z,\nabla^2F_i(x^*)z\rangle]}-1\Big)(>)\geqq$$
$$\frac{1}{\beta(x,x^*)}\langle\xi_i(x,x^*)[\nabla F_i(x^*)+\nabla^2F_i(x^*)z],e^{\beta(x,x^*)\eta(x,x^*)}-\mathbf{1}\rangle+\rho_i(x,x^*)\|\theta(x,x^*)\|^m$$
$$\text{if } \alpha(x,x^*)\neq 0 \text{ and } \beta(x,x^*)\neq 0 \text{ for all } x\in X,$$

$$\frac{1}{\alpha(x,x^*)}\gamma_i(x,x^*)\Big(e^{\alpha(x,x^*)[F_i(x)-F_i(x^*)+\frac{1}{2}\langle z,\nabla^2F_i(x^*)z\rangle]}-1\Big)(>)\geqq$$
$$\big\langle\xi_i(x,x^*)[\nabla F_i(x^*)+\nabla^2F_i(x^*)z],\eta(x,x^*)\big\rangle+\rho_i(x,x^*)\|\theta(x,x^*)\|^m$$
$$\text{if } \alpha(x,x^*)\neq 0 \text{ and } \beta(x,x^*)=0 \text{ for all } x\in X,$$

$$\gamma_i(x,x^*)[F_i(x)-F_i(x^*)+\frac{1}{2}\langle z,\nabla^2F_i(x^*)z\rangle](>)\geqq$$
$$\frac{1}{\beta(x,x^*)}\langle\xi_i(x,x^*)[\nabla F_i(x^*)+\nabla^2F_i(x^*)z],e^{\beta(x,x^*)\eta(x,x^*)}-\mathbf{1}\rangle+\rho_i(x,x^*)\|\theta(x,x^*)\|^m$$
$$\text{if } \alpha(x,x^*)=0 \text{ and } \beta(x,x^*)\neq 0 \text{ for all } x\in X,$$

$$\gamma_i(x,x^*)[F_i(x)-F_i(x^*)+\frac{1}{2}\langle z,\nabla^2F_i(x^*)z\rangle](>)\geqq$$
$$\big\langle\xi_i(x,x^*)[\nabla F_i(x^*)+\nabla^2F_i(x^*)z],\eta(x,x^*)\big\rangle+\rho_i(x,x^*)\|\theta(x,x^*)\|^m$$
$$\text{if } \alpha(x,x^*)=0 \text{ and } \beta(x,x^*)=0 \text{ for all } x\in X,$$

where

$$\left(e^{\beta(x,x^*)\eta(x,x^*)} - \mathbf{1}\right) \equiv \left(e^{\beta(x,x^*)\eta_1(x,x^*)} - 1, \ldots, e^{\beta(x,x^*)\eta_n(x,x^*)} - 1\right)$$

and $\|\cdot\|$ is a norm on $\mathbb{R}^n$.

The function $F$ is said to be (strictly) $HA(\alpha,\beta,\gamma,\xi,\eta,\rho,\theta,m)$-V-sonvex on $X$ if it is (strictly) HA($\alpha$, $\beta$, $\gamma$, $\xi$, $\eta$, $\rho$, $\theta$, $m$) - V-sonvex at each point $x^* \in X$.

**Definition 3.68.** The function $F$ is said to be (strictly) $HA(\alpha,\beta,\gamma,\xi,\eta,\rho,\theta,m)$-V-pseudosonvex at $x^* \in X$ if there exist functions $\alpha : X \times X \to \mathbb{R}$, $\beta : X \times X \to \mathbb{R}$, $\gamma : X \times X \to \mathbb{R}_+$, $\xi_i : X \times X \to \mathbb{R}_+$, $i \in \underline{p}$, $\eta : X \times X \to \mathbb{R}^n$, $\rho : X \times X \to \mathbb{R}$, and $\theta : X \times X \to \mathbb{R}^n$ such that for all $x \in X (x \neq x^*)$ and $z \in \mathbb{R}^n$,

$$\frac{1}{\beta(x,x^*)}\left\langle \sum_{i=1}^{p} \xi_i(x,x^*)[\nabla F_i(x^*) + \nabla^2 F_i(x^*)z], e^{\beta(x,x^*)\eta(x,x^*)} - \mathbf{1}\right\rangle$$
$$\geqq -\rho(x,x^*)\|\theta(x,x^*)\|^m$$
$$\Rightarrow \quad \frac{1}{\alpha(x,x^*)}\gamma(x,x^*)\left(e^{\alpha(x,x^*)\sum_{i=1}^{p}[F_i(x) - F_i(x^*) + \frac{1}{2}\langle z, \nabla^2 F_i(x^*)z\rangle]} - 1\right)(>) \geqq 0$$
$$\text{if } \alpha(x,x^*) \neq 0 \text{ and } \beta(x,x^*) \neq 0 \text{ for all } x \in X,$$

$$\left\langle \sum_{i=1}^{p} \xi_i(x,x^*)[\nabla F_i(x^*) + \nabla^2 F_i(x^*)z], \eta(x,x^*)\right\rangle \geqq -\rho(x,x^*)\|\theta(x,x^*)\|^m$$
$$\Rightarrow \quad \frac{1}{\alpha(x,x^*)}\gamma(x,x^*)\left(e^{\alpha(x,x^*)\sum_{i=1}^{p}[F_i(x) - F_i(x^*) + \frac{1}{2}\langle z, \nabla^2 F_i(x^*)z\rangle]} - 1\right)(>) \geqq 0$$
$$\text{if } \alpha(x,x^*) \neq 0 \text{ and } \beta(x,x^*) = 0 \text{ for all } x \in X,$$

$$\frac{1}{\beta(x,x^*)}\left\langle \sum_{i=1}^{p} \xi_i(x,x^*)[\nabla F_i(x^*) + \nabla^2 F_i(x^*)z], e^{\beta(x,x^*)\eta(x,x^*)} - \mathbf{1}\right\rangle$$
$$\geqq -\rho(x,x^*)\|\theta(x,x^*)\|^m$$
$$\Rightarrow \quad \gamma(x,x^*)\sum_{i=1}^{p}[F_i(x) - F_i(x^*) + \frac{1}{2}\langle z, \nabla^2 F_i(x^*)z\rangle](>) \geqq 0$$
$$\text{if } \alpha(x,x^*) = 0 \text{ and } \beta(x,x^*) \neq 0 \text{ for all } x \in X,$$

$$\left\langle \sum_{i=1}^{p} \xi_i(x,x^*)[\nabla F_i(x^*) + \nabla^2 F_i(x^*)z], \eta(x,x^*)\right\rangle \geqq -\rho(x,x^*)\|\theta(x,x^*)\|^m$$
$$\Rightarrow \quad \gamma(x,x^*)\sum_{i=1}^{p}[F_i(x) - F_i(x^*) + \frac{1}{2}\langle z, \nabla^2 F_i(x^*)z\rangle](>) \geqq 0$$
$$\text{if } \alpha(x,x^*) = 0 \text{ and } \beta(x,x^*) = 0 \text{ for all } x \in X.$$

The function $F$ is said to be (strictly) $HA(\alpha,\beta,\gamma,\xi,\eta,\rho,\theta,m)$-V-pseudosonvex on $X$ if it is (strictly) HA($\alpha$, $\beta$, $\gamma$, $\xi$, $\eta$, $\rho$, $\theta$, $m$) - V-pseudosonvex at each point $x^* \in X$.

**Definition 3.69.** The function $F$ is said to be (prestrictly) HA($\alpha$, $\beta$, $\gamma$, $\xi$, $\eta$, $\rho$, $\theta$, $m$) - V-quasisonvex at $x^* \in X$ if there exist functions $\alpha : X \times X \to \mathbb{R}$, $\beta : X \times X \to \mathbb{R}$, $\gamma : X \times X \to \mathbb{R}_+$, $\xi_i : X \times X \to \mathbb{R}_+$, $i \in \underline{p}$, $\eta : X \times X \to \mathbb{R}^n$, $\rho : X \times X \to \mathbb{R}$, and $\theta : X \times X \to \mathbb{R}^n$ such that for all $x \in X$ and $z \in \mathbb{R}^n$,

$$\frac{1}{\alpha(x,x^*)}\gamma(x,x^*)\Big(e^{\alpha(x,x^*)\sum_{i=1}^{p}[F_i(x)-F_i(x^*)+\frac{1}{2}\langle z,\nabla^2F_i(x^*)z\rangle]}-1\Big)(<)\leqq 0$$
$$\Rightarrow \frac{1}{\beta(x,x^*)}\Big\langle \sum_{i=1}^{p}\xi_i(x,x^*)[\nabla F_i(x^*)+\nabla^2F_i(x^*)z], e^{\beta(x,x^*)\eta(x,x^*)}-\mathbf{1}\Big\rangle$$
$$\leqq -\rho(x,x^*)\|\theta(x,x^*)\|^m$$
$$\text{if } \alpha(x,x^*)\neq 0 \text{ and } \beta(x,x^*)\neq 0 \text{ for all } x\in X,$$

$$\frac{1}{\alpha(x,x^*)}\gamma(x,x^*)\Big(e^{\alpha(x,x^*)\sum_{i=1}^{p}[F_i(x)-F_i(x^*)+\frac{1}{2}\langle z,\nabla^2F_i(x^*)z\rangle]}-1\Big)(<)\leqq 0$$
$$\Rightarrow \Big\langle \sum_{i=1}^{p}\xi_i(x,x^*)[\nabla F_i(x^*)+\nabla^2F_i(x^*)z], \eta(x,x^*)\Big\rangle \leqq -\rho(x,x^*)\|\theta(x,x^*)\|^m$$
$$\text{if } \alpha(x,x^*)\neq 0 \text{ and } \beta(x,x^*)=0 \text{ for all } x\in X,$$

$$\gamma(x,x^*)\sum_{i=1}^{p}[F_i(x)-F_i(x^*)+\frac{1}{2}\langle z,\nabla^2F_i(x^*)z\rangle](<)\leqq 0$$
$$\Rightarrow \frac{1}{\beta(x,x^*)}\Big\langle \sum_{i=1}^{p}\xi_i(x,x^*)[\nabla F_i(x^*)+\nabla^2F_i(x^*)z], e^{\beta(x,x^*)\eta(x,x^*)}-\mathbf{1}\Big\rangle$$
$$\leqq -\rho(x,x^*)\|\theta(x,x^*)\|^m$$
$$\text{if } \alpha(x,x^*)=0 \text{ and } \beta(x,x^*)\neq 0 \text{ for all } x\in X,$$

$$\gamma(x,x^*)\sum_{i=1}^{p}[F_i(x)-F_i(x^*)+\frac{1}{2}\langle z,\nabla^2F_i(x^*)z\rangle](<)\leqq 0$$
$$\Rightarrow \Big\langle \sum_{i=1}^{p}\xi_i(x,x^*)[\nabla F_i(x^*)+\nabla^2F_i(x^*)z], \eta(x,x^*)\Big\rangle \leqq -\rho(x,x^*)\|\theta(x,x^*)\|^m$$
$$\text{if } \alpha(x,x^*)=0 \text{ and } \beta(x,x^*)=0 \text{ for all } x\in X.$$

The function $F$ is said to be (prestrictly) HA($\alpha$, $\beta$, $\gamma$, $\xi$, $\eta$, $\rho$, $\theta$, $m$) - V-quasisonvex on $X$ if it is (prestrictly) HA($\alpha$, $\beta$, $\gamma$, $\xi$, $\eta$, $\rho$, $\theta$, $m$) - V-quasisonvex at each point $x^* \in X$.

**Definition 3.70.** The function $F$ is said to be (strictly) first-order HA($\alpha$, $\beta$, $\gamma$, $\xi$, $\eta$, $\rho$, $\theta$, $m$) - V-invex at $x^* \in X$ if there exist functions $\alpha : X \times X \to \mathbb{R}$, $\beta : X \times X \to \mathbb{R}$, $\gamma_i$ :

$X \times X \to \mathbb{R}_+$, $\xi_i : X \times X \to \mathbb{R}_+ \setminus \{0\}$, $i \in \underline{p}$, $\eta : X \times X \to \mathbb{R}^n$, $\rho_i : X \times X \to \mathbb{R}$, $i \in \underline{p}$, and $\theta : X \times X \to \mathbb{R}^n$ such that for all $x \in X (x \neq x^*)$ and $i \in \underline{p}$,

$$\frac{1}{\alpha(x,x^*)}\gamma_i(x,x^*)\left(e^{\alpha(x,x^*)[F_i(x)-F_i(x^*)]}-1\right)(>)$$
$$\geqq \frac{1}{\beta(x,x^*)}\langle \xi_i(x,x^*)\nabla F_i(x^*), e^{\beta(x,x^*)\eta(x,x^*)}-\mathbf{1}\rangle$$
$$+\rho_i(x,x^*)\|\theta(x,x^*)\|^m \quad \text{if } \alpha(x,x^*) \neq 0 \text{ and } \beta(x,x^*) \neq 0 \text{ for all } x \in X,$$

$$\frac{1}{\alpha(x,x^*)}\gamma_i(x,x^*)\left(e^{\alpha(x,x^*)[F_i(x)-F_i(x^*)]}-1\right)(>) \geqq \langle \xi_i(x,x^*)\nabla F_i(x^*), \eta(x,x^*)\rangle$$
$$+\rho_i(x,x^*)\|\theta(x,x^*)\|^m \quad \text{if } \alpha(x,x^*) \neq 0 \text{ and } \beta(x,x^*) = 0 \text{ for all } x \in X,$$

$$\gamma_i(x,x^*)[F_i(x)-F_i(x^*)](>) \geqq \frac{1}{\beta(x,x^*)}\langle \xi_i(x,x^*)\nabla F_i(x^*), e^{\beta(x,x^*)\eta(x,x^*)}-\mathbf{1}\rangle$$
$$+\rho_i(x,x^*)\|\theta(x,x^*)\|^m \quad \text{if } \alpha(x,x^*) = 0 \text{ and } \beta(x,x^*) \neq 0 \text{ for all } x \in X,$$

$$\gamma_i(x,x^*)[F_i(x)-F_i(x^*)](>) \geqq \langle \xi_i(x,x^*)\nabla F_i(x^*), \eta(x,x^*)\rangle + \rho_i(x,x^*)\|\theta(x,x^*)\|^m$$
$$\text{if } \alpha(x,x^*) = 0 \text{ and } \beta(x,x^*) = 0 \text{ for all } x \in X,$$

where $\|\cdot\|$ is a norm on $\mathbb{R}^n$ and

$$\left(e^{\beta(x,x^*)\eta(x,x^*)}-\mathbf{1}\right) \equiv \left(e^{\beta(x,x^*)\eta_1(x,x^*)}-1,\ldots,e^{\beta(x,x^*)\eta_n(x,x^*)}-1\right).$$

The function $F$ is said to be (strictly) HA($\alpha$, $\beta$, $\gamma$, $\xi$, $\eta$, $\rho$, $\theta$, $m$) - V-invex on $X$ if it is (strictly) $HA(\alpha,\beta,\gamma,\xi,\eta,\rho,\theta,m)$-V-invex at each point $x^* \in X$.

# Bibliography


[1] I. Ahmad and Z. Husain, Second order (F,$\alpha$,$\rho$,d)-convexity and duality in multiobjective programming, *Inform. Sci.* **176** (2006), 3094 - 3103.

[2] T. Antczak, The notion of V-$r$-invexity in differentiable multiobjective programming, *J. Appl. Anal.* **11** (2005), 63 - 79.

[3] X. Chen, Sufficient conditions and duality for a class of multiobjective fractional programming problems with higher-order $(F,\alpha,\rho,d)$-convexity, *J. Appl. Math. Comput.* **28** (2008), 107 - 121.

[4] M. A. Hanson, Second order invexity and duality in mathematical programming, *Opsearch* **30** (1993), 313 - 320.

[5] O. L. Mangasarian, Second- and higher-order duality theorems in nonlinear programming, *J. Math. Anal. Appl.* **51** (1975), 607 - 620.

[6] B. Mond, Second order duality for nonlinear programs, *Opsearch* **11** (1974), 90 - 99.

[7] B. Mond and T. Weir, Generalized concavity and duality, in *Generalized Concavity in Optimization and Economics* (S. Schaible and W. T. Ziemba, eds.), Academic Press, New York, 1981, pp. 263 - 279.

[8] B. Mond and J. Zhang, Duality for multiobjective programming involving second-order V-invex functions, in *Proceedings of the Optimization Miniconference II* (B. M. Glover and V. Jeyakumar, eds.), University of New South Wales, Sydney, Australia, 1995, pp. 89 - 100.

[9] B. Mond and J. Zhang, Higher order invexity and duality in mathematical programming, in *Generalized Convexity, Generalized Monotonicity: Recent Results* (J. P. Crouzeix, et al., eds.), Kluwer Academic Publishers, printed in the Netherlands, 1998, pp. 357 - 372.

[10] R. U. Verma, *Semi-Infinite Fractional Programming*, Springer Nature, Singapore, 2017.

[11] G. J. Zalmai, Generalized second-order $(\mathscr{F},\beta,\phi,\rho,\theta)$-univex functions and parametric duality models in semiinfinite discrete minmax fractional programming, *Adv. Nonlin. Var. Ineq.* **15** (2012), 63 - 91.

[12] G. J. Zalmai, Hanson-Antczak-type generalized $(\alpha,\beta,\gamma,\xi,\eta,\rho,\theta)$-V-invex functions in semiinfinite multiobjective fractional programming, Part I: Sufficient efficiency conditions, *Advances in Nonlinear Variational Inequalities* **16** (1) (2013), 91 - 114.

[13] G. J. Zalmai, Hanson-Antczak-type generalized $(\alpha,\beta,\gamma,\xi,\eta,\rho,\theta)$-V-invex functions in semiinfinite multiobjective fractional programming, Part II: First-order parametric duality models, *Advances in Nonlinear Variational Inequalities* **16** (2) (2017), 61 - 90.

[14] G. J. Zalmai, Hanson-Antczak-type generalized $(\alpha,\beta,\gamma,\xi,\eta,\rho,\theta)$-V-invex functions in semiinfinite multiobjective fractional programming, Part III: Second-order parametric duality models, *Advances in Nonlinear Variational Inequalities* **16** (2) (2017), 91 - 126.

[15] G. J. Zalmai and Q. Zhang, Global semiparametric sufficient efficiency conditions for semiinfinite multiobjective fractional programming problems involving generalized $(\alpha,\eta,\rho)$-V-invex functions, *Southeast Asian Bull. Math.* **32** (2008), 573 - 599.

# Chapter 4

# Fractional Integral Type Programming

This chapter deals with presenting hybrid minmax fractional integral type programming problems by applying a class of $(\phi, b, \beta, \eta, \omega)$- univexities, which is equivalent to minimizing a maximum of several time dependent ratios involving integral expressions. First, some optimality conditions based on $(\phi, b, \beta, \eta, \omega)$- univexities are established, and then some classes of Wolfe type dual models and Mond-Weir type dual models are explored to the context of developing a class of related duality results, which are in terms of weak, strong and strictly converse duality theorems based on $(\phi, b, \beta, \eta, \omega)$- univexity frameworks.

## 1. Introduction

The fractional integral type programming is one of the most application-enhanced fast-expanding areas of optimization which features in several types of real-world problems. It can be applied to different branches of engineering (including multitime multiobjective mechanical engineering problems) as well as to economics to minimize a ratio of functions between given periods of time. Furthermore, it can be utilized as a resource in order to measure the efficiency or productivity of a system. In such type of problems the objective function is given as a ratio of functions. In this chapter, we consider a problem that deals with minimizing a maximum of several time dependent ratios involving integral expressions. Among researchers who have investigated these types of problems are Chandra et al. [1], Crouzeix et al. [2], Mond et al. [5], Mond and Husain [6], and others, while we generalize the results in Mishra et al. [4] to a larger class nonconvex minmax fractional optimization problems, that is, to a class of hybrid minmax fractional integral type problems in which the involved functionals are $(\phi, b, \beta, \eta, \omega)$- univex. The obtained results extend (and unify) the works of Mishra et al. [4], Lai et al. [3], and others. As our problem involves a state function $x(t)$ in the integrand of the integrals, the obtained results also relate to optimal control problems. For more details, we refer the reader to works [1 - 10].

Consider the following hybrid minmax fractional variational problem:

(P) $$\min_x \left\{ \sum_{1\le i\le p} \max \frac{\int_a^b f^i(t,x,\dot{x})dt}{\int_a^b g^i(t,x,\dot{x})dt} \right\}$$

subject to $x \in PS(T,R^n), x(a) = \alpha, x(b) = \beta,$

$$\int_a^b h^j(t,x,\dot{x})dt \le 0, j \in \underline{m} \equiv \{1,2,\ldots,m\},$$

where functions $f^i, g^i, i \in \underline{p}$ and $h^j, j \in \underline{m}$ are continuous in $t$ and have continuous partial derivatives with respect to $x$ and $\dot{x}$; $T = [a,b]$ denotes the time space, and $PS(T,R^n)$ stands for the state space of all piecewise smooth functions $x : T \to R^n$ with norm defined by

$$\|x\| = \|x\|_\infty + \|Dx\|_\infty$$

and
$D$ is the differential operator on $PS(T,R^n)$ defined by

$$y = Dx \text{ if and only if } x(t) = x(a) + \int_a^b y(s)ds.$$

Thus $D = \dfrac{d}{dt}$ except at the point of discontinuity. Without loss of any generality, we may assume throughout the paper

$$\int_a^b g^i(t,x,\dot{x}) > 0, \int_a^b f^i(t,x,\dot{x}) \ge 0 \text{ for each } i \in \underline{p}.$$

Let $F_p$ denote the set of all feasible solutions to $(P)$.

## 2. Preliminaries

For $x \in PS(T,R^n)$, let us define as follows:

$$F_i(x) = \int_a^b f^i(t,x,\dot{x})dt,$$

$$G_i(x) = \int_a^b g^i(t,x,\dot{x})dt$$

for $i \in \underline{p}$ and

$$H_j(x) = \int_a^b h^j(t,x,\dot{x})dt$$

for $j \in \underline{m}$.

Based on the assumptions imposed on the functions $f^i$, $g^i$ and $h^j$ constituting the considered hybrid minmax fractional integral type problem $(P)$, it follows that the functionals $F = (F_1,...,F_p)$, $G = (G_1,..,G_p)$, $H = (H_1,...,H_m)$ are Fréchet differentiable on $PS(T,R^n)$. In order to simplify the symbols in problem $(P)$, we scalarize functionals as follows:

$$\Phi(x,y) = \langle y, F(x)\rangle = \sum_{i=1}^{p} y_i F_i(x) = \sum_{i=1}^{p} y_i \int_a^b f^i(t,x,\dot{x})dt,$$

$$\Psi(x,y) = \langle y, G(x)\rangle = \sum_{i=1}^{p} y_i G_i(x) = \sum_{i=1}^{p} y_i \int_a^b g^i(t,x,\dot{x})dt,$$

$$\Omega(x,y) = \langle z, H(x)\rangle = \sum_{i=1}^{p} z_j H_j(x) = \sum_{i=1}^{p} z_j \int_a^b h^j(t,x,\dot{x})dt,$$

where $y \in I = \left\{ y \in R^p_+ | \sum_{i=1}^{p} y_i = 1 \right\}$ and $z \in R^m_+$. Then for any feasible solution $x$ of $(P)$, the objective fractional function can be represented by

$$\varphi(x) = \max_{i \in \underline{p}} \frac{F_i(x)}{G_i(x)} = \max_{y \in I} \frac{\langle y, F(x)\rangle}{\langle y, G(x)\rangle} = \max_{y \in I} \frac{\Phi(y,x)}{\Psi(y,x)}. \tag{4.1}$$

The problem $(P)$ is equivalent to

$$(\tilde{P}) \min_{x \in PS} \max_{y \in I} \frac{\Phi(y,x)}{\Psi(y,x)} = \min_{x} \left( \max_{i \in \underline{p}} \frac{F_i(x)}{G_i(x)} \right)$$

subject to $x \in PS(T,R^n)$, $x(a) = \alpha$, $x(b) = \beta$ and $H(x) \leq 0$.
It is equivalent to the parametric minimization problem:

$(EP_v)$ minimize $q(v)$

subject to $F_i(x) - vG_i(x) \leq q$, $i \in \underline{p}$, and $H_j(x) \leq 0$, $j \in \underline{m}$,

$$x \in PS(T,R^n), x(a) = \alpha, x(b) = \beta.$$

From $(EP_v)$, one can reduce to the optimal solution $x^*$ for $(P)$ with its optimal value $v^*$ which is given by

$$v^* = \varphi(x^*) = \max_{i \in \underline{p}} \frac{F^i(x^*)}{G^i(x^*)} = \max_{y \in I} \frac{\Phi(x^*, y)}{\Psi(x^*, y)} = \frac{\Phi(x^*, y^*)}{\Psi(x^*, y^*)}. \tag{4.2}$$

The concept used here for solution of $(EP_v)$, coincides with finding the minimax solution $(x^*, y^*)$ of the Lagrangian

$$\begin{aligned} L(x, y; x, z) &= \langle y, F(x) \rangle - v \langle y, G(x) \rangle + \langle z, H(x) \rangle \\ - & \Phi(x, y) - v\Psi(x, y) + \Omega(x, z), \end{aligned} \tag{4.3}$$

with multipliers $v^* \in R_+$ and $z^* \in R_+^m$. The minimax solution $(x^*, y^*)$ is given in the equation

$$L_x^{'}(x, y)\xi = 0 \text{ for all } \xi \in C(T, R^n).$$

We need the following auxiliary results in order to establish our results on hand regarding problem $(P)$:

**Lemma 4.1.** [3] Let $x^*$ be an optimal solution of $(P)$. Then there exist Lagrangian multipliers $y^* \in I \subset R_+^p$ and $z^* \in R_+^m$ such that the Kuhn-Tucker type conditions hold for the Lagrangian (6.3):

$$L_x^{'}(x^*, y^*; v^*, z^*) = 0, \tag{4.4}$$

$$\Omega(x^*, z^*) = 0 \tag{4.5}$$

where $v^* = \dfrac{\Phi(x^*, y^*)}{\Psi(x^*, y^*)}$ given in (6.2) is the optimal value of problem (P), and the equation $L_x^{'}(\cdot, \cdot) = 0$ is the expressed by

$$\Psi(x^*, y^*) \left[\Phi^{'}(x^*, y^*) + \Omega^{'}(x^*, z^*)\right] - \Phi(x^*, y^*)\Psi^{'}(x^*, y^*) = 0. \tag{4.6}$$

**Lemma 4.2.** [3] Let $x^*$ be an optimal solution of $(P)$. Then there exist $y^* \in I \subset R_+^p$, multipliers $v^* \in R$ and $z^* \in R_+^m$ such that $(x^*, y^*, z^*)$ satisfies

$$\Psi(x^*, y^*)\Phi_1^{'}(x^*, y^*) - \Phi(x^*, y^*)\Psi_1^{'}(x^*, y^*) + \Psi(x^*, y^*)\Omega_1^{'}(x^*, z^*) = 0, \tag{4.7}$$

$$\Omega(x^*, z^*) = 0, \tag{4.8}$$

and obtain the optimal value by

$$K(x^*) = \frac{\Phi(x^*, y^*)}{\Psi(x^*, y^*)} = \max_{y \in I} \frac{\Phi(x^*, y)}{\Psi(x^*, y)} \left( = \min_{x} \max_{1 \le i \le p} \frac{F^i(x)}{G^i(x)} \right). \tag{4.9}$$

Now we consider $(\phi, b, \beta, \eta)$- univexity concepts, which encompass most of the concepts available in the literature. Let $x \in PS(T, R^n)$ and $f \in C^1(T \times R^n \times R^n)$. Define a functional $J : PS(T,R^n) \to R$ by $J(x) = \int_a^b f(t,x,\dot{x})dt$. We assume that the boundary points $x(a)$ and $x(b)$ are fixed. Consider the admissible vectors variations $w \in C([a,b],R^n)$ vanishing at the boundary points. Then the differential of $J$ is a linear functional on $C([a,b],R^n)$,

$$\begin{aligned} J'(x)w &= \frac{d}{d\alpha}\int_a^b f(t,x+\alpha w,\dot{x}+\alpha\dot{w})dt|_{\alpha=0} \\ &= \int_a^b [f_x(t,x,\dot{x})w(t)+f_{\dot{x}}(t,x,\dot{x})\dot{w}(t)]dt \\ &= \int_a^b [f_x(t,x,\dot{x})-Df_{\dot{x}}(t,x,\dot{x})]\,w(t)dt + f_{\dot{x}}(t,x,\dot{x})\dot{w}(x)|_a^b \\ &= \int_a^b [f_x(t,x,\dot{x})-Df_{\dot{x}}(t,x,\dot{x})]\,w(t)dt. \end{aligned}$$

Thus, we have

$$J'(x)w = \int_a^b [f_x(t,x,\dot{x})-Df_{\dot{x}}(t,x,\dot{x})]\,w(t)dt$$

for all $w \in C([a,b],R^n), w(a) = 0 = w(b)$, where $D = \frac{d}{dt}$.

We further assume that $b : PS(T,R^n) \times PS(T,R^n) \to R_+$, $\beta : PS(T,R^n) \times PS(T,R^n) \to R_+$, $\phi : R \to R$, and let $\eta : PS(T,R^n) \times PS(T,R^n) \to C(T,R^n)$, ($\eta(x,u) = 0$ only if $x = u$) and $\omega : PS(T,R^n) \times PS(T,R^n) \to C(T,R^n)$, ($\omega(x,u) = 0$ only if $x = u$) be vector functions.

**Definition 4.3.** A differentiable function J is said to be ($\phi$, $b$, $\beta$, $\eta$, $\omega$) - univex at $u \in PS(T,R^n)$ with respect to $b$, $\beta$, $\eta$ and $\omega$ if

$$b(x,u)\phi[J(x)-J(u)] \geq \beta(x,u)\left(\frac{1}{2}\right)[J'(u)\eta(x,u)+J'(u)\omega(x,u)].$$

**Definition 4.4.** A differentiable function $J$ is said to be ($\phi$, $b$, $\beta$, $\eta$, $\omega$)- pseudo-univex at $u \in PS(T,R^n)$ with respect to $b$, $\beta$,$\eta$, $\omega$ and $\phi$ if

$$\beta(x,u)\left(\frac{1}{2}\right)[J'(u)\eta(x,u)+J'(u)\omega(x,u] \geq 0 \Rightarrow b(x,u)\phi[J(x)-J(u)] \geq 0,$$

equivalently,

$$b(x,u)\phi[J(x)-J(u)] < 0 \Rightarrow \beta(x,u)\left(\frac{1}{2}\right)[J'(u)\eta(x,u)+J'(u)\omega(x,u] < 0.$$

**Definition 4.5.** A differentiable function $J$ is said to be strictly ($\phi$, $b$, $\beta$, $\eta$, $\omega$)- pseudo-univex at $u \in PS(T,R^n)$ with respect to $\eta$, $\omega$ and $\phi$ if

$$\beta(x,u)\left(\frac{1}{2}\right)[J'(u)\eta(x,u)+J'(u)\omega(x,u] \geq 0 \Rightarrow b(x,u)\phi[J(x)-J(u)] > 0,$$

equivalently,

$$b(x,u)\phi[J(x)-J(u)] \leq 0 \Rightarrow \beta(x,u)(\frac{1}{2})[J'(u)\eta(x,u)+J'(u)\omega(x,u] < 0.$$

**Definition 4.6.** A differentiable function $J$ is said to be ($\phi$, $b$, $\beta$, $\eta$, $\omega$)- quasi-univex at $u \in PS(T,R^n)$ with respect to $b$, $\beta$, $\eta$, $\omega$ and $\phi$if

$$b(x,u)\phi[J(x)-J(u)] \leq 0 \Rightarrow \beta(x,u)(\frac{1}{2})[J'(u)\eta(x,u)+J'(u)\omega(x,u] \leq 0,$$

equivalently,

$$\beta(x,u)(\frac{1}{2})[J'(u)\eta(x,u)+J'(u)\omega(x,u] > 0 \Rightarrow b(x,u)\phi[J(x)-J(u)] > 0.$$

**Remark 4.7.** For $\omega = \eta$, a differentiable function $J$ is ($\phi$, $b$, $\beta$, $\eta$)- univex if

$$b(x,u)\phi[J(x)-J(u)] \geq \beta(x,u)[J'(u)\eta(x,u)].$$

**Remark 4.8.** For $\beta$ = 1 and $\omega = \eta$, a differentiable function $J$ is ($\phi$, $b$, $\eta$)- univex [4] if

$$b(x,u)\phi[J(x)-J(u)] \geq [J'(u)\eta(x,u)].$$

**Remark 4.9.** For $b = \beta = 1$ and $\omega = \eta$, a differentiable function $J$ said to be ($\phi$, $\eta$)-univex [3] if

$$\phi[J(x)-J(u)] \geq [J'(u)\eta(x,u)].$$

## 3. Sufficient Optimality Conditions

Next, we consider the theorem on the necessary optimality conditions with some extra assumptions. Furthermore, we explore some sufficient optimality conditions regarding problem $(P)$.

**Theorem 4.10.** (Sufficient optimality). Let $x^* \in F_p$, $y^* \in I$, $z^* \in R^m_+$ and $(x^*,y^*,z^*)$ satisfy (6.7)-(6.9). Let us define

$$A(x) = \Psi(x^*,y^*)\Phi(x,y^*) - \Psi(x,y^*)\Phi(x^*,y^*),$$

and

$$B(x) = A(x) + \Psi(x^*,y^*)\Omega(x,z^*).$$

Assume, for the given $y^*$ and $z^*$, that any one of the following conditions holds:

**(a)** $\Phi(.,y^*), -\Psi(.,y^*)$ and $\Omega(.,z^*)$ are ($\phi$, $b$, $\beta$, $\eta$, $\omega$)- univex with respect to $b(.,.)$, $\beta(.,.)$, $\phi$, $\eta(.,.)$ and $\omega(.,.)$, where $b(.,.) > 0$, $\beta(.,.) > 0$ and $\phi$ is superlinear.

**(b)** $A$ is ($\phi$, $b$, $\beta$, $\eta$, $\omega$)- pseudo-invex and $\Omega(.,z^*)$ is ($\phi$, $b$, $\beta$, $\eta$, $\omega$)- quasi-invex with respect to $b(.,.)$, $\beta(.,.)$, $\phi$, $\eta(.,.)$ and $\omega(.,.)$, where $b(.,.) > 0$, $\beta(.,.) > 0$ and $\phi$ is strictly increasing.

**(c)** $A$ is ($\phi$, $b$, $\beta$, $\eta$, $\omega$)- quasi-invex and $\Omega(.,z^*)$ is strictly ($\phi$, $b$, $\beta$, $\eta$, $\omega$)- pseudo-invex with respect to $b(.,.)$, $\beta(.,.)$, $\phi$, $\eta(.,.)$ and $\omega(.,.)$, where $b(.,.) > 0$, $\beta(.,.) > 0$ and $\phi$ is increasing.

**(d)** $B$ is ($\phi$, $b$, $\beta$, $\eta$, $\omega$)- pseudo-invex with respect to $b(.,.)$, $\beta(.,.)$, $\phi$, $\eta(.,.)$ and $\omega(.,.)$, where $b(.,.) > 0$, $\beta(.,.) > 0$ and $\phi$ is increasing.

Then $x^*$ is an optimal solution to $(P)$.

**Proof**

Assume to contrary $x^*$ is not an optimal solution of $(P)$. Then there is a feasible solution $u \in F_p$ such that

$$K(x^*) > K(u). \tag{4.10}$$

Applying (6.9), we have

$$\frac{\Phi(x^*,y^*)}{\Psi(x^*,y^*)} = K(x^*) > K(u) = \max_{y \in I} \frac{\Phi(u,y)}{\Psi(u,y)} \geq \frac{\Phi(u,y^*)}{\Psi(u,y^*)}, \quad y^* \in I.$$

It follows that

$$A(u) = \Phi(u,y^*)\Psi(x^*,y^*) - \Phi(x^*,y^*)\Psi(u,y^*) < 0 = A(x^*), \tag{4.11}$$

i.e.,

$$A(u) < A(x^*).$$

Now by rearranging the above inequality (11), we get

$$\Psi(x^*,y^*)[\Phi(u,y^*) - \Phi(x^*,y^*)] - \Phi(x^*,y^*)[\Psi(u,y^*) - \Psi(x^*,y^*)] < 0. \tag{4.12}$$

Since $u \in F_p$ , by condition (6.8), we have

$$\Omega(u,z^*) \leq 0 = \Omega(x^*,z^*). \tag{4.13}$$

Now (6.29) and (6.31) yield $B(u) < B(x^*)$.

If condition (a) holds, for $y^* \in I, z^* \in R^m_+$, employing the ($\phi$, $b$, $\beta$, $\eta$, $\omega$)- univexity of $\Phi(.,y^*), -\Psi(.,y^*)$ and $\Omega(.,z^*)$, we have

$$\begin{aligned} & b(u,x^*)\phi\left[\Phi(u,y^*) - \Phi(x^*,y^*)\right] \\ \geq\ & \beta(u,x^*)\frac{1}{2}[\Phi_1'(x^*,y^*)\eta(u,x^*) + \Phi_1'(x^*,y^*)\omega(u,x^*)], \end{aligned} \tag{4.14}$$

$$
\begin{aligned}
&-b(u,x^*)\phi\,[\Psi(u,y^*)-\Psi(x^*,y^*)]\\
\geq\;& -\beta(u,x^*)\frac{1}{2}[\Psi_1'(x^*,y^*)\eta(u,x^*)+\Psi_1'(x^*,y^*)\omega(u,x^*)], \qquad (4.15)
\end{aligned}
$$

$$
\begin{aligned}
&b(u,x^*)\phi\,[\Omega(u,z^*)-\Omega(x^*,z^*)]\\
\geq\;& \beta(u,x^*)\frac{1}{2}[\Omega_1'(x^*,z^*)\eta(u,x^*)+\Omega_1'(x^*,z^*)\omega(u,x^*)]. \qquad (4.16)
\end{aligned}
$$

Here for each $y^*$ and $z^*$, $\Phi_1', \Psi_1'$ and $\Omega_1'$ are partial Frechet derivatives of $\Phi(.,y^*), -\Psi(.,y^*)$ and $\Omega(.,z^*)$ at $x^*$, respectively. Since $\Psi(x^*,y^*)>0$ and $\Phi(x^*,y^*)\geq 0$, we multiply (6.32) by $\Psi(x^*,y^*)$, (6.33) by $\Phi(x^*,y^*)$ and (6.34) by $\Psi(x^*,y^*)$. As $\phi$ is superlinear, by adding up the resulting inequalities, it follows from (6.30) and (6.31) that

$$
\begin{aligned}
&0>\beta(u,x^*)\frac{1}{2}[\Big[\Psi(x^*,y^*)\Phi_1'(x^*,y^*)-\Phi(x^*,y^*)\Psi_1'(x^*,y^*)\Big]\\
+\;& \Big[\Psi(x^*,y^*)\Omega_1'(x^*,z^*)\Big]\,\eta(u,x^*)]\\
+\;& \beta(u,x^*)\frac{1}{2}[\Big[\Psi(x^*,y^*)\Phi_1'(x^*,y^*)-\Phi(x^*,y^*)\Psi_1'(x^*,y^*)\Big]\\
+\;& \Big[\Psi(x^*,y^*)\Omega_1'(x^*,z^*)\Big]\,\omega(u,x^*)].
\end{aligned}
$$

This leads to a contradiction in light of (6.7). Hence (6.28) does not hold, and so $x^*$ must be an optimal solution of $(P)$.

For condition (b), $A$ is $(\phi, b, \beta, \eta, \omega)$- pseudo-univex (applying the equivalent form of the pseudo-univexity of $A$ in this situation) and so from (6.29), for $b(.,.)>0$ and $\phi(.)$ is strictly increasing, we have

$$
\begin{aligned}
&b(u,x^*)\phi\,[A(u)-A(x^*)]<0\\
&\Rightarrow\beta(u,x^*)\Big(\frac{1}{2}[A'(x^*)\eta(u,x^*)]+\frac{1}{2}[A'(x^*)\omega(u,x^*)]\Big)<0,
\end{aligned}
$$

that is,

$$
\begin{aligned}
&\beta(u,x^*)\frac{1}{2}\Big([\Big[\Psi(x^*,y^*)\Phi_1'(x^*,y^*)-\Phi(x^*,y^*)\Psi_1'(x^*,y^*)\Big]\,\eta(u,x^*)]\\
+\;& [\Big[\Psi(x^*,y^*)\Phi_1'(x^*,y^*)-\Phi(x^*,y^*)\Psi_1'(x^*,y^*)\Big]\,\omega(u,x^*)]\Big)<0. \qquad (4.17)
\end{aligned}
$$

The identity (6.7) and inequality (6.35) yield

$$\beta(u,x^*)\left([\Omega_1'(x^*,z^*)\Psi(x^*,y^*)\eta(u,x^*)] + [\Omega_1'(x^*,z^*)\Psi(x^*,y^*)\omega(u,x^*)]\right) > 0.$$

Since $\beta(u,x^*) > 0$ and $\Psi(x^*,y^*) > 0$, it follows that

$$\Omega_1'(x^*,z^*)\eta(u,x^*) + \Omega_1'(x^*,z^*)\omega(u,x^*) > 0. \tag{4.18}$$

On the other hand, the function $\Omega$ in condition (b) assumed to be ($\phi$, $b$, $\beta$, $\eta$, $\omega$)-quasi-invex with respect to $\eta$ and $\omega$. Thus it implies

$$\beta(u,x^*)\frac{1}{2}\left([\Omega_1'(x^*,z^*)\eta(u,x^*)] + [\Omega_1'(x^*,z^*)\omega(u,x^*)]\right) \leq 0$$

(by the ($\phi$, $b$, $\beta$, $\eta$, $\omega$)-quasi-univexity) which contradicts (6.36). This shows that $x^*$ is optimal for $(P)$.
Similarly one can prove for conditions (c) and (d).

## 4. Wolfe Type Dual

In this section, we consider the following Wolfe type model:

$$\text{(WD)} \qquad \text{maximize}\ \frac{\Phi(u,y)+\Omega(u,z)}{\Psi(u,y)}$$

subject to $(u,z) \in PS(T,R^n) \times R^m_+$,

$$u(a) = \alpha, u(b) = \beta \text{ and } y \in I \subset R^m_+,$$

$$\Psi(u,y)\Phi_1'(u,y) - [\Phi(u,y)+\Omega(u,z)]\Psi_1'(u,y) + \Psi(u,y)\Omega_1'(u,z) = 0. \tag{4.19}$$

Let the set of all feasible solutions of problem (WD) be denoted by $K_1$. We assume throughout this section that

$$\Phi(u,y) + \Omega(u,z) \geq 0 \text{ and } \Psi(u,y) > 0 \text{ for all } (u,y,z) \in K_1. \tag{4.20}$$

In what follows, we plan to establish weak, strong and strict converse duality theorems for $(P)-(WD)$.

**Theorem 4.11** (Weak duality). Let $x \in F_p$ , $(u,y,z) \in K_1$ and let

$$C(x) = \Psi(u,y)[\Phi(x,y)+\Omega(x,z)] - [\Phi(u,y)+\Omega(u,z)]\Psi(x,y).$$

Suppose that for each, $y$ and $z$, either one of the conditions holds:

**(a)** $\Phi(.,y^*), -\Psi(.,y^*)$ and $\Omega(.,z^*)$ are ($\phi$, $b$, $\beta$, $\eta$ $\omega$)- univex with respect to $b$, $\beta$, $\phi$, $\eta$ and $\omega$, where $b(.,.) > 0$, $\beta(.,.) > 0$ and $\phi$ is superlinear.

**(b)** $C$ is $(\phi, b, \beta, \eta, \omega)$- pseudo-univex with respect to $b$, $\beta$, $\phi$, $\eta$ and $\omega$, where $b > 0$, $\beta(.,.) > 0$ and $\phi$ is strictly increasing.

Then

$$\varphi(x) \geq \frac{\Phi(u,y) + \Omega(u,z)}{\Psi(u,y)}, \tag{4.21}$$

where $\varphi(x)$ is defined by (6.1).

Proof. Suppose, on contrary

$$\varphi(x) < \frac{\Phi(u,y) + \Omega(u,z)}{\Psi(u,y)}. \tag{4.22}$$

It follows from (6.2) that for any $y \in I$,

$$\frac{\Phi(x,y)}{\Psi(x,y)} \leq \max_{\beta \in I} \frac{\Phi(x,\beta)}{\Psi(x,\beta)} = \varphi(x) < \frac{\Phi(u,y) + \Omega(u,z)}{\Psi(u,y)},$$

$$\Phi(x,y)\Psi(u,y) - [\Phi(u,y) + \Omega(u,z)]\Psi(x,y) < 0. \tag{4.23}$$

Rewriting the above inequality, we get

$$[\Phi(x,y) + \Omega(x,z)]\Psi(u,y) - [\Phi(u,y) + \Omega(u,z)]\Psi(x,y) < \Omega(x,z)\Psi(u,y). \tag{4.24}$$

Since $\Omega(x,z) \leq 0$, if $x \in F_p$, $z \in R^m_+$ and $\Psi(u,y) > 0, \Omega(x,z)\Psi(u,y) \leq 0$. It follows from (6.58) that

$$C(x) < 0 = C(u).$$

Since $b(.,.) > 0$, the above inequality yields

$$b(x,u)\phi[C(x) - C(u)] < 0. \tag{4.25}$$

Now if condition (a) holds, for each $y \in I, z \in R^m_+$, the $(\phi, b, \beta, \eta, \omega)$- univexity of $\Phi(.,y^*)$, $-\Psi(.,y^*)$ and $\Omega(.,z^*)$ implies that

$$b(x,u)\phi\left[\Phi(x,y) - \Phi(u,y)\right] \geq \beta(x,u)\frac{1}{2}\Big(\Phi_1'(u,y)\eta(x,u) + \Phi_1'(u,y)\omega(x,u)\Big), \tag{4.26}$$

$$-b(x,u)\phi\left[\Psi(x,y) - \Psi(u,y)\right] \geq -\beta(x,u)\frac{1}{2}\Big(\Psi_1'(u,y)\eta(x,u) + \Psi_1'(u,y)\omega(x,u)\Big), \tag{4.27}$$

$$b(x,u)\phi\left[\Omega(x,z) - \Omega(u,z)\right] \geq \beta(x,u)\frac{1}{2}\Big(\Omega_1'(u,z)\eta(x,u) + \Omega_1'(u,z)\omega(x,u)\Big). \tag{4.28}$$

From (6.38), $\Psi(u,y)$ and $\Phi(u,y)+\Omega(u,z)$ are nonnegative. Multiplying (6.52) by $\Psi(u,y)$, (6.61) by $\Phi(u,y)+\Omega(u,z)$ and (6.62) by $\Psi(u,y)$, and summing up the resulting inequalities in light of the superlinearity of $\phi$, we obtain from (6.57) that

$$\begin{aligned}
&\beta(x,u)\frac{1}{2}\Big(\Big\{\Psi(u,y)\Phi_1'(u,y)-[\Phi(u,y)+\Omega(u,z)]\Psi_1'(u,y)\Big\}\eta(x,u)\\
+&\Big\{\Psi(u,y)\Omega_1'(u,z)\Big\}\eta(x,u)\\
+&\Big\{\Psi(u,y)\Phi_1'(u,y)-[\Phi(u,y)+\Omega(u,z)]\Psi_1'(u,y)\Big\}\omega(x,u)\\
+&\Big\{\Psi(u,y)\Omega_1'(u,z)\Big\}\omega(x,u)\Big)<0.
\end{aligned}$$

This contradicts (6.37), so (6.56) does not hold and (6.55) holds.

For the proof under the assumption (b), $C$ is ($\phi$, $b$, $\beta$, $\eta$, $\omega$)- pseudo-univex and (6.59) implies that

$$\beta(x,u)C'(u)\eta(x,u)<0,$$

that is,

$$\begin{aligned}
&\beta(x,u)\frac{1}{2}\Big(\Big\{\Psi(u,y)\Phi_1'(u,y)-[\Phi(u,y)+\Omega(u,z)]\Psi_1'(u,y)\Big\}\eta(x,u)\\
+&\Big\{\Psi(u,y)\Omega_1'(u,z)\Big\}\eta(x,u)\\
+&\Big\{\Psi(u,y)\Phi_1'(u,y)-[\Phi(u,y)+\Omega(u,z)]\Psi_1'(u,y)\Big\}\omega(x,u)\\
+&\Big\{\Psi(u,y)\Omega_1'(u,z)\Big\}\omega(x,u)\Big)<0,
\end{aligned}$$

which contradicts (6.37).

**Theorem 4.12.** (Strong duality). Let $x^*$ be an optimal solution of $(P)$. Then there exist $y\in I$ and $z\in R_+^m$ such that $(x^*,y^*,z^*)$ is feasible for $(WD)$. If the hypotheses of Lemma 4.2 and Theorem 4.10 are satisfied, then $(x^*,y^*,z^*)$ is an optimal solution of $(WD)$ and, two problems $(P)$ and $(WD)$ have the same optimal values.

**Proof.** Assume that $x^*$ is an optimal solution of $(P)$. Then, by Theorem 4.11, there exist $y\in I$ and $z\in R_+^m$ satisfying the constraints of $(WD)$, so that $(x^*,y^*,z^*)\in K_1$. Furthermore, we have

$$\frac{\Phi(x^*,y^*)+\Omega(x^*,z^*)}{\Psi(x^*,y^*)}=\frac{\Phi(x^*,y^*)}{\Psi(x^*,y^*)}=\varphi(x^*).$$

Since $\Omega(x^*,z^*)=0$, it implies $(x^*,y^*,z^*)$ is an optimal solution of $(WD)$. Consequently, (6.55) demonstrates that $(P)$ and $(WD)$ have the same optimal values.

**Theorem 4.13.** (Strict converse duality). Let $x_1$ and $(x^*,y_0,z_0)$ be optimal solutions for (P) and (WD) that have the same optimal value

$$\varphi(x_1)=\frac{\Phi(x^*,y_0)+\Omega(x^*,z_0)}{\Psi(x^*,y_0)}.$$

Proof. Suppose on contrary that $x_1 \neq x^*$. Then by Theorem 4.12, there exist $y_1 \in I$ and $z_1 \in R^m_+$ such that $(x_1, y_1, z_1)$ is an optimal solution of $(WD)$ and

$$\varphi(x_1) = \frac{\Phi(x_1, y_1) + \Omega(x_1, z_1)}{\Psi(x_1, y_1)}.$$

As in the proof of Theorem 4.11, we replace $x$ by $x_1$, and $(u, y, z)$ by $(x^*, y_0, z_0)$ to derive the inequality

$$\varphi(x_1) > \frac{\Phi(x^*, y_0) + \Omega(x^*, z_0)}{\Psi(x^*, y_0)}.$$

This contradicts the fact that

$$\varphi(x_1) = \frac{\Phi(x_1, y_1) + \Omega(x_1, z_1)}{\Psi(x_1, y_1)} = \frac{\Phi(x^*, y_0) + \Omega(x^*, z_0)}{\Psi(x^*, y_0)}.$$

Hence, we conclude that

$$x_1 = x^* \text{ and } \varphi(x_1) = \frac{\Phi(x^*, y_0) + \Omega(x^*, z_0)}{\Psi(x^*, y_0)}.$$

## 5. Mond - Weir Type Dual

In this section, we consider the Mond-Weir type dual problems under the assumptions of $(\phi, b, \beta, \eta, \omega)$- univexities as follows:

$$\text{(MWD)} \qquad \text{Maximize } \frac{\Phi(u, y)}{\Psi(u, y)}$$

subject to

$$(u, y) \in PS(T, R^n) \times I, \quad u(a) = \alpha, u(b) = \beta,$$

$$\begin{aligned} & \frac{1}{2}\left\{\Psi(u,y)\Phi_1'(u,y) - (\Phi(u,y) + \Omega(u,z))\Psi_1'(u,y)\right\}\eta(x,u) \\ + & \left\{\Psi(u,y)\Omega_1'(u,z)\right\}\eta(x,u) \\ + & \frac{1}{2}\left\{\Psi(u,y)\Phi_1'(u,y) - (\Phi(u,y) + \Omega(u,z))\Psi_1'(u,y)\right\}\omega(x,u) \\ + & \left\{\Psi(u,y)\Omega_1'(u,z)\right\}\omega(x,u) < 0, \end{aligned} \tag{4.29}$$

$$\Omega(u, z) \geq 0, z \in R^m_+. \tag{4.30}$$

Let $K_2$ denote the set of all feasible solutions of problem $(MWD)$. We assume throughout that $\Phi(u, y) \geq 0$ and $\Psi(u, y) > 0$ for all $(u, y, z) \in K_2$.

Now we will establish weak, strong and strict converse duality theorems for $(P)-(MWD)$.

**Theorem 4.14.** (Weak duality) Let $x \in F_p, (u,y,z) \in K_2$ and define

$$D(x) = \Psi(u,y)\Phi(x,y) - \Psi(x,y)\Phi(u,y)$$

and

$$E(x) = D(x) + \Psi(u,y)\Omega(x,z).$$

Suppose, for each $y \in I, z \in R_+^m$, that any one of the following conditions holds:

**(a)** $\Phi(.,y), -\Psi(.,y)$ and $\Omega(.,z)$ are $(\phi, b, \beta, \eta, \omega)$- univex with respect to $b, \beta, \phi, \omega$ and $\eta$ where $b(.,.) > 0, \beta(.,.) > 0$ and $\phi$ is superlinear.

**(b)** $D$ is $(\phi, b, \beta, \eta, \omega)$- pseudo-univex and $\Omega(.,z)$ is $(\phi, b, \beta, \eta, \omega)$- quasi-univex with respect to $b, \beta, \phi, \omega$ and $\eta$, where $b(.,.) > 0, \beta(.,.) > 0$ and $\phi$ is strictly increasing.

**(c)** $D$ is $(\phi, b, \beta, \eta, \omega)$- quasi-univex and $\Omega(.,z)$ is strictly $(\phi, b, \beta, \eta, \omega)$- pseudo-univex with respect to $b, \beta, \phi, \omega$ and $\eta$, where $b(.,.) > 0, \beta(.,.) > 0$ and $\phi$ is increasing.

**(d)** $E$ is $(\phi, b, \beta, \eta, \omega)$- pseudo-univex with respect to $b, \beta, \phi, \omega$ and $\eta$, where $b(.,.) > 0$, $\beta(.,.) > 0$ and $\phi$ is increasing.

Then

$$\varphi(x) \geq \frac{\Phi(u,y)}{\Psi(u,y)}. \tag{4.31}$$

**Proof.** Suppose to contrary. Then we have

$$\frac{\Phi(u,y)}{\Psi(u,y)} > \varphi(x) = \max_{y_1 \in I} \frac{\Phi(x,y_1)}{\Psi(x,y_1)} \geq \frac{\Phi(x,y)}{\Psi(x,y)}.$$

This implies that

$$D(x) = \Phi(x,y)\Psi(u,y) - \Phi(u,y)\Psi(x,y) < 0 = D(u). \tag{4.32}$$

As $x \in F_p, z \in R_+^m$ and $\Omega(x,z) \leq 0$, the constraints inequality (6.64) of $(MWD)$ yields

$$\Omega(x,z) \leq \ 0 \leq \Omega(u,z). \tag{4.33}$$

Consequently, (6.94) and (6.95) yield

$$E(x) < E(u). \tag{4.34}$$

If condition (a) holds, then by the $(\phi, b, \beta, \eta, \omega)$- univexity of $\Phi$, $-\Psi$ and $\Omega$, we have

$$b(x,u)\phi\left[\Phi(x,y) - \Phi(u,y)\right] \geq \beta(x,u)\frac{1}{2}\Big(\Phi_1^{'}(u,y)\eta(x,u) + \Phi_1^{'}(u,y)\omega(x,u)\Big), \tag{4.35}$$

$$-b(x,u)\phi\left[\Psi(x,y) - \Psi(u,y)\right] \geq -\beta(x,u)\frac{1}{2}\Big(\Psi_1^{'}(u,y)\eta(x,u) + \Big(\Psi_1^{'}(u,y)\omega(x,u)\Big), \tag{4.36}$$

$$b(x,u)\phi\left[\Omega(x,z)-\Omega(u,z)\right]\geq\beta(x,u)\frac{1}{2}\Big(\Omega_1^{'}(u,z)\eta(x,u)+\Omega_1^{'}(u,z)\omega(x,u)\Big). \quad (4.37)$$

Since $\Psi(u,y)$ and $\Phi(u,y)$ are nonnegative, multiplying (6.97) by $\Psi(u,y)$, (6.98) by $\Phi(u,y)$ and (6.99) by $\Psi(u,y)$, and summing up the resulting inequalities in light of superlinearity of $\phi$, we obtain from (6.94) and (6.95)

$$\begin{aligned}
0 > \beta(x,u)&\frac{1}{2}\\
&\Big(\Big[\Psi(u,y)\Phi_1^{'}(u,y)-\Phi(u,y)\Psi_1^{'}(u,y)+\Psi(u,y)\Omega_1^{'}(u,z)\Big]\eta(x,u))\\
+\ &\Big[\Psi(u,y)\Phi_1^{'}(u,y)-\Phi(u,y)\Psi_1^{'}(u,y)+\Psi(u,y)\Omega_1^{'}(u,z)\Big]\omega(x,u)\Big).
\end{aligned}$$

which contradicts (6.63). Hence (6.93) holds.
In condition (b), $D$ is ($\phi$, $b$, $\beta$, $\eta$, $\omega$)- pseudo-univex. Since $b(.,.)>0$, (6.94) implies $\beta(x,u)\frac{1}{2}\Big(D'(u)\eta(x,u)+D'(u)\eta(x,u)\Big)<0$, that is,

$$\begin{aligned}
&\beta(x,u)\frac{1}{2}\\
&\Big(\Big[\Psi(u,y)\Phi_1^{'}(u,y)-\Phi(u,y)\Psi_1^{'}(u,y)\Big]\eta(x,u)\Big)\\
&+\Big[\Psi(u,y)\Phi_1^{'}(u,y)-\Phi(u,y)\Psi_1^{'}(u,y)\Big]\omega(x,u)\Big)<0. \quad (4.38)
\end{aligned}$$

It follows from (29) and (38) that

$$-\beta(x,u)\Big(\Psi(u,y)\Omega_1^{'}(u,z)\eta(x,u)+\Psi(u,y)\Omega_1^{'}(u,z)\omega(x,u)\Big)<0$$

or

$$\Omega^{'}(u,z)\eta(x,u)+\Omega^{'}(u,z)\omega(x,u)>0,$$

since $\Psi(u,y)>0$. But $\Omega$ is ($\phi$, $b$, $\beta$, $\eta$, $\omega$)- quasi-univex, so the above inequality implies

$$\Omega(x,z)>\Omega(u,z)$$

which contradicts (6.95). Hence, (6.93) holds.

Similarly, one can prove for the rest of two cases (c) and (d).

**Theorem 4.15.** (Strong duality). Let $x^*$ be an optimal solution of $(P)$. Then there exist $y^*\in I, z^*\in R_+^m$ such that $(x^*,y^*,z^*)$ is feasible for $(MWD)$. If any of the conditions of Theorem 4.14 hold, then $(x^*,y^*,z^*)$ is an optimal solution of $(MWD)$. Furthermore, $(MWD)$ and $(P)$ have the same optimal values.

**Proof.** If $x^*$ is an optimal solution of $(P)$, then by Theorem 4.14, there exist $y^*\in I$ and $z^*\in R_+^m$ such that $(x,y,z)\in K_2$, the feasible solution of $(MWD)$ and

$$\varphi(x^*)=\frac{\Phi(x^*,y^*)}{\Psi(x^*,y^*)}\leq\max_{y\in I}\frac{\Phi(x^*,y)}{\Psi(x^*,y)}.$$

The optimality of $(MWD)$ follows from Theorem 4.14.

**Theorem 4.16.** (Strict converse duality). Let $x_1$ and $(x^*, y_0, z_0)$ be optimal solutions of $(P)$ and $(MWD)$, respectively. Assume that the hypothesis of Theorem 4.14 is fulfilled. Assume that $D(.) = \Psi(x^*, y_0)\Phi(., y_0) - \Psi(., y_0)\Phi(x^*, y_0)$ is strictly $(\phi, b, \beta, \eta, \omega)$- pseudo-univex and $\Omega(., z_0)$ is $(\phi, b, \beta, \eta, \omega)$- quasi-univex with respect to $b$, $\beta$, $\phi$, $\omega$ and $\eta$. Then $x_1 = x^*$ is an optimal solution of $(P)$, and $(P)$ and $(MWD)$ have the same optimal values

$$\varphi(x_1) = \Phi(x^*, y_0)/\Psi(x^*, y_0).$$

**Proof.** Since $x_1$ is an optimal solution of $(P)$, by Theorem 4.14, there exist $y_1 \in I$ and $z_1 \in R^m_+$ such that $(x_1, y_1, z_1)$ is an optimal solution of $(MWD)$ and

$$\varphi(x_1) = \frac{\Phi(x_1, y_1)}{\Psi(x_1, y_1)}.$$

On the other hand, $(x^*, y_0, z_0)$ is an optimal solution of $(MWD)$ with optimal values equal to $\varphi(x_1)$, that is,

$$\left[\varphi(x_1) = \frac{\Phi(x_1, y_1)}{\Psi(x_1, y_1)} = \frac{\Phi(x^*, y_0)}{\Psi(x^*, y_0)}\right]. \tag{4.39}$$

If $x_1 \neq x^*$, we could deduce

$$\left[\varphi(x_1) > \frac{\Phi(x^*, y_0)}{\Psi(x^*, y_0)}\right], \tag{4.40}$$

which contradicts (6.101).

Assume (6.102) does not hold. Then $\varphi(x_1) = \Phi(x^*, y_0)/\Psi(x^*, y_0)$, that is,

$$\frac{\Phi(x^*, y^*)}{\Psi(x^*, y^*)} \geq \varphi(x_1) = \max_{y \in I} \frac{\Phi(x_1, y)}{\Psi(x_1, y)} \geq \frac{\Phi(x_1, y_0)}{\Psi(x_1, y_0)}$$

or

$$D(x_1) = \Phi(x_1, y_0)\Psi(x^*, y_0) - \Phi(x^*, y_0)\Psi(x_1, y_0) \leq 0 = D(x^*),$$

that is,

$$D(x_1) \leq D(x^*).$$

Since $b(.,.) > 0$, the above inequality yields

$$b(x_1, x^*)\phi[D(x_1) - D(x^*)] \leq 0. \tag{4.41}$$

Since $D$ is strictly $(\phi, b, \beta, \eta, \omega)$- pseudo-univex, (41) implies $\beta(x_1, x^*)(D'(x^*)\eta(x_1, x^*)) < 0$,

$$\beta(x_1, x^*)\Big([\Psi(x^*, y_0)\Phi_1'(x^*, y_0) - \Phi(x^*, y_0)\Psi_1'(x^*, y_0)]\eta(x_1, x^*)\Big) < 0. \tag{4.42}$$

By (29) and (42), we get

$$-\beta(x_1, x^*)\Big(\Psi(x^*, y_0)\Omega_1'(x^*, z_0)\eta(x_1, x^*) < 0,$$

i.e.,

$$\beta(x_1,x^*)\Omega_1'(x^*,z_0)\eta(x_1,x^*) > 0. \tag{4.43}$$

Since $\Psi(x^*,y_0) > 0$ by the $(\phi, b, \beta, \eta, \omega)$- quasi-univexity of $\Omega$, (43) implies

$$b(x_1,x^*)\phi\,[\Omega(x_1,z_0) - \Omega(x^*,z_0)] > 0.$$

Since $b(.,.) > 0$, the above inequality yields

$$\Omega(x_1,z_0) > \Omega(x^*,z_0). \tag{4.44}$$

On the other hand, $x_1 \in F_p, z_0 \in R_1^m$ and $\Omega(x_1,z_0) \leq 0$. By (6.64), $\Omega\ (x^*,z_0) \geq 0$. Thus $\Omega(x_1,z_0) \leq \Omega\ (x^*,z_0)$, this contradicts (6.106). Hence (6.102) holds under $x_1 \neq x^*$. Therefore $x_1 = x^*$.

## 6. General Remarks

The results established in this chapter can be generalized, based on the work of Pitea and Postolache [7], to applications in mechanical engineering, where curvilinear integral objectives are used largely driven by their physical meaning as mechanical work. These kinds of objectives have also significant roles in mathematical modeling of certain processes relating robotics, tribology, engines, and others.

# Bibliography


[1] S. Chandra, B.D. Craven and B. Mond, Generalized Fractional programming duality: a ratio game approach, *J. Austral. Math. Soc. Ser. B*, **28** (1986) 170 - 180.

[2] J.P. Crouzeix, J.A. Ferland and S. Schaible, Duality in generalized fractional programming, *Math. Prog.* **27**, (1983) 342 - 354.

[3] H.C. Lai and J.C. Liu, Minimax fractional programming involving generalized invex functions, *ANZIAM J.* **44**, (2003) 339 - 354.

[4] S. K. Mishra, K. Shukla and R. U. Verma, Minmax fractional integral programming problems involving univexity, *DCDIS, Series A: Mathematical Analysis* **18** (2012), 335 - 355.

[5] B. Mond, S. Chandra and I. Husain, Duality for variational problems with invexity, *J. Math. Anal. Appl.* **134** (1988) 322 - 328.

[6] B. Mond and I. Husain, Sufficient optimality criteria and duality for variational problems with generalized invexity, *J. Austral. Math. Soc. Ser. B* **31** (1) (1989) 108 - 121.

[7] A. Pitea and M. Postolache, Duality theorems for a new class of multitime multiobjective variational problems, *J. Global Optimization* **54** (2012), 47 - 58.

[8] I.M. Stancu-Minasian, *Fractional Programming Theory, Methods and Applications*, Kluwer, Dordrecht, 1997.

[9] R. U. Verma, *Semi-Infinite Fractional Programming*, Springer Nature/Springer 2017. https://link.springer.com/book/10.1007/10-978-981-10-6256-8

[10] R. U. Vema and G. J. Zalmai, Parameter-free duality models and applications to semi-infinite minmax fractional programming based on higher-order ($\phi$, $\eta$, $\rho$, $\theta$, $m$)- sonvexities, *OPSEARCH*, Springer Nature DOI: 10.1007/s 12597-017-0323-8 30 Pages.

[11] G. J. Zalmai, Generalized ($\mathscr{F}$, $b$, $\phi$, $\rho$, $\theta$)-univex n-set functions and semiparametric duality models in multiobjective fractional subset programming, *Int. J. Math. Math. Sci.* **2005** (7) (2005), 1105 - 1133.

# Chapter 5

# Integral Type Programming

This research deals with a class of problems of minmax fractional integral type programming (or programs) by applying the generalized version of univexity considered in Mishra, Shukla and Verma [4], which is equivalent to minimizing a maximum of several time dependent ratios involving integral expressions. We establish optimality conditions based on the generalized univexity, and then we explore the Wolfe type dual model, and Mond-Weir type dual model. Furthermore, a class of weak, strong and strictly converse duality theorems based on the generalized univexity frameworks is established.

## 1. Introduction

The fractional integral type programming is one of the most application-enhanced and fast-expanding areas of optimization which features in several types of practical problems. It can be applied to different branches of engineering as well as to economics, including, but not limited to, multiobjective control problems, multiobjective variational control problems, to minimize a ratio of functions between given periods of time. Furthermore it can be utilized as a resource in order to measure the efficiency or productivity of a system. In such type of problems the objective function is given as a ratio of functions. In the present chapter, we consider a problem that deals with minimizing a maximum of several time dependent ratios involving integral expressions. Several researchers have investigated this type of problem, including Chandra et al. [1], Crouzeix et al. [2], Mond et al. [5], and Mond and Husain [6], while we generalize the results in [4] to the case of the generalized univexity. As our problem involves a state function $x(t)$ in the integrand of the integrals, the obtained results also relate to the problems on optimal control. For more details, we refer the reader [1 - 8].

Consider the fractional programming for variational problem of the following form:

$$\text{(P)} \qquad \min_{x} \left\{ \sum_{1\leq i\leq p} \max \frac{\int_a^b f^i(t,x,\dot{x})dt}{\int_a^b g^i(t,x,\dot{x})dt} \right\}$$

$$\text{subject to } x \in PS(T,R^n), x(a) = \alpha, x(b) = \beta,$$

$$\int_a^b h^j(t,x,\dot{x})dt \leq 0, j \in \underline{m} \equiv \{1,2,\ldots,m\},$$

where functions $f^i, g^i, i \in \underline{p}$ and $h^j, j \in \underline{m}$ are continuous in $t$ and have continuous partial derivatives with respect to $x$ and $\dot{x}$; $T = [a,b]$ denotes the time space, and $PS(T,R^n)$ stands for the state space of all piecewise smooth functions $x : T \to R^n$ with norm defined by $\|x\| = \|x\|_\infty + \|Dx\|_\infty$ and $D$ is the differential operator on $PS(T,R^n)$ defined by

$$y = Dx \text{ if and only if } x(t) = x(a) + \int_a^b y(s)ds.$$

Thus $D = \frac{d}{dt}$ except at the point of discontinuity.

$$\int_a^b g^i(t,x,\dot{x}) > 0, \int_a^b f^i(t,x,\dot{x}) \geq 0 \text{ for each } i \in \underline{p}.$$

Let $F_p$ denote the set of all feasible solutions to $(P)$.

## 2. Preliminaries

In order to simplify the symbols in problem $(P)$, we scalarize functionals as follows:

$$\Phi(x,y) = \langle y, F(x) \rangle = \sum_{i=1}^{p} y_i F_i(x) = \sum_{i=1}^{p} y_i \int_a^b f^i(t,x,\dot{x})dt,$$

$$\Psi(x,y) = \langle y, G(x) \rangle - \sum_{i=1}^{p} y_i G_i(x) = \sum_{i=1}^{p} y_i \int_a^b g^i(t,x,\dot{x})dt,$$

$$\Omega(x,z) = \langle z, H(x) \rangle = \sum_{j=1}^{m} z_j H_j(x) = \sum_{j=1}^{m} z_j \int_a^b h^j(t,x,\dot{x})dt,$$

where $y \in I = \left\{ y \in R_+^p | \sum_{i=1}^{p} y_i = 1 \right\}$ and $z \in R_+^m$. Then for any feasible solution $x$ of $(P)$, the objective fractional function can be represented by

$$\varphi(x) = \max_{i \in \underline{p}} \frac{F_i(x)}{G_i(x)} = \max_{y \in I} \frac{\langle y, F(x) \rangle}{\langle y, G(x) \rangle} = \max_{y \in I} \frac{\Phi(y,x)}{\Psi(y,x)}. \tag{5.1}$$

The problem $(P)$ is equivalent to

$$(\tilde{P}) \qquad \min_{x\in PS}\max_{y\in I}\frac{\Phi(y,x)}{\Psi(y,x)} = \min_{x}\left(\max_{i\in \underline{p}}\frac{F_i(x)}{G_i(x)}\right)$$

subject to $x \in PS(T,R^n)$, $x(a) = \alpha$, $x(b) = \beta$ and $H(x) \leq 0$.
It is equivalent to the parametric minimization problem:

$$(EP_v) \qquad \text{minimize} \quad q(v)$$

subject to $F_i(x) - vG_i(x) \leq q$, $i \in \underline{p}$, and $H_j(x) \leq 0$, $j \in \underline{m}$,

$$x \in PS(T,R^n), x(a) = \alpha, x(b) = \beta.$$

From $(EP_v)$, one can reduce to the optimal solution $x^*$ for $(P)$ with its optimal value $v^*$ which is given by

$$v^* = \varphi(x^*) = \max_{i\in \underline{p}}\frac{F^i(x^*)}{G^i(x^*)} = \max_{y\in I}\frac{\Phi(x^*,y)}{\Psi(x^*,y)} = \frac{\Phi(x^*,y^*)}{\Psi(x^*,y^*)}. \tag{5.2}$$

The concept used here for solution of $(EP_v)$, coincides with finding the minimax solution $(x^*,y^*)$ of the Lagrangian

$$\begin{aligned} & L(x,y;x,z) = \langle y,F(x)\rangle - v\langle y,G(x)\rangle + \langle z,H(x)\rangle \\ = \; & \Phi(x,y) - v\Psi(x,y) + \Omega(x,z), \end{aligned} \tag{5.3}$$

with multipliers $v^* \in R_+$ and $z^* \in R^m_+$. The minimax solution $(x^*,y^*)$ is given in the equation

$$L'_x(x,y)\xi = 0 \text{ for all } \xi \in C(T,R^n).$$

We need the following results in order to establish our results on hand regarding problem $(P)$:

**Lemma 5.1.** [3] Let $x^*$ be an optimal solution of $(P)$. Then there exist Lagrangian multipliers $y^* \in I \subset R^p_+$ and $z^* \in R^m_+$ such that the Kuhn-Tucker type conditions hold for the Lagrangian (3):

$$L'_x(x^*,y^*;v^*,z^*) = 0, \tag{5.4}$$

$$\Omega(x^*,z^*) = 0 \tag{5.5}$$

where $v^* = \dfrac{\Phi(x^*,y^*)}{\Psi(x^*,y^*)}$ given in (2) is the optimal value of problem (P), and the equation $L'_x(\cdot,\cdot) = 0$ is the expressed by

$$\Psi(x^*,y^*)\left[\Phi'(x^*,y^*) + \Omega'(x^*,z^*)\right] - \Phi(x^*,y^*)\Psi'(x^*,y^*) = 0. \tag{5.6}$$

**Lemma 5.2.** [3] Let $x^*$ be an optimal solution of $(P)$. Then there exist $y^* \in I \subset R^p_+$, multipliers $v^* \in R$ and $z^* \in R^m_+$ such that $(x^*, y^*, z^*)$ satisfies

$$\Psi(x^*,y^*)\Phi_1'(x^*,y^*) - \Phi(x^*,y^*)\Psi_1'(x^*,y^*) + \Psi(x^*,y^*)\Omega_1'(x^*,z^*) = 0, \tag{5.7}$$

$$\Omega(x^*,z^*) = 0, \tag{5.8}$$

and obtain the optimal value by

$$K(x^*) = \frac{\Phi(x^*,y^*)}{\Psi(x^*,y^*)} = \max_{y \in I} \frac{\Phi(x^*,y)}{\Psi(x^*,y)} \left( -\min_{x} \max_{1 \le i \le p} \frac{F^i(x)}{G^i(x)} \right). \tag{5.9}$$

Now we introduce a new generalization to the univexity concept applied in [4], which encompasses most of the concepts available in the literature. Let $x \in PS(T,R^n)$ and $f \in C^1(T \times R^n \times R^n)$. Define a functional $J : PS(T,R^n) \to R$ by $J(x) = \int_a^b f(t,x,\dot{x})dt$. We assume that the boundary points $x(a)$ and $x(b)$ are fixed. Consider the admissible vectors variations $w \in C([a,b],R^n)$ vanishing at the boundary points. Then the differential of $J$ is a linear functional on $C([a,b],R^n)$,

$$\begin{aligned} J'(x)w &= \frac{d}{d\alpha}\int_a^b f(t,x+\alpha w,\dot{x}+\alpha\dot{w})dt|_{\alpha=0} \\ &= \int_a^b [f_x(t,x,\dot{x})w(t) + f_{\dot{x}}(t,x,\dot{x})\dot{w}(t)]dt \\ &= \int_a^b [f_x(t,x,\dot{x}) - Df_{\dot{x}}(t,x,\dot{x})]\, w(t)dt + f_{\dot{x}}(t,x,\dot{x})\dot{w}(x)|_a^b \\ &= \int_a^b [f_x(t,x,\dot{x}) - Df_{\dot{x}}(t,x,\dot{x})]\, w(t)dt \end{aligned}$$

$$\text{i.e., } J'(x)w = \int_a^b [f_x(t,x,\dot{x}) - Df_{\dot{x}}(t,x,\dot{x})]\, w(t)dt$$

for all $w \in C([a,b],R^n), w(a) = 0 = w(b)$, where $D = \frac{d}{dt}$.

We assume $b : PS(T,R^n) \times PS(T,R^n) \to R_+$, $\beta : PS(T,R^n) \times PS(T,R^n) \to R_+$, $\phi : R \to R$ and let $\eta : PS(T,R^n) \times PS(T,R^n) \to C(T,R^n)$, ($\eta(x,u) = 0$ only if $x = u$) and $\zeta : PS(T,R^n) \times PS(T,R^n) \to C(T,R^n)$, ($\zeta(x,u) = 0$ only if $x = u$) be vector functions.

**Definition 5.3.** A differentiable function J is said to be generalized univex at $u \in PS(T,R^n)$ with respect to $\eta$ and $\zeta$ if

$$b(x,u)\phi[J(x) - J(u)] \ge \beta(x,u)[J'(u)\eta(x,u) - \frac{1}{2}J'(u)\zeta(x,u)].$$

**Definition 5.4.** A differentiable function $J$ is said to be generalized pseudo-univex at $u \in PS(T,R^n)$ with respect to $\eta$ and $\zeta$ if

$$\beta(x,u)[J'(u)\eta(x,u) - \frac{1}{2}J'(u)\zeta(x,u)] \geq 0 \Rightarrow b(x,u)\phi[J(x) - J(u)] \geq 0.$$

**Definition 5.5.** differentiable function $J$ is said to be strictly generalized pseudo-univex at $u \in PS(T,R^n)$ with respect to $\eta$ and $\zeta$ if

$$\beta(x,u)[J'(u)\eta(x,u) - \frac{1}{2}J'(u)\zeta(x,u)] \geq 0 \Rightarrow b(x,u)\phi[J(x) - J(u)] > 0.$$

**Definition 5.6.** A differentiable function $J$ is said to be generalized quasi-univex at $u \in PS(T,R^n)$ with respect to $\eta$ and $\zeta$ if

$$b(x,u)\phi[J(x) - J(u)] \leq 0 \Rightarrow \beta(x,u)[J'(u)\eta(x,u) - \frac{1}{2}J'(u)\zeta(x,u)] \leq 0.$$

## 3. Sufficient Optimality Conditions

Now we consider the theorem on the necessary optimality conditions with some extra assumptions. Furthermore, we consider some sufficient optimality conditions regarding problem $(P)$.

**Theorem 5.7.** (Sufficient optimality). Let $x^* \in F_p$, $y^* \in I$, $z^* \in R^m_+$ and $(x^*,y^*,z^*)$ satisfy (7)-(9). Let us define

$$A(x) = \Psi(x^*,y^*)\Phi(x,y^*) - \Psi(x,y^*)\Phi(x^*,y^*),$$

and

$$B(x) = A(x) + \Psi(x^*,y^*)\Omega(x,z^*).$$

Assume, for the given $y^*$ and $z^*$, that any one of the following conditions holds:

**(a)** $\Phi(.,y^*), -\Psi(.,y^*)$ and $\Omega(.,z^*)$ are generalized univex with respect to $b(.,.)$, $\beta(.,.)$, $\phi$, $\eta(.,.)$ and $\zeta(.,.)$, where $b(.,.) > 0$, $\beta(.,.) > 0$ and $\phi$ is superlinear.

**(b)** $A$ is generalized pseudo-univex and $\Omega(.,z^*)$ is generalized quasi-univex with respect to $b(.,.)$, $\beta(.,.)$, $\phi$, $\eta(.,.)$ and $\zeta(.,.)$, where $b(.,.) > 0$, $\beta(.,.) > 0$ and $\phi$ is increasing.

**(c)** $A$ is generalized quasi-univex and $\Omega(.,z^*)$ is strictly generalized pseudo-univex with respect to $b(.,.)$, $\beta(.,.) > 0$, $\phi$, $\eta(.,.)$ and $\zeta(.,.)$, where $b(.,.) > 0$, $\beta(.,.) > 0$ and $\phi$ is increasing.

**(d)** $B$ is generalized pseudo-univex with respect to $b(.,.)$, $\beta(.,.)$, $\phi$, $\eta(.,.)$ and $\zeta(.,.)$, where $b(.,.) > 0$, $\beta(.,.) > 0$ and $\phi$ is increasing.

Then $x^*$ is an optimal solution to $(P)$.

**Proof.** Assume to contrary $x^*$ is not an optimal solution of $(P)$. Then there is a feasible solution $u \in F_p$ such that

$$K(x^*) > K(u). \tag{5.10}$$

Applying (9), we have

$$\frac{\Phi(x^*,y^*)}{\Psi(x^*,y^*)} = K(x^*) > K(u) = \max_{y\in I}\frac{\Phi(u,y)}{\Psi(u,y)} \geq \frac{\Phi(u,y^*)}{\Psi(u,y^*)}, \quad y^* \in I.$$

It follows that

$$A(u) = \Phi(u,y^*)\Psi(x^*,y^*) - \Phi(x^*,y^*)\Psi(u,y^*) < 0 = A(x^*), \tag{5.11}$$

i.e.,

$$A(u) < A(x^*).$$

Now by rearranging the above inequality (11), we get

$$\Psi(x^*,y^*)[\Phi(u,y^*) - \Phi(x^*,y^*)] - \Phi(x^*,y^*)[\Psi(u,y^*) - \Psi(x^*,y^*)] < 0. \tag{5.12}$$

Since $u \in F_p$ , by condition (8), we have

$$\Omega(u,z^*) \leq 0 = \Omega(x^*,z^*). \tag{5.13}$$

Now (11) and (13) yield $B(u) < B(x^*)$.

If condition (a) holds, for $y^* \in I, z^* \in R^m_+$, employing the generalized univexity of $\Phi(.,y^*), -\Psi(.,y^*)$ and $\Omega(.,z^*)$, we have

$$b(u,x^*)\phi\,[\Phi(u,y^*) - \Phi(x^*,y^*)]$$

$$\geq \beta(u,x^*)[\Phi_1'(x^*,y^*)\eta(u,x^*) - \frac{1}{2}\Phi_1'(x^*,y^*)\zeta(u,x^*)], \tag{5.14}$$

$$-b(u,x^*)\phi\,[\Psi(u,y^*) - \Psi(x^*,y^*)] \geq -\beta(u,x^*)[\Psi_1'(x^*,y^*)\eta(u,x^*)$$

$$- \frac{1}{2}\Psi_1'(x^*,y^*)\zeta(u,x^*)], \tag{5.15}$$

$$b(u,x^*)\phi\,[\Omega(u,z^*) - \Omega(x^*,z^*)]$$

$$\geq \beta(u,x^*)[\Omega_1'(x^*,z^*)\eta(u,x^*) - \frac{1}{2}\Omega_1'(x^*,z^*)\zeta(u,x^*)]. \tag{5.16}$$

Here for each $y^*$ and $z^*$, $\Phi_1^{'}, \Psi_1^{'}$ and $\Omega_1^{'}$ are partial Fréchet derivatives of $\Phi(.,y^*), -\Psi(.,y^*)$ and $\Omega(.,z^*)$ at $x^*$, respectively. Since $\Psi(x^*,y^*) > 0$ and $\Phi(x^*,y^*) \geq 0$, we multiply (14) by $\Psi(x^*,y^*)$, (15) by $\Phi(x^*,y^*)$ and (16) by $\Psi(x^*,y^*)$. As $\phi$ is superlinear, by adding up the resulting inequalities, it follows from (12) and (13) that

$$\begin{aligned} 0 > & \left[\Psi(x^*,y^*)\Phi_1^{'}(x^*,y^*) - \Phi(x^*,y^*)\Psi_1^{'}(x^*,y^*)\right] \eta(u,x^*) \\ - & \frac{1}{2}\left[\Psi(x^*,y^*)\Phi_1^{'}(x^*,y^*) - \Phi(x^*,y^*)\Psi_1^{'}(x^*,y^*)\right] \zeta(u,x^*) \\ + & \left[\Psi(x^*,y^*)\Omega_1^{'}(x^*,z^*)\right] \eta(u,x^*) - \frac{1}{2}\left[\Psi(x^*,y^*)\Omega_1^{'}(x^*,z^*)\right] \zeta(u,x^*). \end{aligned}$$

This contradicts an upgraded version of (7). Hence (10) does not hold, and so $x^*$ must be optimal for $(P)$.

For condition (b), $A$ is generalized pseudo-univex and so from (11), $b(.,.) > 0$, $\beta(.,.) > 0$ and $\phi(.)$ is increasing, we get

$$b(u,x^*)\phi\left[A(u) - A(x^*)\right] < 0 \Rightarrow \beta(u,x^*)[A'(x^*)\eta(u,x^*) - \frac{1}{2}A'(x^*)\zeta(u,x^*)] < 0,$$

that is,

$$\begin{aligned} & \beta(u,x^*)[\left[\Psi(x^*,y^*)\Phi_1^{'}(x^*,y^*) - \Phi(x^*,y^*)\Psi_1^{'}(x^*,y^*)\right] \eta(u,x^*) \\ & -\frac{1}{2}\left[\Psi(x^*,y^*)\Phi_1^{'}(x^*,y^*) - \Phi(x^*,y^*)\Psi_1^{'}(x^*,y^*)\right] \zeta(u,x^*) < 0]. \end{aligned} \tag{5.17}$$

The identity (7) and inequality (17) yield

$$\beta(u,x^*)[\Omega_1^{'}(x^*,z^*)\Psi(x^*,y^*)\eta(u,x^*) - \frac{1}{2}\Omega_1^{'}(x^*,z^*)\Psi(x^*,y^*)\zeta(u,x^*)] > 0.$$

Since $\Psi(x^*,y^*) > 0$ and $\beta(u,x^*) > 0$, it follows that

$$\Omega_1^{'}(x^*,z^*)\eta(u,x^*) - \frac{1}{2}\Omega_1^{'}(x^*,z^*)\zeta(u,x^*) > 0. \tag{5.18}$$

On the other hand, the function $\Omega$ in condition (b) assumed to be generalized quasi-univex with respect to $\eta$ and $\zeta$. Thus it implies $\Omega_1^{'}(x^*,z^*)\eta(u,x^*) - \frac{1}{2}\Omega_1^{'}(x^*,z^*)\zeta(u,x^*) \leq 0$, (by the quasi-univexity) which contradicts (6.36). This shows that $x^*$ is optimal for $(P)$. Similarly one can prove for conditions (c) and (d).

# 4. Wolfe Type Dual

In this section, we consider the following Wolfe type model:

$$\text{(WD)} \qquad \text{maximize} \frac{\Phi(u,y)+\Omega(u,z)}{\Psi(u,y)}$$

subject to $(u,z) \in PS(T,R^n) \times R^m_+$,

$$u(a)=\alpha, u(b)=\beta \text{ and } y \in I \subset R^m_+,$$

$$\Psi(u,y)\Phi_1'(u,y) - [\Phi(u,y)+\Omega(u,z)]\Psi_1'(u,y) + \Psi(u,y)\Omega_1'(u,z) = 0. \tag{5.19}$$

Let the set of all feasible solutions of problem (WD) be denoted by $K_1$. We assume throughout this section that

$$\Phi(u,y)+\Omega(u,z) \geq 0 \text{ and } \Psi(u,y) > 0 \text{ for all } (u,y,z) \in K_1. \tag{5.20}$$

In what follows, we plan to establish weak, strong and strict converse duality theorems for $(P)-(WD)$.

**Theorem 5.8.** (Weak duality). Let $x \in F_p$ , $(u,y,z) \in K_1$ and let

$$C(x) = \Psi(u,y)[\Phi(x,y)+\Omega(x,z)] - [\Phi(u,y)+\Omega(u,z)]\Psi(x,y).$$

Suppose that for each, $y$ and $z$, either one of the conditions holds:

**(a)** $\Phi(.,y^*), -\Psi(.,y^*)$ and $\Omega(.,z^*)$ are generalized univex with respect to $b$, $\beta$, $\phi$, $\eta$ and $\zeta$, where $b>0$, $\beta>0$ and $\phi$ is superlinear.

**(b)** $C$ is generalized pseudo-univex with respect to $b$, $\beta$, $\phi$, $\eta$ and $\zeta$, where $b>0$, $\beta>0$ and $\phi$ is increasing.

Then

$$\varphi(x) \geq \frac{\Phi(u,y)+\Omega(u,z)}{\Psi(u,y)}, \tag{5.21}$$

where $\varphi(x)$ is defined by (1).

**Proof.** Suppose, on contrary

$$\varphi(x) < \frac{\Phi(u,y)+\Omega(u,z)}{\Psi(u,y)}. \tag{5.22}$$

It follows from (2) that for any $y \in I$,

$$\frac{\Phi(x,y)}{\Psi(x,y)} \leq \max_{\beta \in I} \frac{\Phi(x,\beta)}{\Psi(x,\beta)} = \varphi(x) < \frac{\Phi(u,y)+\Omega(u,z)}{\Psi(u,y)},$$

$$\Phi(x,y)\Psi(u,y) - [\Phi(u,y)+\Omega(u,z)]\Psi(x,y) < 0. \tag{5.23}$$

Rewriting the above inequality, we get

$$[\Phi(x,y)+\Omega(x,z)]\Psi(u,y)-[\Phi(u,y)+\Omega(u,z)]\Psi(x,y)<\Omega(x,z)\Psi(u,y). \qquad (5.24)$$

Since $\Omega(x,z)\leq 0$, if $x\in F_p$, $z\in R_+^m$ and $\Psi(u,y)>0, \Omega(x,z)\Psi(u,y)\leq 0$. It follows from (24) that

$$C(x)<0=C(u).$$

Since $b(.,.)>0$, $\beta(.,.)>0$ and $(.)\leq 0\Rightarrow \phi(.)<0$, the above inequality yields

$$b(x,u)\phi[C(x)-C(u)]<0. \qquad (5.25)$$

Now if condition (a) holds, for each $y\in I, z\in R_+^m$, the univexity of $\Phi(.,y^*)$, $-\Psi(.,y^*)$ and $\Omega(.,z^*)$ implies that

$$b(x,u)\phi\,[\Phi(x,y)-\Phi(u,y)]\geq \beta(x,u)[\Phi_1'(u,y)\eta(x,u)-\frac{1}{2}\Phi_1'(u,y)\zeta(x,u)], \qquad (5.26)$$

$$-b(x,u)\phi\,[\Psi(x,y)-\Psi(u,y)]\geq \beta(x,u)[-\Psi_1'(u,y)\eta(x,u)-(-\frac{1}{2}\Psi_1'(u,y)\zeta(x,u))], \quad (5.27)$$

$$b(x,u)\phi\,[\Omega(x,z)-\Omega(u,z)]\geq \beta(x,u)[\Omega_1'(u,z)\eta(x,u)-\frac{1}{2}\Omega_1'(u,z)\zeta(x,u)]. \qquad (5.28)$$

From (20), $\Psi(u,y)$ and $\Phi(u,y)+\Omega(u,z)$ are nonnegative. Multiplying (26) by $\Psi(u,y)$, (27) by $\Phi(u,y)+\Omega(u,z)$ and (28) by $\Psi(u,y)$, and summing up the resulting inequalities in light of the superlinearity of $\phi$, we obtain from (23) that

$$\begin{aligned}
&\beta(x,u)[\left\{\Psi(u,y)\Phi_1'(u,y)-[\Phi(u,y)+\Omega(u,z)]\Psi_1'(u,y)\right\}\eta(x,u)\\
-\;&\frac{1}{2}\left\{\Psi(u,y)\Phi_1'(u,y)-[\Phi(u,y)+\Omega(u,z)]\Psi_1'(u,y)\right\}\zeta(x,u)\\
+\;&\left\{\Psi(u,y)\Omega_1'(u,z)\right\}\eta(x,u)-\frac{1}{2}\left\{\Psi(u,y)\Omega_1'(u,z)\right\}\zeta(x,u)]<0.
\end{aligned}$$

This contradicts (19), so (22) does not hold and (6.55) holds.

For the proof under the assumption (b), $C$ is pseudo-univex and (25) implies that

$$\beta(x,u)[C'(u)\eta(x,u)-\frac{1}{2}C'(u)\zeta(x,u)]<0,$$

that is,

$$\beta(x,u)[\Big\{\Psi(u,y)\Phi_1'(u,y)-[\Phi(u,y)+\Omega(u,z)]\Psi_1'(u,y)\Big\}\eta(x,u)$$

$$-\quad \frac{1}{2}\Big\{\Psi(u,y)\Phi_1'(u,y)-[\Phi(u,y)+\Omega(u,z)]\Psi_1'(u,y)\Big\}\zeta(x,u)$$

$$+\quad \Big\{\Psi(u,y)\Omega_1'(u,z)\Big\}\eta(x,u)]-\frac{1}{2}\Big\{\Psi(u,y)\Omega_1'(u,z)\Big\}\zeta(x,u)]<0,$$

which contradicts (19).

**Theorem 5.9.** (Strong duality). Let $x^*$ be an optimal solution of $(P)$. Then there exist $y \in I$ and $z \in R_+^m$ such that $(x^*, y^*, z^*)$ is feasible for $(WD)$. If the hypotheses of Lemma 5.2 and Theorem 5.8 are satisfied, then $(x^*, y^*, z^*)$ is an optimal solution of $(WD)$ and, two problems $(P)$ and $(WD)$ have the same optimal values.

**Proof.** Assume that $x^*$ is an optimal solution of $(P)$. Then, by Lemma 5.2, there exist $y \in I$ and $z \in R_+^m$ satisfying the constraints of $(WD)$, so that $(x^*, y^*, z^*) \in K_1$. Furthermore, we have

$$\frac{\Phi(x^*,y^*)+\Omega(x^*,z^*)}{\Psi(x^*,y^*)}=\frac{\Phi(x^*,y^*)}{\Psi(x^*,y^*)}=\varphi(x^*).$$

Since $\Omega(x^*, z^*) = 0$, it implies $(x^*, y^*, z^*)$ is an optimal solution of $(WD)$. Consequently, (21) demonstrates that $(P)$ and $(WD)$ have the same optimal values.

**Theorem 5.10.** (Strict converse duality). Let $x_1$ and $(x^*, y_0, z_0)$ be optimal solutions for (P) and (WD) that have the same optimal value

$$\varphi(x_1)=\frac{\Phi(x^*,y_0)+\Omega(x^*,z_0)}{\Psi(x^*,y_0)}.$$

**Proof.** Suppose on contrary that $x_1 \neq x^*$. Then by Theorem 5.9, there exist $y_1 \in I$ and $z_1 \in R_+^m$ such that $(x_1, y_1, z_1)$ is an optimal solution of $(WD)$ and

$$\varphi(x_1)=\frac{\Phi(x_1,y_1)+\Omega(x_1,z_1)}{\Psi(x_1,y_1)}.$$

As in the proof of Theorem 5.8, we replace $x$ by $x_1$, and $(u,y,z)$ by $(x^*, y_0, z_0)$ to derive the inequality

$$\varphi(x_1)>\frac{\Phi(x^*,y_0)+\Omega(x^*,z_0)}{\Psi(x^*,y_0)}.$$

This contradicts the fact that

$$\varphi(x_1)=\frac{\Phi(x_1,y_1)+\Omega(x_1,z_1)}{\Psi(x_1,y_1)}=\frac{\Phi(x^*,y_0)+\Omega(x^*,z_0)}{\Psi(x^*,y_0)}.$$

Hence, we conclude that

$$x_1=x^* \text{ and } \varphi(x_1)=\frac{\Phi(x^*,y_0)+\Omega(x^*,z_0)}{\Psi(x^*,y_0)}.$$

## 5. Mond - Weir Type Dual

In this section, we consider the Mond-Weir type dual problems under the assumptions of generalized univex functions as follows:

$$(\text{MWD}) \qquad \text{Maximize } \frac{\Phi(u,y)}{\Psi(u,y)}$$

subject to

$$(u,y) \in PS(T,R^n) \times I, \quad u(a) = \alpha, u(b) = \beta,$$

$$\begin{aligned} &\beta(x,u)[\left\{\Psi(u,y)\Phi_1'(u,y) - (\Phi(u,y) + \Omega(u,z))\Psi_1'(u,y)\right\}\eta(x,u) \\ + &\left\{\Psi(u,y)\Omega_1'(u,z)\right\}\eta(x,u)] \\ - &\frac{1}{2}\beta(x,u)[\left\{\Psi(u,y)\Phi_1'(u,y) - (\Phi(u,y) + \Omega(u,z))\Psi_1'(u,y)\right\}\zeta(x,u) \\ + &\left\{\Psi(u,y)\Omega_1'(u,z)\right\}\zeta(x,u)] < 0, \end{aligned} \tag{5.29}$$

$$\Omega(u,z) \geq 0, z \in R_+^m. \tag{5.30}$$

Let $K_2$ denote the set of all feasible solutions of problem $(MWD)$. We assume throughout that $\Phi(u,y) \geq 0$ and $\Psi(u,y) > 0$ for all $(u,y,z) \in K_2$.

Now we will establish weak, strong and strict converse duality theorems for $(P) - (MWD)$.

**Theorem 5.11.** (Weak duality). Let $x \in F_p, (u,y,z) \in K_2$ and define

$$D(x) = \Psi(u,y)\Phi(x,y) - \Psi(x,y)\Phi(u,y)$$

and

$$E(x) = D(x) + \Psi(u,y)\Omega(x,z).$$

Suppose, for each $y \in I, z \in R_+^m$, that any one of the following conditions holds:

**(a)** $\Phi(.,y), -\Psi(.,y)$ and $\Omega(.,z)$ are generalized univex with respect to $b$, $\beta$, $\phi$, $\eta$ and $\zeta$, where $b > 0$, $\beta > 0$ and $\phi$ is superlinear.

**(b)** $D$ is generalized pseudo-univex and $\Omega(.,z)$ is quasi-univex with respect to $b$, $\beta$, $\phi$, $\eta$ and $\zeta$, where $b > 0$, $\beta > 0$ and $\phi$ is increasing.

**(c)** $D$ is generalized quasi-univex and $\Omega(.,z)$ is strictly pseudo-univex with respect to $b$, $\beta$, $\phi$, $\eta$ and $\zeta$, where $b > 0$, $\beta > 0$ and $\phi$ is increasing.

**(d)** $E$ is generalized pseudo-univex with respect to $b$, $\beta$, $\phi$, $\eta$ and $\zeta$, where $b > 0$, $\beta > 0$ and $\phi$ is increasing.

Then

$$\varphi(x) \geq \frac{\Phi(u,y)}{\Psi(u,y)}. \tag{5.31}$$

**Proof.** Suppose to contrary. Then we have

$$\frac{\Phi(u,y)}{\Psi(u,y)} > \varphi(x) = \max_{y_1 \in I} \frac{\Phi(x,y_1)}{\Psi(x,y_1)} \geq \frac{\Phi(x,y)}{\Psi(x,y)}.$$

This implies that

$$D(x) = \Phi(x,y)\Psi(u,y) - \Phi(u,y)\Psi(x,y) < 0 = D(u). \tag{5.32}$$

As $x \in F_p, z \in R_+^m$ and $\Omega(x,z) \leq 0$, the constraints inequality (30) of $(MWD)$ yields

$$\Omega(x,z) \leq 0 \leq \Omega(u,z). \tag{5.33}$$

Consequently, (32) and (33) yield

$$E(x) < E(u). \tag{5.34}$$

If condition (a) holds, then by the generalized univexity of $\Phi$, $-\Psi$ and $\Omega$, we have

$$b(x,u)\phi\,[\Phi(x,y) - \Phi(u,y)] \geq \beta(x,u)[\Phi_1'(u,y)\eta(x,u) - \frac{1}{2}\Phi_1'(u,y)\zeta(x,u)], \tag{5.35}$$

$$-b(x,u)\phi\,[\Psi(x,y) - \Psi(u,y)] \geq \beta(x,u)[-\Psi_1'(u,y)\eta(x,u) - (-\frac{1}{2}\Psi_1'(u,y)\zeta(x,u))], \tag{5.36}$$

$$b(x,u)\phi\,[\Omega(x,z) - \Omega(u,z)] \geq \beta(x,u)[\Omega_1'(u,z)\eta(x,u) - \frac{1}{2}\Omega_1'(u,z)\zeta(x,u)]. \tag{5.37}$$

Since $\Psi(u,y)$ and $\Phi(u,y)$ are nonnegative, multiplying (35) by $\Psi(u,y)$, (36) by $\Phi(u,y)$ and (37) by $\Psi(u,y)$, and summing up the resulting inequalities in light of superlinearity of $\phi$, we obtain from (32) and (33)

$$0 > \beta(x,u)[\Big[\Psi(u,y)\Phi_1'(u,y) - \Phi(u,y)\Psi_1'(u,y) + \Psi(u,y)\Omega_1'(u,z)\Big]\eta(x,u)$$

$$- \frac{1}{2}\Big[\Psi(u,y)\Phi_1'(u,y) - \Phi(u,y)\Psi_1'(u,y) + \Psi(u,y)\Omega_1'(u,z)\Big]\zeta(x,u)],$$

which contradicts (29). Hence (31) holds.

In condition (b), $D$ is generalized pseudo-univex. Since $b(.,.) > 0$, $\beta(.,.) > 0$ and $(.) \leq 0 \Rightarrow \phi(.) \leq 0$, (32) implies $D'(u)\eta(x,u) < 0$, that is,

$$\beta(x,u)[\left[\Psi(u,y)\Phi_1'(u,y)-\Phi(u,y)\Psi_1'(u,y)\right]\eta(x,u)$$
$$-\quad\frac{1}{2}\left[\Psi(u,y)\Phi_1'(u,y)-\Phi(u,y)\Psi_1'(u,y)\right]\zeta(x,u)]<0. \tag{5.38}$$

It follows from (29) and (38) that

$$-\Psi(u,y)\Omega_1'(u,z)\eta(x,u)-(-\frac{1}{2}\Psi(u,y)\Omega_1'(u,z)\zeta(x,u)<0$$

or

$$\Omega'(u,z)\eta(x,u)-\frac{1}{2}\Omega'(u,z)\zeta(x,u)>0,$$

since $\Psi(u,y)>0$. On the other hand, $\Omega$ is quasi-univex, so the above inequality implies

$$\Omega(x,z)>\Omega(u,z)$$

which contradicts (33). Hence, (31) holds.

Similarly, one can prove for the rest of two cases (c) and (d).

**Theorem 5.12.** (Strong duality). Let $x^*$ be an optimal solution of $(P)$. Then there exist $y^*\in I, z^*\in R_+^m$ such that $(x^*,y^*,z^*)$ is feasible for $(MWD)$. If any of the conditions of Theorem 5.11 hold, then $(x^*,y^*,z^*)$ is an optimal solution of $(MWD)$. Furthermore, $(MWD)$ and $(P)$ have the same optimal values.

**Proof.** If $x^*$ is an optimal solution of $(P)$, then by Theorem 5.11, there exist $y^*\in I$ and $z^*\in R_+^m$ such that $(x,y,z)\in K_2$, the feasible solution of $(MWD)$ and

$$\varphi(x^*)=\frac{\Phi(x^*,y^*)}{\Psi(x^*,y^*)}\leq\max_{y\in I}\frac{\Phi(x^*,y)}{\Psi(x^*,y)}.$$

The optimality of $(MWD)$ follows from Theorem 5.11.

**Theorem 5.13.** (Strict converse duality). Let $x_1$ and $(x^*,y_0,z_0)$ be optimal solutions of $(P)$ and $(MWD)$, respectively. Assume that the hypothesis of Theorem 5.12 is fulfilled. Assume that $D(.)=\Psi(x^*,y_0)\Phi(.,y_0)-\Psi(.,y_0)\Phi(x^*,y_0)$ is strictly pseudo-univex and $\Omega(.,z_0)$ is quasi-univex with respect to $b$, $\beta$, $\phi$, $\eta$ and $\zeta$. Then $x_1=x^*$ is an optimal solution of $(P)$, and $(P)$ and $(MWD)$ have the same optimal values

$$\varphi(x_1)=\Phi(x^*,y_0)/\Psi(x^*,y_0).$$

**Proof.** Since $x_1$ is an optimal solution of $(P)$, by Theorem 5.12, there exist $y_1\in I$ and $z_1\in R_+^m$ such that $(x_1,y_1,z_1)$ is an optimal solution of $(MWD)$ and

$$\varphi(x_1)=\frac{\Phi(x_1,y_1)}{\Psi(x_1,y_1)}.$$

On the other hand, $(x^*, y_0, z_0)$ is an optimal solution of $(MWD)$ with optimal values equal to $\varphi(x_1)$, that is,

$$\left[\varphi(x_1) = \frac{\Phi(x_1, y_1)}{\Psi(x_1, y_1)} = \frac{\Phi(x^*, y_0)}{\Psi(x^*, y_0)}\right]. \tag{5.39}$$

If $x_1 \neq x^*$, we could deduce

$$\left[\varphi(x_1) > \frac{\Phi(x^*, y_0)}{\Psi(x^*, y_0)}\right], \tag{5.40}$$

which contradicts (39).

Assume (40) does not hold. Then $\varphi(x_1) = \Phi(x^*, y_0)/\Psi(x^*, y_0)$, that is,

$$\frac{\Phi(x^*, y^*)}{\Psi(x^*, y^*)} \geq \varphi(x_1) = \max_{y \in I} \frac{\Phi(x_1, y)}{\Psi(x_1, y)} \geq \frac{\Phi(x_1, y_0)}{\Psi(x_1, y_0)}$$

or

$$D(x_1) = \Phi(x_1, y_0)\Psi(x^*, y_0) - \Phi(x^*, y_0)\Psi(x_1, y_0) \leq 0 = D(x^*),$$

that is,

$$D(x_1) \leq D(x^*).$$

Since $b(.,.) \geq 0$ and $(.) \leq 0 \Rightarrow \phi(.) \leq 0$, the above inequality yields

$$b(x_1, x^*)\phi[D(x_1) - D(x^*)] \leq 0. \tag{5.41}$$

Since $D$ is strictly pseudo-univex, (41) implies
$D'(x^*)\eta(x_1, x^*) - \frac{1}{2}D'(x^*)\zeta(x_1, x^*) < 0,$

$$\begin{aligned} & [\Psi(x^*, y_0)\Phi_1'(x^*, y_0) - \Phi(x^*, y_0)\Psi_1'(x^*, y_0)]\eta(x_1, x^*) \\ - & \frac{1}{2}[\Psi(x^*, y_0)\Phi_1'(x^*, y_0) - \Phi(x^*, y_0)\Psi_1'(x^*, y_0)]\zeta(x_1, x^*) < 0. \end{aligned} \tag{5.42}$$

By (29) and (42), we get

$$-\Psi(x^*, y_0)\Omega_1'(x^*, z_0)\eta(x_1, x^*) - (-\frac{1}{2}\Psi(x^*, y_0)\Omega_1'(x^*, z_0)\zeta(x_1, x^*) < 0,$$

i.e.,

$$\Omega_1'(x^*, z_0)\eta(x_1, x^*) - \frac{1}{2}\Omega_1'(x^*, z_0)\zeta(x_1, x^*) > 0. \tag{5.43}$$

Since $\Psi(x^*, y_0) > 0$ by the quasi-univexity of $\Omega$, (43) implies

$$b(x_1, x^*)\phi\left[\Omega(x_1, z_0) - \Omega(x^*, z_0)\right] > 0.$$

Since $b(.,.) \geq 0$ and $\phi(.) \geq 0 \Rightarrow (.) \geq 0$, the above inequality yields

$$\Omega(x_1, z_0) > \Omega(x^*, z_0). \tag{5.44}$$

On the other hand, $x_1 \in F_p, z_0 \in R_+^m$ and $\Omega(x_1, z_0) \leq 0$. By (30), $\Omega\ (x^*, z_0) \geq 0$. Thus $\Omega(x_1, z_0) \leq \Omega\ (x^*, z_0)$, this contradicts (44). Hence (40) holds under $x_1 \neq x^*$. Therefore $x_1 = x^*$.

## 6. General Remarks

The obtained results can be generalized not only to the case of the $(\mathscr{F}, b, \phi, \eta, \zeta)-$*invexity*, but also to the $(b, c, \phi, \eta, \zeta)-$*invexity* and beyond. For more details, we refer the reader to [8].

# Bibliography


[1] S. Chandra, B.D. Craven and B. Mond, Generalized Fractional programming duality: a ratio game approach, *J. Austral. Math. Soc. Ser. B*, **28** (1986) 170 - 180.

[2] J.P. Crouzeix, J.A. Ferland and S. Schaible, Duality in generalized fractional programming, *Math. Prog.* **27** (1983) 342 - 354.

[3] H.C. Lai and J.C. Liu, Minimax fractional programming involving generalized invex functions, *ANZIAM J.* **44**, (2003) 339 - 354.

[4] S.K. Mishra, K. Shukla and R.U. Verma, Minmax fractional integral programming problems involving univexity, *DCDIS, Series A: Mathematical Analysis* **18** (2012), 335 - 355.

[5] B. Mond, S. Chandra and I. Husain, Duality for variational problems with invexity, *J. Math. Anal. Appl.* **134** (1988) 322 - 328.

[6] B. Mond and I. Husain, Sufficient optimality criteria and duality for variational problems with generalized invexity, *J. Austral. Math. Soc. Ser. B* **31** (1) (1989) 108 - 121.

[7] I.M. Stancu-Minasian, *Fractional Programming Theory, Methods and Applications*, Kluwer, Dordrecht, 1997.

[8] R.U. Verma, *Semi-Infinite Fractional Programming*, Springer Nature, Singapore, 2017.

# Chapter 6

# Generalized Fractional Integral Programming

In this chapter, we present a class of generalized minmax fractional integral type programming problems by applying a class of ($\phi$, $c$, $\beta$, $\gamma$, $\eta$, $\zeta$, $\omega$)- univexities, which is equivalent to minimizing a maximum of several time dependent ratios involving integral expressions. First, some optimality conditions based on ($\phi$, $c$, $\beta$, $\gamma$, $\zeta$, $\eta$, $\omega$)- univexities are established, and then some classes of Wolfe type dual models and Mond-Weir type dual models are explored to the context of developing a class of related duality results, which are in terms of weak, strong and strictly converse duality theorems based on ($\phi$, $c$, $\beta$, $\gamma$, $\eta$, $\zeta$, $\omega$)- univexity frameworks. This chapter would prove extremely advancing further research in light of interdisciplinary research collaboration worldwide.

## 1. Introduction

Recently, the role of fractional integral type programming is one of the most research - enhanced fast-expanding areas of optimization control which features in several types of real-world problems. It can be applied to different branches of engineering (including multitime multiobjective mechanical engineering problems) as well as to economics to minimize a ratio of functions between given periods of time. Furthermore, it can be utilized as a resource in order to measure the efficiency or productivity of a system. In such type of problems the objective function is given as a ratio of functions. In this chapter, we consider a problem that deals with minimizing a maximum of several time dependent ratios involving integral expressions. Among researchers who have investigated this type of problems are Chandra et al. [1], Crouzeix et al. [2], Mond et al. [5], Mond and Husain [6], and others, while we generalize the results in Mishra et al. [4] to a larger class of nonconvex minmax fractional optimization problems, that is, to a class of hybrid minmax fractional integral type problems in which the involved functionals are ($\phi$, $b$, $\beta$, $\eta$, $\omega$)- univex. The obtained results extend (and unify) the works of Mishra et al. [4], Lai et al. [3], and others. As our problem involves a state function $x(t)$ in the integrand of the integrals, the obtained results also relate to optimal control problems. For more details, we refer the reader to publications [1 - 10].

Consider the following hybrid minmax fractional variational problem:

$$\text{(P)} \qquad \min_{x} \left\{ \sum_{1\le i\le p} \max \frac{\int_a^b f^i(t,x,\dot{x})dt}{\int_a^b g^i(t,x,\dot{x})dt} \right\}$$

subject to $x \in PS(T,R^n), x(a) = \alpha, x(b) = \beta$,

$$\int_a^b h^j(t,x,\dot{x})dt \le 0, j \in \underline{m} \equiv \{1,2,\ldots,m\},$$

where functions $f^i, g^i, i \in \underline{p}$ and $h^j, j \in \underline{m}$ are continuous in $t$ and have continuous partial derivatives with respect to $x$ and $\dot{x}$; $T = [a,b]$ denotes the time space, and $PS(T,R^n)$ stands for the state space of all piecewise smooth functions $x : T \to R^n$ with norm defined by

$$\|x\| = \|x\|_\infty + \|Dx\|_\infty$$

and $D$ is the differential operator on $PS(T,R^n)$ defined by

$$y = Dx \text{ if and only if } x(t) = x(a) + \int_a^b y(s)ds.$$

Thus $D = \frac{d}{dt}$ except at the point of discontinuity. Without loss of any generality, we may assume throughout the paper

$$\int_a^b g^i(t,x,\dot{x}) > 0, \int_a^b f^i(t,x,\dot{x}) \ge 0 \text{ for each } i \in \underline{p}.$$

Let $F_p$ denote the set of all feasible solutions to $(P)$.

## 2. Preliminaries

This section deals with the fundamentals to this chapter.
For $x \in PS(T,R^n)$, let us define as follows:

$$F_i(x) = \int_a^b f^i(t,x,\dot{x})dt,$$

$$G_i(x) = \int_a^b g^i(t,x,\dot{x})dt$$

for $i \in \underline{p}$ and

$$H_j(x) = \int_a^b h^j(t,x,\dot{x})dt$$

for $j \in \underline{m}$.

Based on the assumptions imposed on the functions $f^i$, $g^i$ and $h^j$ constituting the considered hybrid minmax fractional integral type problem $(P)$, it follows that the functionals $F = (F_1,...,F_p)$, $G = (G_1,..,G_p)$, $H = (H_1,...,H_m)$ are Fréchet differentiable on $PS(T,R^n)$. In order to simplify the symbols in problem $(P)$, we scalarize functionals as follows:

$$\Phi(x,y) = \langle y, F(x)\rangle = \sum_{i=1}^{p} y_i F_i(x) = \sum_{i=1}^{p} y_i \int_a^b f^i(t,x,\dot{x})dt,$$

$$\Psi(x,y) = \langle y, G(x)\rangle = \sum_{i=1}^{p} y_i G_i(x) = \sum_{i=1}^{p} y_i \int_a^b g^i(t,x,\dot{x})dt,$$

$$\Omega(x,y) = \langle z, H(x)\rangle = \sum_{i=1}^{m} z_j H_j(x) = \sum_{i=1}^{m} z_j \int_a^b h^j(t,x,\dot{x})dt,$$

where $y \in I = \left\{ y \in R_+^p | \sum_{i=1}^{p} y_i = 1 \right\}$ and $z \in R_+^m$. Then for any feasible solution $x$ of $(P)$, the objective fractional function can be represented by

$$\varphi(x) = \max_{i \in \underline{p}} \frac{F_i(x)}{G_i(x)} = \max_{y \in I} \frac{\langle y, F(x)\rangle}{\langle y, G(x)\rangle} = \max_{y \in I} \frac{\Phi(y,x)}{\Psi(y,x)}. \tag{6.1}$$

The problem $(P)$ is equivalent to

$$(\tilde{P}) \min_{x \in PS} \max_{y \in I} \frac{\Phi(y,x)}{\Psi(y,x)} = \min_{x} \left( \max_{i \in \underline{p}} \frac{F_i(x)}{G_i(x)} \right)$$

subject to $x \in PS(T,R^n)$, $x(a) = \alpha$, $x(b) = \beta$ and $H(x) \leq 0$.
It is equivalent to the parametric minimization problem:

$(EP_v)$ minimize $q(v)$

subject to $F_i(x) - vG_i(x) \leq q$, $i \in \underline{p}$, and $H_j(x) \leq 0$, $j \in \underline{m}$,

$$x \in PS(T,R^n), x(a) = \alpha, x(b) = \beta.$$

From $(EP_v)$, one can reduce to the optimal solution $x^*$ for $(P)$ with its optimal value $v^*$ which is given by

$$v^* = \varphi(x^*) = \max_{i \in \underline{p}} \frac{F^i(x^*)}{G^i(x^*)} = \max_{y \in I} \frac{\Phi(x^*, y)}{\Psi(x^*, y)} = \frac{\Phi(x^*, y^*)}{\Psi(x^*, y^*)}. \tag{6.2}$$

The concept used here for solution of $(EP_v)$, coincides with finding the minimax solution $(x^*, y^*)$ of the Lagrangian

$$\begin{aligned} L(x,y;x,z) &= \langle y, F(x)\rangle - v\langle y, G(x)\rangle + \langle z, H(x)\rangle \\ &= \Phi(x,y) - v\Psi(x,y) + \Omega(x,z), \end{aligned} \tag{6.3}$$

with multipliers $v^* \in R_+$ and $z^* \in R^m_+$. The minimax solution $(x^*, y^*)$ is given in the equation

$$L_x'(x,y)\xi = 0 \text{ for all } \xi \in C(T, R^n).$$

We recall some of the following auxiliary results in order to establish our main results on hand regarding problem $(P)$:

**Lemma 6.1.** [3] Let $x^*$ be an optimal solution of $(P)$. Then there exist Lagrangian multipliers $y^* \in I \subset R^p_+$ and $z^* \in R^m_+$ such that the Kuhn-Tucker type conditions hold for the Lagrangian (6.3):

$$L_x'(x^*, y^*; v^*, z^*) = 0, \tag{6.4}$$

$$\Omega(x^*, z^*) = 0 \tag{6.5}$$

where $v^* = \dfrac{\Phi(x^*, y^*)}{\Psi(x^*, y^*)}$ given in (6.2) is the optimal value of problem (P), and the equation $L_x'(\cdot,\cdot) = 0$ is the expressed by

$$\Psi(x^*, y^*)\left[\Phi'(x^*, y^*) + \Omega'(x^*, z^*)\right] - \Phi(x^*, y^*)\Psi'(x^*, y^*) = 0. \tag{6.6}$$

**Lemma 6.2.** [3] Let $x^*$ be an optimal solution of $(P)$. Then there exist $y^* \in I \subset R^p_+$, multipliers $v^* \in R$ and $z^* \in R^m_+$ such that $(x^*, y^*, z^*)$ satisfies

$$\Psi(x^*, y^*)\Phi_1'(x^*, y^*) - \Phi(x^*, y^*)\Psi_1'(x^*, y^*) + \Psi(x^*, y^*)\Omega_1'(x^*, z^*) = 0, \tag{6.7}$$

$$\Omega(x^*, z^*) = 0, \tag{6.8}$$

and obtain the optimal value by

$$K(x^*) = \frac{\Phi(x^*, y^*)}{\Psi(x^*, y^*)} = \max_{y \in I} \frac{\Phi(x^*, y)}{\Psi(x^*, y)} \left( = \min_x \max_{1 \le i \le p} \frac{F^i(x)}{G^i(x)} \right). \tag{6.9}$$

Now we consider ($\phi$, $b$, $\beta$, $\eta$)- univexity concepts, which encompass most of the concepts available in the literature. Let $x \in PS(T, R^n)$ and $f \in C^1(T \times R^n \times R^n)$. Define a functional $J : PS(T,R^n) \to R$ by $J(x) = \int_a^b f(t,x,\dot{x})dt$. We assume that the boundary points $x(a)$ and $x(b)$ are fixed. Consider the admissible vectors variations $w \in C([a,b],R^n)$ vanishing at the boundary points. Then the differential of $J$ is a linear functional on $C([a,b],R^n)$,

$$\begin{aligned} J'(x)w &= \frac{d}{d\alpha}\int_a^b f(t,x+\alpha w,\dot{x}+\alpha\dot{w})dt|_{\alpha=0} \\ &= \int_a^b [f_x(t,x,\dot{x})w(t)+f_{\dot{x}}(t,x,\dot{x})\dot{w}(t)]\,dt \\ &= \int_a^b [f_x(t,x,\dot{x})-Df_{\dot{x}}(t,x,\dot{x})]\,w(t)dt+f_{\dot{x}}(t,x,\dot{x})\dot{w}(x)|_a^b \\ &= \int_a^b [f_x(t,x,\dot{x})-Df_{\dot{x}}(t,x,\dot{x})]\,w(t)dt. \end{aligned}$$

Thus, we have

$$J'(x)w = \int_a^b [f_x(t,x,\dot{x})-Df_{\dot{x}}(t,x,\dot{x})]\,w(t)dt$$

for all $w \in C([a,b],R^n), w(a)=0=w(b)$, where $D = \frac{d}{dt}$.

We further assume that $c : PS(T,R^n)\times PS(T,R^n) \to R_+$, $\beta : PS(T,R^n)\times PS(T,R^n) \to R_+$, $\phi : R \to R$, and let $\eta : PS(T,R^n)\times PS(T,R^n) \to C(T,R^n)$, ($\eta(x,u) = 0$ only if $x = u$), $\omega : PS(T,R^n)\times PS(T,R^n) \to C(T,R^n)$, ($\omega(x,u) = 0$ only if $x = u$) and $\zeta : PS(T,R^n)\times PS(T,R^n) \to C(T,R^n)$, ($\zeta(x,u) = 0$ only if $x = u$) be vector functions.

**Definition 6.3.** A differentiable function J is said to be ($\phi$, $c$, $\beta$, $\gamma$, $\eta$, $\omega$, $\zeta$) - univex at $u \in PS(T,R^n)$ with respect to $c$, $\beta$, $\gamma$, $\eta$, $\omega$ and $\zeta$ if

$$\begin{aligned} & c(x,u)\phi[J(x)-J(u)] \geq \beta(x,u)[J'(u)\eta(x,u)-\frac{1}{2}J'(u)\omega(x,u)] \\ + & \frac{1}{2}\gamma(x,u)[J'(u)\zeta(x,u)]. \end{aligned}$$

**Definition 6.4.** A differentiable function $J$ is said to be ($\phi$, $b$, $\beta$, $\eta$, $\omega$ $\zeta$)- pseudo-univex at $u \in PS(T,R^n)$ with respect to $c$, $\beta$, $\gamma$, $\eta$, $\zeta$, $\omega$ and $\phi$ if

$$\begin{aligned} & \beta(x,u)[J'(u)\eta(x,u)-\frac{1}{2}J'(u)\omega(x,u)]+\frac{1}{2}\gamma(x,u)[J'(u)\zeta(x,u)] \geq 0 \\ & \Rightarrow c(x,u)\phi[J(x)-J(u)] \geq 0, \end{aligned}$$

equivalently,

$$\begin{aligned}
&c(x,u)\phi[J(x)-J(u)]<0 \Rightarrow \\
&\beta(x,u)[J'(u)\eta(x,u)-\frac{1}{2}J'(u)\omega(x,u] \\
&+\frac{1}{2}\gamma(x,u)[J'(u)\zeta(x,u]<0.
\end{aligned}$$

**Definition 6.5.** A differentiable function $J$ is said to be strictly ($\phi$, $c$, $\beta$, $\gamma$, $\eta$, $\zeta$, $\omega$)-pseudo-univex at $u \in PS(T,R^n)$ with respect to $\eta$, $\omega$ and $\phi$ if

$$\begin{aligned}
&\beta(x,u)[J'(u)\eta(x,u)-\frac{1}{2}J'(u)\omega(x,u)]+\gamma(x,u)\frac{1}{2}[J'(u)\zeta(x,u)]\geq 0 \\
&\Rightarrow c(x,u)\phi[J(x)-J(u)]>0,
\end{aligned}$$

equivalently,

$$\begin{aligned}
&c(x,u)\phi[J(x)-J(u)]\leq 0 \\
&\Rightarrow \beta(x,u)[J'(u)\eta(x,u)-\frac{1}{2}J'(u)\zeta(x,u)]<0.
\end{aligned}$$

**Definition 6.6.** A differentiable function $J$ is said to be ($\phi$, $c$, $\beta$, $\gamma$, $\eta$, $\zeta$, $\omega$)- quasi-univex at $u \in PS(T,R^n)$ with respect to $b$, $\beta$, $\eta$, $\omega$ and $\phi$if

$$\begin{aligned}
&c(x,u)\phi[J(x)-J(u)]\leq 0 \Rightarrow \\
&\beta(x,u))[J'(u)\eta(x,u)-\frac{1}{2}J'(u)\omega(x,u)] \\
&+\gamma(x,u)\frac{1}{2}J'(u)\zeta(x,u)]\leq 0,
\end{aligned}$$

equivalently,

$$\begin{aligned}
&\beta(x,u)[J'(u)\eta(x,u)-\frac{1}{2}J'(u)\omega(x,u)]+\gamma(x,u)\frac{1}{2}[J'(u)\zeta(x,u)]>0 \\
&\Rightarrow c(x,u)\phi[J(x)-J(u)]>0.
\end{aligned}$$

**Remark, 6.7.** For $\omega=\eta$, a differentiable function $J$ is ($\phi$, $c$, $\beta$, $\gamma$, $\zeta$, $\eta$)- univex if

$$c(x,u)\phi[J(x)-J(u)]\geq\beta(x,u)\frac{1}{2}[J'(u)\eta(x,u)]+\frac{1}{2}\gamma(x,u)[J'(u)\zeta(x,u)].$$

**Remark 6.8.** For $\zeta=\omega=\eta$ and $\gamma=\beta$, a differentiable function $J$ is ($\phi$, $c$, $\beta$, $\eta$)- univex if

$$c(x,u)\phi[J(x)-J(u)] \geq \beta(x,u)[J'(u)\eta(x,u)].$$

**Remark 6.9.** For $\beta = \gamma = 1$ and $\zeta = \omega = \eta$, a differentiable function $J$ is $(\phi, c, \eta)$- univex [4] if

$$c(x,u)\phi[J(x)-J(u)] \geq [J'(u)\eta(x,u)].$$

**Remark 6.10.** For $c = \beta = \gamma = 1$ and $\zeta = \omega = \eta$, a differentiable function $J$ said to be $(\phi, \eta)$- univex [3] if

$$\phi[J(x)-J(u)] \geq [J'(u)\eta(x,u)].$$

# 3. Sufficient Optimality Conditions

Next, we consider the theorem on the necessary optimality conditions with some extra assumptions. Furthermore, we explore some sufficient optimality conditions regarding problem $(P)$.

**Theorem 6.11.** (Sufficient optimality). Let $x^* \in F_p$, $y^* \in I$, $z^* \in R_+^m$ and $(x^*, y^*, z^*)$ satisfy (6.7) - (6.9). Let us define

$$A(x) = \Psi(x^*, y^*)\Phi(x, y^*) - \Psi(x, y^*)\Phi(x^*, y^*),$$

and

$$B(x) = A(x) + \Psi(x^*, y^*)\Omega(x, z^*).$$

Assume, for the given $y^*$ and $z^*$, that any one of the following conditions holds:

**(a)** $A$ is $\Phi(., y^*)$, $-\Psi(., y^*)$ and $\Omega(., z^*)$ are $(\phi, c, \beta, \gamma, \zeta, \eta, \omega)$- univex with respect to $c(.,.), \beta(.,.), \gamma(.,.), \phi, \zeta(.,.), \eta(.,.)$ and $\omega(.,.)$, where $c(.,.) > 0$, $\beta(.,.) > 0$, $\gamma(.,.) > 0$ and $\phi$ is superlinear.

**(b)** $A$ is $(\phi, c, \beta, \gamma, \zeta, \eta, \omega)$- pseudo-invex, and $\Omega(., z^*)$ is $(\phi, c, \beta, \gamma, \eta, \omega, \zeta)$- quasi-invex with respect to $c(.,.)$, $\beta(.,.)$, $\gamma(.,.)$, $\phi$, $\zeta(.,.)$, $\eta(.,.)$ and $\omega(.,.)$, where $c(.,.) > 0$, $\beta(.,.) > 0$, $\gamma(.,.) > 0$ and $\phi$ is strictly increasing.

**(c)** $A$ is $(\phi, c, \beta, \gamma, \zeta, \eta, \omega)$- quasi-invex and $\Omega(., z^*)$ is strictly $(\phi, c, \beta, \gamma, \zeta, \eta, \omega)$- pseudo-invex with respect to $c(.,.)$, $\beta(.,.)$, $\gamma(.,.)$, $\phi$, $\zeta(.,.)$, $\eta(.,.)$ and $\omega(.,.)$, where $c(.,.) > 0$, $\beta(.,.) > 0$, $\gamma(.,.) > 0$, and $\phi$ is increasing.

**(d)** $B$ is $(\phi, c, \beta, \gamma, \zeta, \eta, \omega)$- pseudo-invex with respect to $c(.,.)$, $\beta(.,.)$, $\gamma(.,.)$, $\phi$, $\eta(.,.)$, $\zeta(.,.)$ and $\omega(.,.)$, where $c(.,.) > 0$, $\beta(.,.) > 0$ $\gamma(.,.) > 0$ and $\phi$ is increasing.

Then $x^*$ is an optimal solution to $(P)$.

**Proof.** Assume to contrary $x^*$ is not an optimal solution of $(P)$. Then there is a feasible solution $u \in F_p$ such that (in light of (9))

$$K(x^*) > K(u). \tag{6.10}$$

Applying (6.9), we have

$$\frac{\Phi(x^*,y^*)}{\Psi(x^*,y^*)} = K(x^*) > K(u) = \max_{y \in I} \frac{\Phi(u,y)}{\Psi(u,y)} \geq \frac{\Phi(u,y^*)}{\Psi(u,y^*)}, \quad y^* \in I.$$

It follows that

$$A(u) = \Phi(u,y^*)\Psi(x^*,y^*) - \Phi(x^*,y^*)\Psi(u,y^*) < 0 = A(x^*), \tag{6.11}$$

i.e.,

$$A(u) < A(x^*).$$

Now by rearranging the above inequality (11), we get

$$\Psi(x^*,y^*)[\Phi(u,y^*) - \Phi(x^*,y^*)] - \Phi(x^*,y^*)[\Psi(u,y^*) - \Psi(x^*,y^*)] < 0. \tag{6.12}$$

Since $u \in F_p$ , by condition (6.8), we have

$$\Omega(u,z^*) \leq 0 = \Omega(x^*,z^*). \tag{6.13}$$

Now (11) and (13) yield $B(u) < B(x^*)$.

If condition (a) holds, for $y^* \in I, z^* \in R^m_+$, employing the $(\phi, c, \beta, \gamma, \eta, \zeta, \omega)$- univexity of $\Phi(.,y^*), -\Psi(.,y^*)$ and $\Omega(.,z^*)$, we have

$$\begin{aligned} & c(u,x^*)\phi\,[\Phi(u,y^*) - \Phi(x^*,y^*)] \\ \geq\ & \beta(u,x^*)[\Phi_1'(x^*,y^*)\eta(u,x^*) - \frac{1}{2}\Phi_1'(x^*,y^*)\omega(u,x^*)] \\ & + \frac{1}{2}\gamma(u,x^*)[\Phi_1'(x^*,y^*)\zeta(u,x^*)], \end{aligned} \tag{6.14}$$

$$\begin{aligned} & -c(u,x^*)\phi\,[\Psi(u,y^*) - \Psi(x^*,y^*)] \\ \geq\ & -\beta(u,x^*)[\Psi_1'(x^*,y^*)\eta(u,x^*) - \frac{1}{2}\Psi_1'(x^*,y^*)\omega(u,x^*)] \\ & - \frac{1}{2}\gamma(u,x^*)[\Psi_1'(x^*,y^*)\zeta(u,x^*)], \end{aligned} \tag{6.15}$$

$$\begin{aligned} & c(u,x^*)\phi\,[\Omega(u,z^*) - \Omega(x^*,z^*)] \\ \geq\ & \beta(u,x^*)[\Omega_1'(x^*,z^*)\eta(u,x^*) - \frac{1}{2}\Omega_1'(x^*,z^*)\omega(u,x^*)] \\ & + \frac{1}{2}\gamma(u,x^*)[\Omega_1'(x^*,z^*)\zeta(u,x^*)]. \end{aligned} \tag{6.16}$$

Here for each $y^*$ and $z^*$, $\Phi_1^{'}, \Psi_1^{'}$ and $\Omega_1^{'}$ are partial Frechet derivatives of $\Phi(.,y^*)$, -$\Psi(.,y^*)$ and $\Omega(.,z^*)$ at $x^*$, respectively. Since $\Psi(x^*,y^*) > 0$ and $\Phi(x^*,y^*) \geq 0$, we multiply (14) by $\Psi(x^*,y^*)$, (15) by $\Phi(x^*,y^*)$ and (16) by $\Psi(x^*,y^*)$. As $\phi$ is superlinear, by adding up the resulting inequalities, it follows from (12) and (13) that

$$
\begin{aligned}
&0 > \beta(u,x^*)\big[[\Psi(x^*,y^*)\Phi_1^{'}(x^*,y^*) - \Phi(x^*,y^*)\Psi_1^{'}(x^*,y^*)] \\
&+[\Psi(x^*,y^*)\Omega_1^{'}(x^*,z^*)]\big]\eta(u,x^*) \\
- \quad &\beta(u,x^*)\frac{1}{2}\big[\Psi(x^*,y^*)\Phi_1^{'}(x^*,y^*) + \Phi(x^*,y^*)\Psi_1^{'}(x^*,y^*)] \\
+ \quad &[\Psi(x^*,y^*)\Omega_1^{'}(x^*,z^*)\omega(u,x^*)]\big] \\
&+\frac{1}{2}\gamma(u.x^*)\big[\Psi(x^*,y^*)\Phi_1^{'}(x^*,y^*) - \Phi(x^*,y^*)\Psi_1^{'}(x^*,y^*)] \\
&+[\Psi(x^*,y^*)\Omega_1^{'}(x^*,z^*)]\big]\zeta(u,x^*).
\end{aligned}
$$

This leads to a contradiction in light of (6.7). Hence (10) does not hold, and so $x^*$ must be an optimal solution of $(P)$.

For condition (b), $A$ is ($\phi$, $c$, $\beta$, $\gamma$, $\zeta$, $\eta$, $\omega$)- pseudo-univex (applying the equivalent form of the pseudo-univexity of $A$ in this situation) and so from (11), for $c(.,.) > 0$, $\beta(.,.) > 0$, $\gamma(.,.) > 0$ and $\phi(.)$ is strictly increasing, we have

$$
\begin{aligned}
&c(u,x^*)\phi[A(u) - A(x^*)] < 0 \Rightarrow \\
&\beta(u,x^*)\Big([A'(x^*)\eta(u,x^*)] - \frac{1}{2}[A'(x^*)\omega(u,x^*)]\Big) \\
&+\frac{1}{2}\gamma(u,x^*)\big([A'(x^*)\zeta(u,x^*)]\big) < 0,
\end{aligned}
$$

that is,

$$
\begin{aligned}
&\beta(u,x^*)\Big[\Big[\Psi(x^*,y^*)\Phi_1^{'}(x^*,y^*) - \Phi(x^*,y^*)\Psi_1^{'}(x^*,y^*)\Big]\eta(u,x^*)] \\
&-\frac{1}{2}[\Big[\Psi(x^*,y^*)\Phi_1^{'}(x^*,y^*) - \Phi(x^*,y^*)\Psi_1^{'}(x^*,y^*)\Big]\omega(u,x^*)] \\
+ \quad &\frac{1}{2}\gamma(u,x^*)\Big[\Big[\Psi(x^*,y^*)\Phi_1^{'}(x^*,y^*) - \Phi(x^*,y^*)\Psi_1^{'}(x^*,y^*)\Big] \\
&\zeta(u,x^*)] < 0. \qquad (6.17)
\end{aligned}
$$

The identity (7) and inequality (17) yield

$$\beta(u,x^*)\Big[[\Omega_1'(x^*,z^*)\Psi(x^*,y^*)\eta(u,x^*)]$$

$$-\frac{1}{2}[\Omega_1'(x^*,z^*)\Psi(x^*,y^*)\omega(u,x^*)]$$

$$+\frac{1}{2}\gamma(u,x^*)[\Omega_1'(x^*,z^*)\Psi(x^*,y^*)\zeta(u,x^*)] > 0.$$

Since $\beta(u,x^*) > 0$, $\gamma(u,x^*) > 0$ and $\Psi(x^*,y^*) > 0$, it follows that

$$\Omega_1'(x^*,z^*)\eta(u,x^*) - \Omega_1'(x^*,z^*)\omega(u,x^*) + \Omega_1'(x^*,z^*)\zeta(u,x^*) > 0. \tag{6.18}$$

On the other hand, the function $\Omega$ in condition (b) assumed to be ($\phi$, $c$, $\beta$, $\gamma$, $\zeta$, $\eta$, $\omega$) - quasi-invex with respect to $\eta$, $\zeta$ and $\omega$. Thus it implies

$$\beta(u,x^*)\Big[[\Omega_1'(x^*,z^*)\eta(u,x^*)]$$

$$-\frac{1}{2}[\Omega_1'(x^*,z^*)\omega(u,x^*)]\Big]$$

$$+\frac{1}{2}\gamma(u,x^*)[\Omega_1'(x^*,z^*)\zeta(u,x^*)]\Big] \leq 0$$

(by the ($\phi$, $c$, $\beta$, $\gamma$, $\eta$, $\zeta$, $\omega$)-quasi-univexity) which contradicts (18). This shows that $x^*$ is optimal for $(P)$.
Similarly one can prove for conditions (c) and (d).

**Theorem 6.12.** (Sufficient optimality). Let $x^* \in F_P$, $y^* \in I$, $z^* \in R_+^m$ and $(x^*,y^*,z^*)$ satisfy (6.7) - (6.9). Let us define

$$A(x) = \Psi(x^*,y^*)\Phi(x,y^*) - \Psi(x,y^*)\Phi(x^*,y^*),$$

and

$$B(x) = A(x) + \Psi(x^*,y^*)\Omega(x,z^*).$$

Assume, for the given $y^*$ and $z^*$, that any one of the following conditions holds:

**(a)** $A$ is $\Phi(.,y^*), -\Psi(.,y^*)$ and $\Omega(.,z^*)$ are ($\phi$, $c$, $\beta$, $\gamma$, $\zeta$, $\eta$)- univex with respect to $c(.,.)$, $\beta(.,.)$, $\gamma(.,.)$, $\phi$, $\zeta(.,.)$, $\eta(.,.)$ , where $c(.,.) > 0$, $\beta(.,.) > 0$, $\gamma(.,.) > 0$ and $\phi$ is superlinear.

**(b)** $A$ is ($\phi$, $c$, $\beta$, $\gamma$, $\zeta$, $\eta$)- pseudo-invex, and $\Omega(.,z^*)$ is ($\phi$, $c$, $\beta$, $\gamma$, $\eta$, $\zeta$)- quasi-invex with respect to $c(.,.)$, $\beta(.,.)$, $\gamma(.,.)$, $\phi$, $\zeta(.,.)$, $\eta(.,.)$, where $c(.,.) > 0$, $\beta(.,.) > 0$, $\gamma(.,.) > 0$ and $\phi$ is strictly increasing.

**(c)** $A$ is $(\phi, c, \beta, \gamma, \zeta, \eta)$- quasi-invex and $\Omega(., z^*)$ is strictly $(\phi, c, \beta, \gamma, \zeta, \eta)$- pseudo-invex with respect to $c(.,.)$, $\beta(.,.)$, $\gamma(.,.)$, $\phi$, $\zeta(.,.)$ and $\eta(.,.)$, where $c(.,.) > 0$, $\beta(.,.) > 0$, $\gamma(.,.) > 0$, and $\phi$ is increasing.

**(d)** $B$ is $(\phi, c, \beta, \gamma, \zeta, \eta)$- pseudo-invex with respect to $c(.,.)$, $\beta(.,.)$, $\gamma(.,.)$, $\phi$, $\eta(.,.)$ and $\zeta(.,.)$, where $c(.,.) > 0$, $\beta(.,.) > 0$ $\gamma(.,.) > 0$ and $\phi$ is increasing.

Then $x^*$ is an optimal solution to $(P)$.

**Proof.** Assume to contrary $x^*$ is not an optimal solution of $(P)$. Then there is a feasible solution $u \in F_p$ such that (in light of (9))

$$K(x^*) > K(u). \tag{6.19}$$

Applying (6.9), we have

$$\frac{\Phi(x^*, y^*)}{\Psi(x^*, y^*)} = K(x^*) > K(u) = \max_{y \in I} \frac{\Phi(u, y)}{\Psi(u, y)} \geq \frac{\Phi(u, y^*)}{\Psi(u, y^*)}, \quad y^* \in I.$$

It follows that

$$A(u) = \Phi(u, y^*)\Psi(x^*, y^*) - \Phi(x^*, y^*)\Psi(u, y^*) < 0 = A(x^*), \tag{6.20}$$

i.e.,

$$A(u) < A(x^*).$$

Now by rearranging the above inequality (20), we get

$$\Psi(x^*, y^*)[\Phi(u, y^*) - \Phi(x^*, y^*)] - \Phi(x^*, y^*)[\Psi(u, y^*) - \Psi(x^*, y^*)] < 0. \tag{6.21}$$

Since $u \in F_p$ , by condition (8), we have

$$\Omega(u, z^*) \leq 0 = \Omega(x^*, z^*). \tag{6.22}$$

Now (20) and (22) yield $B(u) < B(x^*)$.

If condition (a) holds, for $y^* \in I, z^* \in R^m_+$, employing the $(\phi, c, \beta, \gamma, \eta, \zeta)$- univexity of $\Phi(., y^*), -\Psi(., y^*)$, and $\Omega(., z^*)$, we have

$$\begin{aligned} & c(u, x^*)\phi\,[\Phi(u, y^*) - \Phi(x^*, y^*)] \\ \geq\ & \beta(u, x^*)\frac{1}{2}[\Phi_1'(x^*, y^*)\eta(u, x^*)] \\ & + \frac{1}{2}\gamma(u, x^*)[\Phi_1'(x^*, y^*)\zeta(u, x^*)], \end{aligned} \tag{6.23}$$

$$\begin{aligned} & -c(u, x^*)\phi\,[\Psi(u, y^*) - \Psi(x^*, y^*)] \\ \geq\ & -\frac{1}{2}\beta(u, x^*)[\Psi_1'(x^*, y^*)\eta(u, x^*)] \\ & -\frac{1}{2}\gamma(u, x^*)[\Psi_1'(x^*, y^*)\zeta(u, x^*)], \end{aligned} \tag{6.24}$$

$$\begin{aligned} & c(u,x^*)\phi\,[\Omega(u,z^*)-\Omega(x^*,z^*)] \\ \geq\; & \frac{1}{2}\beta(u,x^*)[\Omega_1'(x^*,z^*)\eta(u,x^*)] \\ & +\frac{1}{2}\gamma(u,x^*)[\Omega_1'(x^*,z^*)\zeta(u,x^*)]. \end{aligned} \tag{6.25}$$

Here for each $y^*$ and $z^*$, $\Phi_1', \Psi_1'$ and $\Omega_1'$ are partial Frechet derivatives of $\Phi(.,y^*)$, - $\Psi(.,y^*)$ and $\Omega(.,z^*)$ at $x^*$, respectively. Since $\Psi(x^*,y^*)>0$ and $\Phi(x^*,y^*)\geq 0$, we multiply (23) by $\Psi(x^*,y^*)$, (24) by $\Phi(x^*,y^*)$ and (25) by $\Psi(x^*,y^*)$. As $\phi$ is superlinear, by adding up the resulting inequalities, it follows from (21) and (22) that

$$\begin{aligned} 0 > & \frac{1}{2}\beta(u,x^*)\left[[\Psi(x^*,y^*)\Phi_1'(x^*,y^*)-\Phi(x^*,y^*)\Psi_1'(x^*,y^*)]\right. \\ & \left.+[\Psi(x^*,y^*)\Omega_1'(x^*,z^*)]\right]\eta(u,x^*) \\ & +\frac{1}{2}\gamma(u,x^*)\left[\Psi(x^*,y^*)\Phi_1'(x^*,y^*)-\Phi(x^*,y^*)\Psi_1'(x^*,y^*)]\right. \\ & \left.+[\Psi(x^*,y^*)\Omega_1'(x^*,z^*)]\right]\zeta(u,x^*). \end{aligned}$$

This leads to a contradiction in light of (7). Hence (19) does not hold, and so $x^*$ must be an optimal solution of $(P)$.

For condition (b), $A$ is ($\phi$, $c$, $\beta$, $\gamma$, $\zeta$, $\eta$)- pseudo-univex (applying the equivalent form of the pseudo-univexity of $A$ in this situation) and so from (6.29), for $c(.,.)>0$, $\beta(.,.)>0$, $\gamma(.,.)>0$ and $\phi(.)$ is strictly increasing, we have

$$\begin{aligned} & c(u,x^*)\phi[A(u)-A(x^*)]<0\Rightarrow \\ & \frac{1}{2}\beta(u,x^*)\big([A'(x^*)\eta(u,x^*)]\big) \\ & +\frac{1}{2}\gamma(u,x^*)\big([A'(x^*)\zeta(u,x^*)]\big)<0, \end{aligned}$$

that is,

$$\begin{aligned} & \frac{1}{2}\beta(u,x^*)\Big[\Big[\Psi(x^*,y^*)\Phi_1'(x^*,y^*)-\Phi(x^*,y^*)\Psi_1'(x^*,y^*)\Big]\eta(u,x^*)] \\ + \; & \frac{1}{2}\gamma(u,x^*) \\ & \Big[\Big[\Psi(x^*,y^*)\Phi_1'(x^*,y^*)-\Phi(x^*,y^*)\Psi_1'(x^*,y^*)\Big]\zeta(u,x^*)]<0. \end{aligned} \tag{6.26}$$

The identity (6.7) and inequality (26) yield

$$\frac{1}{2}\beta(u,x^*)\Big[[\Omega_1^{'}(x^*,z^*)\Psi(x^*,y^*)\eta(u,x^*)]$$

$$+\frac{1}{2}\gamma(u,x^*)[\Omega_1^{'}(x^*,z^*)\Psi(x^*,y^*)\zeta(u,x^*)] > 0.$$

Since $\beta(u,x^*) > 0$, $\gamma(u,x^*) > 0$ and $\Psi(x^*,y^*) > 0$, it follows that

$$\Omega_1^{'}(x^*,z^*)\eta(u,x^*) + \Omega_1^{'}(x^*,z^*)\zeta(u,x^*) > 0. \tag{6.27}$$

On the other hand, the function $\Omega$ in condition (b) assumed to be ($\phi$, $c$, $\beta$, $\gamma$, $\zeta$, $\eta$) - quasi-invex with respect to $\eta$, $\zeta$. Thus it implies

$$\frac{1}{2}\beta(u,x^*)\Big[[\Omega_1^{'}(x^*,z^*)\eta(u,x^*)]$$

$$+\frac{1}{2}\gamma(u,x^*)[\Omega_1^{'}(x^*,z^*)\zeta(u,x^*)]\Big] \leq 0$$

(by the ($\phi$, $c$, $\beta$, $\gamma$, $\eta$, $\zeta$)-quasi-univexity) which contradicts (27). This shows that $x^*$ is optimal for $(P)$.
Similarly one can prove for conditions (c) and (d).

**Theorem 6.13.** (Sufficient optimality). Let $x^* \in F_p$, $y^* \in I$, $z^* \in R_+^m$ and $(x^*,y^*,z^*)$ satisfy (6.7) - (6.9). Let us define

$$A(x) = \Psi(x^*,y^*)\Phi(x,y^*) - \Psi(x,y^*)\Phi(x^*,y^*),$$

and

$$B(x) = A(x) + \Psi(x^*,y^*)\Omega(x,z^*).$$

Assume, for the given $y^*$ and $z^*$, that any one of the following conditions holds.

**(a)** $A$ is $\Phi(.,y^*)$, $-\Psi(.,y^*)$ and $\Omega(.,z^*)$ are ($\phi$, $c$, $\eta$)- univex with respect to $c(.,.)$, $\phi$, and $\eta(.,.)$ , where $c(.,.) > 0$, and $\phi$ is superlinear.

**(b)** $A$ is ($\phi$, $c$, $\eta$)- pseudo-invex, and $\Omega(.,z^*)$ is ($\phi$, $c$, $\eta$)- quasi-invex with respect to $c(.,.)$, $\phi$, $\eta(.,.)$, where $c(.,.) > 0$, and $\phi$ is strictly increasing.

**(c)** $A$ is ($\phi$, $c$, $\eta$)- quasi-invex and $\Omega(.,z^*)$ is strictly ($\phi$, $c$, $\eta$)- pseudo-invex with respect to $c(.,.)$, $\phi$, and $\eta(.,.)$, where $c(.,.) > 0$, and $\phi$ is increasing.

**(d)** $B$ is ($\phi$, $c$, $\eta$)- pseudo-invex with respect to $c(.,.)$,, $\phi$, $\eta(.,.)$, where $c(.,.) > 0$, and $\phi$ is increasing.

Then $x^*$ is an optimal solution to $(P)$.

**Proof.** Assume to contrary $x^*$ is not an optimal solution of $(P)$. Then there is a feasible solution $u \in F_p$ such that (in light of (9))

$$K(x^*) > K(u). \tag{6.28}$$

Applying (6.9), we have

$$\frac{\Phi(x^*,y^*)}{\Psi(x^*,y^*)} = K(x^*) > K(u) = \max_{y \subset I} \frac{\Phi(u,y)}{\Psi(u,y)} \geq \frac{\Phi(u,y^*)}{\Psi(u,y^*)}, \quad y^* \in I.$$

It follows that

$$A(u) = \Phi(u,y^*)\Psi(x^*,y^*) - \Phi(x^*,y^*)\Psi(u,y^*) < 0 = A(x^*), \tag{6.29}$$

i.e.,

$$A(u) < A(x^*).$$

Now by rearranging the above inequality (29), we get

$$\Psi(x^*,y^*)[\Phi(u,y^*) - \Phi(x^*,y^*)] - \Phi(x^*,y^*)[\Psi(u,y^*) - \Psi(x^*,y^*)] < 0. \tag{6.30}$$

Since $u \in F_p$ , by condition (6.8), we have

$$\Omega(u,z^*) \leq 0 = \Omega(x^*,z^*). \tag{6.31}$$

Now (29) and (31) yield $B(u) < B(x^*)$.

If condition (a) holds, for $y^* \in I, z^* \in R^m_+$, employing the $(\phi,\ c,\ \eta)$- univexity of $\Phi(.,y^*), -\Psi(.,y^*), \Omega(.,y^*)$, we have

$$\begin{aligned} & c(u,x^*)\phi\,[\Phi(u,y^*) - \Phi(x^*,y^*)] \\ \geq\ & [\Phi_1'(x^*,y^*)\eta(u,x^*)] \end{aligned} \tag{6.32}$$

$$\begin{aligned} & -c(u,x^*)\phi\,[\Psi(u,y^*) - \Psi(x^*,y^*)] \\ \geq\ & -[\Psi_1'(x^*,y^*)\eta(u,x^*)] \end{aligned} \tag{6.33}$$

$$\begin{aligned} & c(u,x^*)\phi\,[\Omega(u,z^*) - \Omega(x^*,z^*)] \\ \geq\ & [\Omega_1'(x^*,z^*)\eta(u,x^*)] \end{aligned} \tag{6.34}$$

Here for each $y^*$ and $z^*$, $\Phi_1', \Psi_1'$ and $\Omega_1'$ are partial Frechet derivatives of $\Phi(.,y^*)$, -$\Psi(.,y^*)$ and $\Omega(.,z^*)$ at $x^*$, respectively. Since $\Psi(x^*,y^*) > 0$ and $\Phi(x^*,y^*) \geq 0$, we multiply (32) by $\Psi(x^*,y^*)$, (33) by $\Phi(x^*,y^*)$ and (34) by $\Psi(x^*,y^*)$. As $\phi$ is superlinear, by adding up the resulting inequalities, it follows from (30) and (31) that

$$0 > \left[[\Psi(x^*,y^*)\Phi_1'(x^*,y^*) - \Phi(x^*,y^*)\Psi_1'(x^*,y^*)]\right.$$

$$\left.+[\Psi(x^*,y^*)\Omega_1'(x^*,z^*)]\right]\eta(u,x^*).$$

This leads to a contradiction in light of (7). Hence (28) does not hold, and so $x^*$ must be an optimal solution of $(P)$.

For condition (b), $A$ is ($\phi$, $c$, $\eta$)- pseudo-univex (applying the equivalent form of the pseudo-univexity of $A$ in this situation) and so from (29), for $c(.,.) > 0$, and $\phi(.)$ is strictly increasing, we have

$$c(u,x^*)\phi[A(u) - A(x^*)] < 0 \Rightarrow$$
$$\left([A'(x^*)\eta(u,x^*)]\right) < 0,$$

that is,

$$\left[\Psi(x^*,y^*)\Phi_1'(x^*,y^*) - \Phi(x^*,y^*)\Psi_1'(x^*,y^*)\right]\eta(u,x^*)] < 0. \tag{6.35}$$

The identity (7) and inequality (35) yield

$$\beta(u,x^*)\left[[\Omega_1'(x^*,z^*)\Psi(x^*,y^*)\eta(u,x^*)] > 0.\right.$$

Since $\Psi(x^*,y^*) > 0$, it follows that

$$\Omega_1'(x^*,z^*)\eta(u,x^*) > 0. \tag{6.36}$$

On the other hand, the function $\Omega$ in condition (b) assumed to be ($\phi$, $c$, $\eta$) - quasi-invex with respect to $\eta$. Thus it implies

$$\left[[\Omega_1'(x^*,z^*)\eta(u,x^*)] \leq 0\right.$$

(by the ($\phi$, $c$, $\eta$)-quasi-univexity) which contradicts (36). This shows that $x^*$ is optimal for $(P)$.

Similarly one can prove for conditions (c) and (d).

## 4. Wolfe Type Dual

In this section, we consider the following Wolfe type model:

$$\text{(WD)} \qquad \text{maximize } \frac{\Phi(u,y) + \Omega(u,z)}{\Psi(u,y)}$$

subject to $(u,z) \in PS(T,R^n) \times R^m_+$,

$$u(a) = \alpha, u(b) = \beta \text{ and } y \in I \subset R^m_+,$$

$$\Psi(u,y)\Phi_1'(u,y) - [\Phi(u,y) + \Omega(u,z)]\Psi_1'(u,y) + \Psi(u,y)\Omega_1'(u,z) = 0. \tag{6.37}$$

Let the set of all feasible solutions of problem (WD) be denoted by $K_1$. We further assume throughout this section that

$$\Phi(u,y) + \Omega(u,z) \geq 0 \text{ and } \Psi(u,y) > 0 \text{ for all } (u,y,z) \in K_1. \tag{6.38}$$

In what follows, we plan to establish weak, strong and strict converse duality theorems for $(P) - (WD)$.

**Theorem 6.14.** (Weak duality). Let $x \in F_p$ , $(u,y,z) \in K_1$ and let

$$C(x) = \Psi(u,y)[\Phi(x,y) + \Omega(x,z)] - [\Phi(u,y) + \Omega(u,z)]\Psi(x,y).$$

Suppose that for each, $y$ and $z$, either one of the conditions holds:

**(a)** $\Phi(.,y^*), -\Psi(.,y^*)$ and $\Omega(.,z^*)$ are ($\phi$, $c$, $\beta$, $\gamma$, $\zeta$, $\eta$ $\omega$)- univex with respect to $c$, $\beta$, $\gamma$,$\phi$, $\zeta$, $\eta$ and $\omega$, where $c(.,.) > 0$, $\beta(.,.) > 0$, $\gamma(.,.) > 0$ and $\phi$ is superlinear.

**(b)** $C$ is ($\phi$, $c$, $\beta$, $\gamma$, $\zeta$, $\eta$, $\omega$)- pseudo-univex with respect to $c$, $\beta$, $\gamma$,$\phi$, $\zeta$, $\eta$ and $\omega$, where $c > 0$, $\beta(.,.) > 0$ $\gamma(.,.) > 0$ and $\phi$ is strictly increasing.

Then

$$\varphi(x) \geq \frac{\Phi(u,y) + \Omega(u,z)}{\Psi(u,y)}, \tag{6.39}$$

where $\varphi(x)$ is defined by (6.1).

**Proof.** Suppose, on contrary

$$\varphi(x) < \frac{\Phi(u,y) + \Omega(u,z)}{\Psi(u,y)}. \tag{6.40}$$

It follows from (6.2) that for any $y \in I$,

$$\frac{\Phi(x,y)}{\Psi(x,y)} \leq \max_{\beta \in I} \frac{\Phi(x,\beta)}{\Psi(x,\beta)} = \varphi(x) < \frac{\Phi(u,y) + \Omega(u,z)}{\Psi(u,y)},$$

$$\Phi(x,y)\Psi(u,y) - [\Phi(u,y) + \Omega(u,z)]\Psi(x,y) < 0. \tag{6.41}$$

Rewriting the above inequality, we get

$$[\Phi(x,y) + \Omega(x,z)]\Psi(u,y) - [\Phi(u,y) + \Omega(u,z)]\Psi(x,y) < \Omega(x,z)\Psi(u,y). \tag{6.42}$$

Since $\Omega(x,z) \leq 0$, if $x \in F_p$, $z \in R_+^m$ and $\Psi(u,y) > 0, \Omega(x,z)\Psi(u,y) \leq 0$. It follows from (42) that

$$C(x) < 0 = C(u).$$

Since $c(.,.) > 0$, the above inequality yields

$$c(x,u)\phi[C(x) - C(u)] < 0. \tag{6.43}$$

Now if condition (a) holds, for each $y \in I, z \in R_+^m$, the ($\phi$, $c$, $\beta$, $\eta$, $\gamma$, $\omega$, $\zeta$)- univexity of $\Phi(.,y^*)$, $-\Psi(.,y^*)$ and $\Omega(.,z^*)$ implies that

$$\begin{aligned} & c(x,u)\phi\left[\Phi(x,y) - \Phi(u,y)\right] \\ & \geq \beta(x,u)\left(\Phi_1'(u,y)\eta(x,u) - \frac{1}{2}\Phi_1'(u,y)\omega(x,u)\right) \\ + & \frac{1}{2}\gamma(x,u)\left(\Phi_1'(u,y)\zeta(x,u)\right), \end{aligned} \tag{6.44}$$

$$\begin{aligned} & -c(x,u)\phi\left[\Psi(x,y) - \Psi(u,y)\right] \\ & \geq -\beta(x,u)\left(\Psi_1'(u,y)\eta(x,u) - \frac{1}{2}\Psi_1'(u,y)\omega(x,u)\right) \\ + & \frac{1}{2}\gamma(x,u)\left(\Psi_1'(u,y)\zeta(x,u)\right), \end{aligned} \tag{6.45}$$

$$\begin{aligned} & c(x,u)\phi\left[\Omega(x,z) - \Omega(u,z)\right] \\ & \geq \beta(x,u)\left(\Omega_1'(u,z)\eta(x,u) - \frac{1}{2}\Omega_1'(u,z)\omega(x,u)\right) \\ + & \frac{1}{2}\gamma(x,u)\left(\Omega_1'(u,z)\zeta(x,u)\right). \end{aligned} \tag{6.46}$$

From (38), $\Psi(u,y)$ and $\Phi(u,y) + \Omega(u,z)$ are nonnegative. Multiplying (44) by $\Psi(u,y)$, (45) by $\Phi(u,y) + \Omega(u,z)$ and (46) by $\Psi(u,y)$, and summing up the resulting inequalities in light of the superlinearity of $\phi$, we obtain from (41) that

$$\begin{aligned} & \beta(x,u)\Big(\Big\{\Psi(u,y)\Phi_1'(u,y) - [\Phi(u,y) + \Omega(u,z)]\Psi_1'(u,y)\Big\}\eta(x,u) \\ + & \Big\{\Psi(u,y)\Omega_1'(u,z)\Big\}\eta(x,u) \\ - & \frac{1}{2}\Big\{\Psi(u,y)\Phi_1'(u,y) - [\Phi(u,y) + \Omega(u,z)]\Psi_1'(u,y)\Big\}\omega(x,u) \\ + & \Big\{\Psi(u,y)\Omega_1'(u,z)\Big\}\omega(x,u)\Big) \\ + & \frac{1}{2}\gamma(x,u)\Big(\Big\{\Psi(u,y)\Phi_1'(u,y) - [\Phi(u,y) + \Omega(u,z)]\Psi_1'(u,y)\Big\}\zeta(x,u) < 0. \end{aligned}$$

This contradicts (37), so (40) does not hold and (39 holds.

For the proof under the assumption (b), $C$ is ($\phi$, $c$, $\gamma$, $\beta$, $\eta$, $\omega$, $\zeta$)- pseudo-univex and (43) implies that

$$\beta(x,u)[C'(u)\eta(x,u) - \frac{1}{2}C'(u)\omega(x,u)]$$
$$+\frac{1}{2}\gamma(x,u)[C'(u)\zeta(x,u)] < 0,$$

that is,

$$\begin{aligned}
& \beta(x,u)\Big(\{\Psi(u,y)\Phi_1'(u,y) - [\Phi(u,y)+\Omega(u,z)]\Psi_1'(u,y)\}\eta(x,u) \\
+ & \{\Psi(u,y)\Omega_1'(u,z)\}\eta(x,u) \\
- & \frac{1}{2}\{\Psi(u,y)\Phi_1'(u,y) - [\Phi(u,y)+\Omega(u,z)]\Psi_1'(u,y)\}\omega(x,u) \\
+ & \{\Psi(u,y)\Omega_1'(u,z)\}\omega(x,u)\Big) \\
+ & \frac{1}{2}\gamma(x,u)\Big(\{\Psi(u,y)\Phi_1'(u,y) - [\Phi(u,y)+\Omega(u,z)]\Psi_1'(u,y) \\
+ & \Psi(u,y)\Omega_1'(u,z)\}\Big)\zeta(x,u) < 0,
\end{aligned}$$

which contradicts (37).

**Theorem 6.15.** (Strong duality). Let $x^*$ be an optimal solution of $(P)$. Then there exist $y \in I$ and $z \in R^m_+$ such that $(x^*, y^*, z^*)$ is feasible for $(WD)$. If the hypotheses of Lemma 6.2 and Theorem 6.14 are satisfied, then $(x^*, y^*, z^*)$ is an optimal solution of $(WD)$ and, two problems $(P)$ and $(WD)$ have the same optimal values.

**Proof.** Assume that $x^*$ is an optimal solution of $(P)$. Then, by Theorem 6.14, there exist $y \in I$ and $z \in R^m_+$ satisfying the constraints of $(WD)$, so that $(x^*, y^*, z^*) \in K_1$. Furthermore, we have

$$\frac{\Phi(x^*,y^*)+\Omega(x^*,z^*)}{\Psi(x^*,y^*)} = \frac{\Phi(x^*,y^*)}{\Psi(x^*,y^*)} = \varphi(x^*).$$

Since $\Omega(x^*, z^*) = 0$, it implies $(x^*, y^*, z^*)$ is an optimal solution of $(WD)$. Consequently, (39) demonstrates that $(P)$ and $(WD)$ have the same optimal values.

**Theorem 6.16.** (Strict converse duality). Let $x_1$ and $(x^*, y_0, z_0)$ be optimal solutions for (P) and (WD) that have the same optimal value

$$\varphi(x_1) = \frac{\Phi(x^*,y_0)+\Omega(x^*,z_0)}{\Psi(x^*,y_0)}.$$

**Proof.** Suppose on contrary that $x_1 \neq x^*$. Then by Theorem 6.15, there exist $y_1 \in I$ and $z_1 \in R^m_+$ such that $(x_1, y_1, z_1)$ is an optimal solution of $(WD)$ and

$$\varphi(x_1) = \frac{\Phi(x_1,y_1)+\Omega(x_1,z_1)}{\Psi(x_1,y_1)}.$$

As in the proof of Theorem 6.14, we replace $x$ by $x_1$, and $(u,y,z)$ by $(x^*,y_0,z_0)$ to derive the inequality

$$\varphi(x_1) > \frac{\Phi(x^*,y_0)+\Omega(x^*,z_0)}{\Psi(x^*,y_0)}.$$

This contradicts the fact that

$$\varphi(x_1) = \frac{\Phi(x_1,y_1)+\Omega(x_1,z_1)}{\Psi(x_1,y_1)} = \frac{\Phi(x^*,y_0)+\Omega(x^*,z_0)}{\Psi(x^*,y_0)}.$$

Hence, we conclude that

$$x_1 = x^* \text{ and } \varphi(x_1) = \frac{\Phi(x^*,y_0)+\Omega(x^*,z_0)}{\Psi(x^*,y_0)}.$$

**Theorem 6.17.** (Weak duality). Let $x \in F_p$ , $(u,y,z) \in K_1$ and let

$$C(x) = \Psi(u,y)[\Phi(x,y)+\Omega(x,z)] - [\Phi(u,y)+\Omega(u,z)]\Psi(x,y).$$

Suppose that for each, $y$ and $z$, either one of the conditions holds:

**(a)** $\Phi(.,y^*), -\Psi(.,y^*)$ and $\Omega(.,z^*)$ are ($\phi$, $c$, $\beta$, $\gamma$, $\zeta$, $\eta$)- univex with respect to $c$, $\beta$, $\gamma$,$\phi$, $\zeta$ and $\eta$, where $c(.,.) > 0$, $\beta(.,.) > 0$, $\gamma(.,.) > 0$ and $\phi$ is superlinear.

**(b)** $C$ is ($\phi$, $c$, $\beta$, $\gamma$, $\zeta$, $\eta$)- pseudo-univex with respect to $c$, $\beta$, $\gamma$,$\phi$, $\zeta$ and $\eta$, where $c > 0$, $\beta(.,.) > 0$ $\gamma(.,.) > 0$ and $\phi$ is strictly increasing.

Then

$$\varphi(x) \geq \frac{\Phi(u,y)+\Omega(u,z)}{\Psi(u,y)}, \tag{6.47}$$

where $\varphi(x)$ is defined by (6.1).

**Proof.** Suppose, on contrary

$$\varphi(x) < \frac{\Phi(u,y)+\Omega(u,z)}{\Psi(u,y)}. \tag{6.48}$$

It follows from (6.2) that for any $y \in I$,

$$\frac{\Phi(x,y)}{\Psi(x,y)} \leq \max_{\beta \in I} \frac{\Phi(x,\beta)}{\Psi(x,\beta)} = \varphi(x) < \frac{\Phi(u,y)+\Omega(u,z)}{\Psi(u,y)},$$

$$\Phi(x,y)\Psi(u,y) - [\Phi(u,y)+\Omega(u,z)]\Psi(x,y) < 0. \tag{6.49}$$

Rewriting the above inequality, we get

$$[\Phi(x,y)+\Omega(x,z)]\,\Psi(u,y) - [\Phi(u,y)+\Omega(u,z)]\Psi(x,y) < \Omega(x,z)\Psi(u,y). \tag{6.50}$$

Since $\Omega(x,z) \leq 0$, if $x \in F_p$, $z \in R_+^m$ and $\Psi(u,y) > 0, \Omega(x,z)\Psi(u,y) \leq 0$. It follows from (50) that

$$C(x) < 0 = C(u).$$

Since $c(.,.) > 0$, the above inequality yields

$$c(x,u)\phi[C(x) - C(u)] < 0. \tag{6.51}$$

Now if condition (a) holds, for each $y \in I, z \in R_+^m$, the ($\phi$, $c$, $\beta$, $\eta$, $\gamma$, $\zeta$)- univexity of $\Phi(.,y^*)$, $-\Psi(.,y^*)$ and $\Omega(.,z^*)$ implies that

$$c(x,u)\phi\left[\Phi(x,y) - \Phi(u,y)\right] \geq \beta(x,u)\frac{1}{2}\Big(\Phi_1'(u,y)\eta(x,u)\Big)$$
$$+ \quad \frac{1}{2}\gamma(x,u)\Big(\Phi_1'(u,y)\zeta(x,u)\Big), \tag{6.52}$$

$$-c(x,u)\phi\left[\Psi(x,y) - \Psi(u,y)\right] \geq -\beta(x,u)\frac{1}{2}\Big(\Psi_1'(u,y)\eta(x,u)\Big)$$
$$+ \quad \frac{1}{2}\gamma(x,u)\Big(\Psi_1'(u,y)\zeta(x,u)\Big), \tag{6.53}$$

$$c(x,u)\phi\left[\Omega(x,z) - \Omega(u,z)\right] \geq \beta(x,u)\frac{1}{2}\Big(\Omega_1'(u,z)\eta(x,u)\Big)$$
$$+ \quad \frac{1}{2}\gamma(x,u)\frac{1}{2}\Big(\Omega_1'(u,z)\zeta(x,u)\Big). \tag{6.54}$$

From (38), $\Psi(u,y)$ and $\Phi(u,y) + \Omega(u,z)$ are nonnegative. Multiplying (52) by $\Psi(u,y)$, (53) by $\Phi(u,y) + \Omega(u,z)$ and (54) by $\Psi(u,y)$, and summing up the resulting inequalities in light of the superlinearity of $\phi$, we obtain from (49) that

$$\beta(x,u)\frac{1}{2}\Big(\Big\{\Psi(u,y)\Phi_1'(u,y) - [\Phi(u,y) + \Omega(u,z)]\Psi_1'(u,y)\Big\}\eta(x,u)$$
$$+ \quad \Big\{\Psi(u,y)\Omega_1'(u,z)\Big\}\eta(x,u)\Big)$$
$$+ \quad \frac{1}{2}\gamma(x,u)\Big(\Big\{\Psi(u,y)\Phi_1'(u,y) - [\Phi(u,y) + \Omega(u,z)]\Psi_1'(u,y)\Big\}$$
$$+ \quad \Big\{\Psi(u,y)\Omega_1'(u,z)\Big\}\Big)\zeta(x,u) < 0.$$

This contradicts (37), so (48) does not hold and (47) holds.

For the proof under the assumption (b), $C$ is ($\phi$, $c$, $\gamma$, $\beta$, $\eta$, $\zeta$)- pseudo-univex and (51) implies that

$$\frac{1}{2}\beta(x,u)[C'(u)\eta(x,u)]$$

$$+\frac{1}{2}\gamma(x,u)[C'(u)\zeta(x,u)] < 0,$$

that is,

$$\beta(x,u)\frac{1}{2}\Big(\Big\{\Psi(u,y)\Phi_1^{'}(u,y)-[\Phi(u,y)+\Omega(u,z)]\Psi_1^{'}(u,y)\Big\}\eta(x,u)$$

$$+\quad\Big\{\Psi(u,y)\Omega_1^{'}(u,z)\Big\}\eta(x,u)$$

$$+\quad\Big\{\Psi(u,y)\Omega_1^{'}(u,z)\Big\}\Big)\zeta(x,u)<0,$$

which contradicts (37).

**Theorem 6.18.** (Strong duality). Let $x^*$ be an optimal solution of $(P)$. Then there exist $y\in I$ and $z\in R_+^m$ such that $(x^*,y^*,z^*)$ is feasible for $(WD)$. If the hypotheses of Lemma 6.2 and Theorem 6.17 are satisfied, then $(x^*,y^*,z^*)$ is an optimal solution of $(WD)$ and, two problems $(P)$ and $(WD)$ have the same optimal values.

**Proof.** Assume that $x^*$ is an optimal solution of $(P)$. Then, by Theorem 6.17, there exist $y\in I$ and $z\in R_+^m$ satisfying the constraints of $(WD)$, so that $(x^*,y^*,z^*)\in K_1$. Furthermore, we have

$$\frac{\Phi(x^*,y^*)+\Omega(x^*,z^*)}{\Psi(x^*,y^*)}=\frac{\Phi(x^*,y^*)}{\Psi(x^*,y^*)}=\varphi(x^*).$$

Since $\Omega(x^*,z^*)=0$, it implies $(x^*,y^*,z^*)$ is an optimal solution of $(WD)$. Consequently, (47) demonstrates that $(P)$ and $(WD)$ have the same optimal values.

**Theorem 6.19.** (Strict converse duality). Let $x_1$ and $(x^*,y_0,z_0)$ be optimal solutions for (P) and (WD) that have the same optimal value

$$\varphi(x_1)=\frac{\Phi(x^*,y_0)+\Omega(x^*,z_0)}{\Psi(x^*,y_0)}.$$

**Proof.** Suppose on contrary that $x_1\neq x^*$. Then by Theorem 6.18, there exist $y_1\in I$ and $z_1\in R_+^m$ such that $(x_1,y_1,z_1)$ is an optimal solution of $(WD)$ and

$$\varphi(x_1)=\frac{\Phi(x_1,y_1)+\Omega(x_1,z_1)}{\Psi(x_1,y_1)}.$$

As in the proof of Theorem 6.17, we replace $x$ by $x_1$, and $(u,y,z)$ by $(x^*,y_0,z_0)$ to derive the inequality

$$\varphi(x_1)>\frac{\Phi(x^*,y_0)+\Omega(x^*,z_0)}{\Psi(x^*,y_0)}.$$

This contradicts the fact that

$$\varphi(x_1)=\frac{\Phi(x_1,y_1)+\Omega(x_1,z_1)}{\Psi(x_1,y_1)}=\frac{\Phi(x^*,y_0)+\Omega(x^*,z_0)}{\Psi(x^*,y_0)}.$$

Hence, we conclude that

$$x_1=x^*\text{ and }\varphi(x_1)=\frac{\Phi(x^*,y_0)+\Omega(x^*,z_0)}{\Psi(x^*,y_0)}.$$

**Theorem 6.20** (Weak duality). Let $x \in F_p$ , $(u,y,z) \in K_1$ and let

$$C(x) = \Psi(u,y)[\Phi(x,y)+\Omega(x,z)] - [\Phi(u,y)+\Omega(u,z)]\Psi(x,y).$$

Suppose that for each, $y$ and $z$, either one of the conditions holds:

**(a)** $\Phi(.,y^*), -\Psi(.,y^*)$ and $\Omega(.,z^*)$ are $(\phi, c, \eta)$- univex with respect to $c$, $\phi$ and $\eta$, where $c(.,.) > 0$ and $\phi$ is superlinear.

**(b)** $C$ is $(\phi, c, \eta)$ pseudo univex with respect to $c$, $\phi$, and $\eta$, where $c > 0$, and $\phi$ is strictly increasing.

Then

$$\varphi(x) \geq \frac{\Phi(u,y)+\Omega(u,z)}{\Psi(u,y)}, \tag{6.55}$$

where $\varphi(x)$ is defined by (6.1).

**Proof.** Suppose, on contrary

$$\varphi(x) < \frac{\Phi(u,y)+\Omega(u,z)}{\Psi(u,y)}. \tag{6.56}$$

It follows from (6.2) that for any $y \in I$,

$$\frac{\Phi(x,y)}{\Psi(x,y)} \leq \max_{\beta \in I} \frac{\Phi(x,\beta)}{\Psi(x,\beta)} = \varphi(x) < \frac{\Phi(u,y)+\Omega(u,z)}{\Psi(u,y)},$$

$$\Phi(x,y)\Psi(u,y) - [\Phi(u,y)+\Omega(u,z)]\Psi(x,y) < 0. \tag{6.57}$$

Rewriting the above inequality, we get

$$[\Phi(x,y)+\Omega(x,z)]\,\Psi(u,y) - [\Phi(u,y)+\Omega(u,z)]\Psi(x,y) < \Omega(x,z)\Psi(u,y). \tag{6.58}$$

Since $\Omega(x,z) \leq 0$, if $x \in F_p$, $z \in R^m_+$ and $\Psi(u,y) > 0, \Omega(x,z)\Psi(u,y) \leq 0$. It follows from (6.58) that

$$C(x) < 0 = C(u).$$

Since $c(.,.) > 0$, the above inequality yields

$$c(x,u)\phi[C(x) - C(u)] < 0. \tag{6.59}$$

Now if condition (a) holds, for each $y \in I, z \in R^m_+$, the $(\phi, c, \eta)$- univexity of $\Phi(.,y^*)$, $-\Psi(.,y^*)$ and $\Omega(.,z^*)$ implies that

$$c(x,u)\phi\left[\Phi(x,y)-\Phi(u,y)\right]$$

$$\geq \beta(x,u)\frac{1}{2}\left(\Phi_1'(u,y)\eta(x,u)\right), \tag{6.60}$$

$$-c(x,u)\phi\left[\Psi(x,y)-\Psi(u,y)\right]$$

$$\geq -\beta(x,u)\frac{1}{2}\left(\Psi_1'(u,y)\eta(x,u)\right), \tag{6.61}$$

$$c(x,u)\phi\left[\Omega(x,z)-\Omega(u,z)\right]$$

$$\geq \beta(x,u)\frac{1}{2}\left(\Omega_1'(u,z)\eta(x,u)\right). \tag{6.62}$$

From (38), $\Psi(u,y)$ and $\Phi(u,y)+\Omega(u,z)$ are nonnegative. Multiplying (60) by $\Psi(u,y)$, (61) by $\Phi(u,y)+\Omega(u,z)$ and (62) by $\Psi(u,y)$, and summing up the resulting inequalities in light of the superlinearity of $\phi$, we obtain from (41) that

$$\left(\left\{\Psi(u,y)\Phi_1'(u,y)-\left[\Phi(u,y)+\Omega(u,z)\right]\Psi_1'(u,y)\right\}\eta(x,u)\right.$$

$$\left.+\ \left\{\Psi(u,y)\Omega_1'(u,z)\right\}\eta(x,u)\right)<0.$$

This contradicts (6.37), so (56) does not hold and (55) holds.

For the proof under the assumption (b), $C$ is ($\phi$, $c$, $\eta$)- pseudo-univex and (59) implies that

$$[C'(u)\eta(x,u)<0,$$

that is,

$$\left(\left\{\Psi(u,y)\Phi_1'(u,y)-\left[\Phi(u,y)+\Omega(u,z)\right]\Psi_1'(u,y)\right\}\eta(x,u)\right.$$

$$\left.+\ \left\{\Psi(u,y)\Omega_1'(u,z)\right\}\eta(x,u)\right)<0,$$

which contradicts (6.37).

**Theorem 6.21.** (Strong duality). Let $x^*$ be an optimal solution of $(P)$. Then there exist $y \in I$ and $z \in R_+^m$ such that $(x^*, y^*, z^*)$ is feasible for $(WD)$. If the hypotheses of Lemma 6.2 and Theorem 6.20 are satisfied, then $(x^*, y^*, z^*)$ is an optimal solution of $(WD)$ and, two problems $(P)$ and $(WD)$ have the same optimal values.

**Proof.** Assume that $x^*$ is an optimal solution of $(P)$. Then, by Lemma 6.2, there exist $y \in I$ and $z \in R_+^m$ satisfying the constraints of $(WD)$, so that $(x^*, y^*, z^*) \in K_1$. Furthermore, we have

$$\frac{\Phi(x^*,y^*)+\Omega(x^*,z^*)}{\Psi(x^*,y^*)} = \frac{\Phi(x^*,y^*)}{\Psi(x^*,y^*)} = \varphi(x^*).$$

Since $\Omega(x^*, z^*) = 0$, it implies $(x^*, y^*, z^*)$ is an optimal solution of $(WD)$. Consequently, (55) demonstrates that $(P)$ and $(WD)$ have the same optimal values.

**Theorem 6.22.** (Strict converse duality). Let $x_1$ and $(x^*, y_0, z_0)$ be optimal solutions for (P) and (WD) that have the same optimal value

$$\varphi(x_1) = \frac{\Phi(x^*,y_0)+\Omega(x^*,z_0)}{\Psi(x^*,y_0)}.$$

**Proof.** Suppose on contrary that $x_1 \neq x^*$. Then by Theorem 6.21, there exist $y_1 \in I$ and $z_1 \in R_+^m$ such that $(x_1, y_1, z_1)$ is an optimal solution of $(WD)$ and

$$\varphi(x_1) = \frac{\Phi(x_1,y_1)+\Omega(x_1,z_1)}{\Psi(x_1,y_1)}.$$

As in the proof of Theorem 6.20, we replace $x$ by $x_1$, and $(u, y, z)$ by $(x^*, y_0, z_0)$ to derive the inequality

$$\varphi(x_1) > \frac{\Phi(x^*,y_0)+\Omega(x^*,z_0)}{\Psi(x^*,y_0)}.$$

This contradicts the fact that

$$\varphi(x_1) = \frac{\Phi(x_1,y_1)+\Omega(x_1,z_1)}{\Psi(x_1,y_1)} = \frac{\Phi(x^*,y_0)+\Omega(x^*,z_0)}{\Psi(x^*,y_0)}.$$

Hence, we conclude that

$$x_1 = x^* \text{ and } \varphi(x_1) = \frac{\Phi(x^*,y_0)+\Omega(x^*,z_0)}{\Psi(x^*,y_0)}.$$

## 5. Mond - Weir Type Dual

In this section, we consider the Mond-Weir type dual problems under the assumptions of $(\phi, c, \beta, \gamma, \zeta, \eta, \omega)$- univexities as follows:

(MWD) $\qquad$ Maximize $\dfrac{\Phi(u,y)}{\Psi(u,y)}$

subject to

$$(u,y) \in PS(T,R^n) \times I, \quad u(a) = \alpha, u(b) = \beta,$$

$$\begin{aligned}
&[\{\Psi(u,y)\Phi_1'(u,y) - (\Phi(u,y) + \Omega(u,z))\Psi_1'(u,y)\}\eta(x,u) \\
+ \; &\{\Psi(u,y)\Omega_1'(u,z)\}\eta(x,u)] \\
- \; &\frac{1}{2}[\{\Psi(u,y)\Phi_1'(u,y) - (\Phi(u,y) + \Omega(u,z))\Psi_1'(u,y)\}\omega(x,u) \\
+ \; &\Psi(u,y)\Omega_1'(u,z)\}\omega(x,u)] \\
+ \; &\frac{1}{2}[\Psi(u,y)\Phi_1'(u,y) - (\Phi(u,y) + \Omega(u,z))\Psi_1'(u,y)\}\zeta(x,u) \\
+ \; &\Psi(u,y)\Omega_1'(u,z)\}\zeta(x,u)] < 0,
\end{aligned} \tag{6.63}$$

$$\Omega(u,z) \geq 0, z \in R_+^m. \tag{6.64}$$

Let $K_2$ denote the set of all feasible solutions of problem $(MWD)$. We assume throughout that $\Phi(u,y) \geq 0$ and $\Psi(u,y) > 0$ for all $(u,y,z) \in K_2$.

Now we will establish weak, strong and strict converse duality theorems for $(P)-(MWD)$.

**Theorem 6.23.** (Weak duality). Let $x \in F_p, (u,y,z) \in K_2$ and define

$$D(x) = \Psi(u,y)\Phi(x,y) - \Psi(x,y)\Phi(u,y)$$

and

$$E(x) = D(x) + \Psi(u,y)\Omega(x,z).$$

Suppose, for each $y \in I, z \in R_+^m$, that any one of the following conditions holds:

**(a)** $\Phi(.,y), -\Psi(.,y)$ and $\Omega(.,z)$ are ($\phi$, $c$, $\beta$, $\gamma$, $\zeta$, $\eta$, $\omega$)- univex with respect to $c$, $\beta$, $\gamma$, $\phi$, $\omega$, $\zeta$ and $\eta$ where $c(.,.) > 0$, $\beta(.,.) > 0$, $\gamma(.,.) > 0$ and $\phi$ is superlinear.

**(b)** $D$ is ($\phi$, $c$, $\beta$, $\gamma$, $\zeta$, $\eta$, $\omega$)- pseudo-univex and $\Omega(.,z)$ is ($\phi$, $c$, $\beta$, $\gamma$, $\eta$, $\zeta$, $\omega$)-quasi-univex with respect to $c$, $\beta$, $\gamma$, $\phi$, $\omega$, $\zeta$ and $\eta$, where $c(.,.) > 0$, $\beta(.,.) > 0$, $\gamma(.,.) > 0$ and $\phi$ is strictly increasing.

**(c)** $D$ is ($\phi$, $c$, $\beta$, $\gamma$, $\eta$, $\zeta$, $\omega$)- quasi-univex and $\Omega(.,z)$ is strictly ($\phi$, $c$, $\beta$, $\gamma$, $\eta$, $\zeta$, $\omega$)-pseudo-univex with respect to $c$, $\beta$, $\gamma$, $\phi$, $\omega$, $\zeta$ and $\eta$, where $c(.,.) > 0$, $\beta(.,.) > 0$, $\gamma(.,.) > 0$ and $\phi$ is increasing.

**(d)** $E$ is ($\phi$, $c$, $\beta$, $\gamma$, $\eta$, $\zeta$, $\omega$)- pseudo-univex with respect to $c$, $\beta$, $\gamma$, $\phi$, $\omega$ and $\eta$, where $c(.,.) > 0$, $\beta(.,.) > 0$, $\gamma(.,.) > 0$ and $\phi$ is increasing.

Then

$$\varphi(x) \geq \frac{\Phi(u,y)}{\Psi(u,y)}. \tag{6.65}$$

**Proof.** Suppose to contrary. Then we have

$$\frac{\Phi(u,y)}{\Psi(u,y)} > \varphi(x) = \max_{y_1 \in I} \frac{\Phi(x,y_1)}{\Psi(x,y_1)} \geq \frac{\Phi(x,y)}{\Psi(x,y)}.$$

This implies that

$$D(x) = \Phi(x,y)\Psi(u,y) - \Phi(u,y)\Psi(x,y) < 0 = D(u). \tag{6.66}$$

As $x \in F_p, z \in R_+^m$ and $\Omega(x,z) \leq 0$, the constraints inequality (64) of $(MWD)$ yields

$$\Omega(x,z) \leq 0 \leq \Omega(u,z). \tag{6.67}$$

Consequently, (66) and (67) yield

$$E(x) < E(u). \tag{6.68}$$

If condition (a) holds, then by the ($\phi$, $c$, $\beta$, $\gamma$, $\eta$, $\zeta$, $\omega$)- univexity of $\Phi$, $-\Psi$ and $\Omega$, we have

$$\begin{aligned} & c(x,u)\phi\,[\Phi(x,y) - \Phi(u,y)] \geq \beta(x,u)[\Phi_1'(u,y)\eta(x,u) \\ & -\frac{1}{2}[\Phi_1'(u,y)\omega(x,u)] \\ & +\frac{1}{2}\gamma(x,u)[\Phi_1'(u,y)\zeta(x,u)], \end{aligned} \tag{6.69}$$

$$\begin{aligned} & -c(x,u)\phi\,[\Psi(x,y) - \Psi(u,y)] \geq -\beta(x,u)[\Psi_1'(u,y)\eta(x,u) \\ & -\frac{1}{2}[\Psi_1'(u,y)\omega(x,u)] \\ & +\frac{1}{2}\gamma(x,u)[\Psi_1'(u,y)\zeta(x,u)], \end{aligned} \tag{6.70}$$

$$\begin{aligned} & c(x,u)\phi\,[\Omega(x,z) - \Omega(u,z)] \geq \beta(x,u)[\Omega_1'(u,z)\eta(x,u) \\ & -\frac{1}{2}\Omega_1'(u,z)\omega(x,u)] \\ & +\frac{1}{2}\gamma(x,u)[\Omega_1'(u,z)\zeta(x,u)]. \end{aligned} \tag{6.71}$$

Since $\Psi(u,y)$ and $\Phi(u,y)$ are nonnegative, multiplying (69) by $\Psi(u,y)$, (70) by $\Phi(u,y)$ and (71) by $\Psi(u,y)$, and summing up the resulting inequalities in light of superlinearity of $\phi$, we obtain from (66) and (67)

$$\begin{aligned} & 0 > \beta(x,u)\big[[\Psi(u,y)\Phi_1'(u,y) - \Phi(u,y)\Psi_1'(u,y) + \Psi(u,y)\Omega_1'(u,z)]\eta(x,u)) \\ - \quad & \frac{1}{2}[\Psi(u,y)\Phi_1'(u,y) - \Phi(u,y)\Psi_1'(u,y) + \Psi(u,y)\Omega_1'(u,z)]\omega(x,u)] \\ + \quad & \frac{1}{2}\gamma(x,u)\big[[\Psi(u,y)\Phi_1'(u,y) - \Phi(u,y)\Psi_1'(u,y) + \Psi(u,y)\Omega_1'(u,z)]\zeta(x,u)\big], \end{aligned}$$

which contradicts (63). Hence (65) holds.
In condition (b), $D$ is ($\phi$, $c$, $\beta$, $\gamma$, $\zeta$, $\eta$, $\omega$)- pseudo-univex. Since $c(.,.) > 0$, (66) implies $\beta(x,u)\left[D'(u)\eta(x,u) + D'(u)\eta(x,u)\right) < 0$, that is,

$$\beta(x,u)\left[[\Psi(u,y)\Phi_1'(u,y) - \Phi(u,y)\Psi_1'(u,y)]\eta(x,u)\right]$$

$$-\frac{1}{2}[\Psi(u,y)\Phi_1'(u,y) - \Phi(u,y)\Psi_1'(u,y)]\omega(x,u)]]$$

$$+\frac{1}{2}\gamma(x,u)[\Psi(u,y)\Phi_1'(u,y) - \Phi(u,y)\Psi_1'(u,y)]\zeta(x,u)] < 0. \tag{6.72}$$

It follows from (63) and (72) that

$$-\left[\beta(x,u)\left(\Psi(u,y)\Omega_1'(u,z)\eta(x,u) - \frac{1}{2}\Psi(u,y)\Omega_1'(u,z)\omega(x,u)\right)\right.$$
$$\left.+\gamma(x,u)[\Psi(u,y)\Omega_1'(u,z)\zeta(x,u)]\right] < 0$$

or

$$\Omega'(u,z)\eta(x,u) - \frac{1}{2}\Omega'(u,z)\omega(x,u) + \frac{1}{2}\Omega'(u,z)\zeta(x,u) > 0,$$

since $\Psi(u,y) > 0$. But $\Omega$ is ($\phi$, $c$, $\beta$, $\gamma$, $\zeta$, $\eta$, $\omega$)- quasi-univex, so the above inequality implies

$$\Omega(x,z) > \Omega(u,z)$$

which contradicts (67). Hence, (65) holds.

Similarly, one can prove for the rest of two cases (c) and (d).

**Theorem 6.24.** (Strong duality). Let $x^*$ be an optimal solution of $(P)$. Then there exist $y^* \in I, z^* \in R_+^m$ such that $(x^*, y^*, z^*)$ is feasible for $(MWD)$. If any of the conditions of Theorem 6.23 holds, then $(x^*, y^*, z^*)$ is an optimal solution of $(MWD)$. Furthermore, $(MWD)$ and $(P)$ have the same optimal values.

**Proof.** If $x^*$ is an optimal solution of $(P)$, then by Theorem 6.23, there exist $y^* \in I$ and $z^* \in R_+^m$ such that $(x,y,z) \in K_2$, the feasible solution of $(MWD)$ and

$$\varphi(x^*) = \frac{\Phi(x^*,y^*)}{\Psi(x^*,y^*)} \leq \max_{y \in I} \frac{\Phi(x^*,y)}{\Psi(x^*,y)}.$$

The optimality of $(MWD)$ follows from Theorem 6.23.

**Theorem 6.25.** (Strict converse duality). Let $x_1$ and $(x^*, y_0, z_0)$ be optimal solutions of $(P)$ and $(MWD)$, respectively. Assume that the hypothesis of Theorem 6.24 is fulfilled. Assume that $D(.) = \Psi(x^*,y_0)\Phi(.,y_0) - \Psi(.,y_0)\Phi(x^*,y_0)$ is strictly ($\phi$, $c$, $\beta$, $\gamma$, $\zeta$, $\eta$, $\omega$)- pseudo-univex and $\Omega(.,z_0)$ is ($\phi$, $c$, $\beta$, $\gamma$, $\zeta$, $\eta$, $\omega$)- quasi-univex with respect to $c$, $\beta$, $\gamma$, $\phi$, $\zeta$, $\omega$ and $\eta$. Then $x_1 = x^*$ is an optimal solution of $(P)$, and $(P)$ and $(MWD)$ have the same optimal values

$$\varphi(x_1) = \Phi(x^*,y_0)/\Psi(x^*,y_0).$$

**Proof.** Since $x_1$ is an optimal solution of $(P)$, by Theorem 6.24, there exist $y_1 \in I$ and $z_1 \in R^m_+$ such that $(x_1, y_1, z_1)$ is an optimal solution of $(MWD)$ and

$$\varphi(x_1) = \frac{\Phi(x_1, y_1)}{\Psi(x_1, y_1)}.$$

On the other hand, $(x^*, y_0, z_0)$ is an optimal solution of $(MWD)$ with optimal values equal to $\varphi(x_1)$, that is,

$$\left[\varphi(x_1) = \frac{\Phi(x_1, y_1)}{\Psi(x_1, y_1)} = \frac{\Phi(x^*, y_0)}{\Psi(x^*, y_0)}\right]. \tag{6.73}$$

If $x_1 \neq x^*$, we could deduce

$$\left[\varphi(x_1) > \frac{\Phi(x^*, y_0)}{\Psi(x^*, y_0)}\right], \tag{6.74}$$

which contradicts (73).

Assume (74) does not hold. Then $\varphi(x_1) = \Phi(x^*, y_0)/\Psi(x^*, y_0)$, that is,

$$\frac{\Phi(x^*, y^*)}{\Psi(x^*, y^*)} \geq \varphi(x_1) = \max_{y \in I} \frac{\Phi(x_1, y)}{\Psi(x_1, y)} \geq \frac{\Phi(x_1, y_0)}{\Psi(x_1, y_0)}$$

or

$$D(x_1) = \Phi(x_1, y_0)\Psi(x^*, y_0) - \Phi(x^*, y_0)\Psi(x_1, y_0) \leq 0 = D(x^*),$$

that is,

$$D(x_1) \leq D(x^*).$$

Since $c(.,.) > 0$, the above inequality yields

$$c(x_1, x^*)\phi[D(x_1) - D(x^*)] \leq 0. \tag{6.75}$$

Since $D$ is strictly ($\phi$, $c$, $\beta$, $\gamma$, $\zeta$, $\eta$, $\omega$)- pseudo-univex, (75) implies

$$\beta(x_1, x^*)\left[[D'(x^*)\eta(x_1, x^*) - \frac{1}{2}D'(x^*)\omega(x_1, x^*)]\right.$$

$$\left. + \frac{1}{2}\gamma(x_1, x^*)[D'(x^*)\zeta(x_1, x^*)]\right] < 0,$$

$$\beta(x_1, x^*)\Big[[\Psi(x^*, y_0)\Phi_1'(x^*, y_0) - \Phi(x^*, y_0)\Psi_1'(x^*, y_0)\eta(x_1, x^*)]$$

$$- \quad \frac{1}{2}[\Psi(x^*, y_0)\Phi_1'(x^*, y_0) - \Phi(x^*, y_0)\Psi_1'(x^*, y_0)]\omega(x_1, x^*)]$$

$$+ \quad \frac{1}{2}\gamma(x_1, x^*)$$

$$[[\Psi(x^*, y_0)\Phi_1'(x^*, y_0) - \Phi(x^*, y_0)\Psi_1'(x^*, y_0)]\zeta(x_1, x^*)] < 0. \tag{6.76}$$

By (63) and (76), we get

$$-\Big[\beta(x_1,x^*)\Big[[\Psi(x^*,y_0)\Omega_1^{'}(x^*,z_0)\eta(x_1,x^*)$$

$$-\frac{1}{2}\Psi(x^*,y_0)\Omega_1^{'}(x^*,z_0)\omega(x_1,x^*)]$$

$$+\frac{1}{2}\gamma(x_1,x^*)[\Psi(x^*,y_0)\Omega_1^{'}(x^*,z_0)\zeta(x_1,x^*)]\Big] < 0,$$

i.e.,

$$\begin{aligned}&\beta(x_1,x^*)\big[[\Omega_1^{'}(x^*,z_0)\eta(x_1,x^*)]\\&-\frac{1}{2}\Omega_1^{'}(x^*,z_0)\omega(x_1,x^*)\big]\\&+\frac{1}{2}\gamma(x_1,x^*)[\Omega_1^{'}(x^*,z_0)\zeta(x_1,x^*)\big] > 0.\end{aligned} \tag{6.77}$$

Since $\Psi(x^*,y_0) > 0$ by the ($\phi$, $c$, $\beta$, $\gamma$, $\eta$, $\omega$, $\zeta$) - quasi-univexity of $\Omega$, (77) implies

$$c(x_1,x^*)\phi\left[\Omega(x_1,z_0) - \Omega(x^*,z_0)\right] > 0.$$

Since $c(.,.) > 0$, the above inequality yields

$$\Omega(x_1,z_0) > \Omega(x^*,z_0). \tag{6.78}$$

On the other hand, $x_1 \in F_p, z_0 \in R_+^m$ and $\Omega(x_1,z_0) \leq 0$. By (6.64), $\Omega\ (x^*,z_0) \geq 0$. Thus $\Omega(x_1,z_0) \leq \Omega\ (x^*,z_0)$, this contradicts (78). Hence (74) holds under $x_1 \neq x^*$. Therefore $x_1 = x^*$.

**Theorem 6.26.** (Weak duality). Let $x \in F_p, (u,y,z) \in K_2$ and define

$$D(x) = \Psi(u,y)\Phi(x,y) - \Psi(x,y)\Phi(u,y)$$

and

$$E(x) = D(x) + \Psi(u,y)\Omega(x,z).$$

Suppose, for each $y \in I, z \in R_+^m$, that any one of the following conditions holds:

**(a)** $\Phi(.,y), -\Psi(.,y)$ and $\Omega(.,z)$ are ($\phi$, $c$, $\beta$, $\gamma$, $\zeta$, $\eta$)- univex with respect to $c$, $\beta$, $\gamma$, $\phi$, $\zeta$ and $\eta$, where $c(.,.) > 0$, $\beta(.,.) > 0$, $\gamma(.,.) > 0$ and $\phi$ is superlinear.

**(b)** $D$ is ($\phi$, $c$, $\beta$, $\gamma$, $\zeta$, $\eta$)- pseudo-univex and $\Omega(.,z)$ is ($\phi$, $c$, $\beta$, $\gamma$, $\eta$, $\zeta$)- quasi-univex with respect to $c$, $\beta$, $\gamma$, $\phi$, $\zeta$ and $\eta$, where $c(.,.) > 0$, $\beta(.,.) > 0$, $\gamma(.,.) > 0$ and $\phi$ is strictly increasing.

**(c)** $D$ is ($\phi$, $c$, $\beta$, $\gamma$, $\eta$, $\zeta$)- quasi-univex and $\Omega(.,z)$ is strictly ($\phi$, $c$, $\beta$, $\gamma$, $\eta$, $\zeta$)- pseudo-univex with respect to $c$, $\beta$, $\gamma$, $\phi$, $\zeta$ and $\eta$, where $c(.,.) > 0$, $\beta(.,.) > 0$, $\gamma(.,.) > 0$ and $\phi$ is increasing.

**(d)** $E$ is ($\phi$, $c$, $\beta$, $\gamma$, $\eta$, $\zeta$)- pseudo-univex with respect to $c$, $\beta$, $\gamma$, $\phi$, $\zeta$ and $\eta$, where $c(.,.) > 0$, $\beta(.,.) > 0$, $\gamma(.,.) > 0$ and $\phi$ is increasing.

Then

$$\varphi(x) \geq \frac{\Phi(u,y)}{\Psi(u,y)}. \tag{6.79}$$

**Proof.** Suppose to contrary. Then we have

$$\frac{\Phi(u,y)}{\Psi(u,y)} > \varphi(x) = \max_{y_1 \in I} \frac{\Phi(x,y_1)}{\Psi(x,y_1)} \geq \frac{\Phi(x,y)}{\Psi(x,y)}.$$

This implies that

$$D(x) = \Phi(x,y)\Psi(u,y) - \Phi(u,y)\Psi(x,y) < 0 = D(u). \tag{6.80}$$

As $x \in F_p, z \in R_+^m$ and $\Omega(x,z) \leq 0$, the constraints inequality (6.64) of $(MWD)$ yields

$$\Omega(x,z) \leq 0 \leq \Omega(u,z). \tag{6.81}$$

Consequently, (80) and (81) yield

$$E(x) < E(u). \tag{6.82}$$

If condition (a) holds, then by the ($\phi$, $c$, $\beta$, $\gamma$, $\eta$, $\zeta$)- univexity of $\Phi$, $-\Psi$ and $\Omega$, we have

$$c(x,u)\phi\left[\Phi(x,y) - \Phi(u,y)\right] \geq \frac{1}{2}\beta(x,u)[\Phi_1'(u,y)\eta(x,u)]$$

$$+\frac{1}{2}\gamma(x,u)[\Phi_1'(u,y)\zeta(x,u)], \tag{6.83}$$

$$-c(x,u)\phi\left[\Psi(x,y) - \Psi(u,y)\right] \geq -\frac{1}{2}\beta(x,u)[\Psi_1'(u,y)\eta(x,u)]$$

$$+\frac{1}{2}[\Psi_1'(u,y)\zeta(x,u)], \tag{6.84}$$

$$c(x,u)\phi\left[\Omega(x,z) - \Omega(u,z)\right] \geq \frac{1}{2}\beta(x,u)[\Omega_1'(u,z)\eta(x,u)]$$

$$+\frac{1}{2}\gamma(x,u)[\Omega_1'(u,z)\zeta(x,u)]. \tag{6.85}$$

Since $\Psi(u,y)$ and $\Phi(u,y)$ are nonnegative, multiplying (83) by $\Psi(u,y)$, (84) by $\Phi(u,y)$ and (85) by $\Psi(u,y)$, and summing up the resulting inequalities in light of superlinearity of $\phi$, we obtain from (80) and (81)

$$\begin{aligned} 0 > &\frac{1}{2}\beta(x,u)\left[[\Psi(u,y)\Phi_1'(u,y) - \Phi(u,y)\Psi_1'(u,y) + \Psi(u,y)\Omega_1'(u,z)]\eta(x,u)\right) \\ + &\frac{1}{2}\gamma(x,u)\left[[\Psi(u,y)\Phi_1'(u,y) - \Phi(u,y)\Psi_1'(u,y) + \Psi(u,y)\Omega_1'(u,z)]\zeta(x,u)\right], \end{aligned}$$

which contradicts (6.63). Hence (79) holds.
In condition (b), $D$ is ($\phi$, $c$, $\beta$, $\gamma$, $\zeta$, $\eta$)- pseudo-univex. Since $c(.,.) > 0$, (80) implies $\beta(x,u)\Big[D'(u)\eta(x,u) + D'(u)\eta(x,u)\Big) < 0$, that is,

$$\frac{1}{2}\beta(x,u)\left[[\Psi(u,y)\Phi_1'(u,y) - \Phi(u,y)\Psi_1'(u,y)]\eta(x,u)\right]$$

$$+\frac{1}{2}\gamma(x,u)[\Psi(u,y)\Phi_1'(u,y) - \Phi(u,y)\Psi_1'(u,y)]\zeta(x,u)] < 0. \tag{6.86}$$

It follows from (63) and (86) that

$$-[\frac{1}{2}\beta(x,u)\Big(\Psi(u,y)\Omega_1'(u,z)\eta(x,u)$$
$$+\frac{1}{2}\gamma(x,u)[\Psi(u,y)\Omega_1'(u,z)\zeta(x,u)]] < 0$$

or

$$\frac{1}{2}\Omega'(u,z)\eta(x,u) + \frac{1}{2}\Omega'(u,z)\zeta(x,u) > 0,$$

since $\Psi(u,y) > 0$. But $\Omega$ is ($\phi$, $c$, $\beta$, $\gamma$, $\zeta$, $\eta$)- quasi-univex, so the above inequality implies

$$\Omega(x,z) > \Omega(u,z)$$

which contradicts (81). Hence, (79) holds.

Similarly, one can prove for the rest of two cases (c) and (d).

**Theorem 6.27.** (Strong duality). Let $x^*$ be an optimal solution of $(P)$. Then there exist $y^* \in I, z^* \in R_+^m$ such that $(x^*, y^*, z^*)$ is feasible for $(MWD)$. If any of the conditions of Theorem 6.26 holds, then $(x^*, y^*, z^*)$ is an optimal solution of $(MWD)$. Furthermore, $(MWD)$ and $(P)$ have the same optimal values.

**Proof.** If $x^*$ is an optimal solution of $(P)$, then by Theorem 6.26, there exist $y^* \in I$ and $z^* \in R_+^m$ such that $(x,y,z) \in K_2$, the feasible solution of $(MWD)$ and

$$\varphi(x^*) = \frac{\Phi(x^*,y^*)}{\Psi(x^*,y^*)} \le \max_{y\in I}\frac{\Phi(x^*,y)}{\Psi(x^*,y)}.$$

The optimality of $(MWD)$ follows from Theorem 6.26.

**Theorem 6.28.** (Strict converse duality). Let $x_1$ and $(x^*, y_0, z_0)$ be optimal solutions of $(P)$ and $(MWD)$, respectively. Assume that the hypothesis of Theorem 6.26 is fulfilled. Assume that $D(.) = \Psi(x^*,y_0)\Phi(.,y_0) - \Psi(.,y_0)\Phi(x^*,y_0)$ is strictly ($\phi$, $c$, $\beta$, $\gamma$, $\zeta$, $\eta$)- pseudo-univex and $\Omega(.,z_0)$ is ($\phi$, $c$, $\beta$, $\gamma$, $\zeta$, $\eta$)- quasi-univex with respect to $c$, $\beta$, $\gamma$, $\phi$, $\zeta$, and $\eta$. Then $x_1 = x^*$ is an optimal solution of $(P)$, and $(P)$ and $(MWD)$ have the same optimal values

$$\varphi(x_1) = \Phi(x^*,y_0)/\Psi(x^*,y_0).$$

**Proof.** Since $x_1$ is an optimal solution of $(P)$, by Theorem 6.26, there exist $y_1 \in I$ and $z_1 \in R^m_+$ such that $(x_1, y_1, z_1)$ is an optimal solution of $(MWD)$ and

$$\varphi(x_1) = \frac{\Phi(x_1, y_1)}{\Psi(x_1, y_1)}.$$

On the other hand, $(x^*, y_0, z_0)$ is an optimal solution of $(MWD)$ with optimal values equal to $\varphi(x_1)$, that is,

$$\left[\varphi(x_1) = \frac{\Phi(x_1, y_1)}{\Psi(x_1, y_1)} = \frac{\Phi(x^*, y_0)}{\Psi(x^*, y_0)}\right]. \tag{6.87}$$

If $x_1 \neq x^*$, we could deduce

$$\left[\varphi(x_1) > \frac{\Phi(x^*, y_0)}{\Psi(x^*, y_0)}\right], \tag{6.88}$$

which contradicts (87).

Assume (88) does not hold. Then $\varphi(x_1) = \Phi(x^*, y_0)/\Psi(x^*, y_0)$, that is,

$$\frac{\Phi(x^*, y^*)}{\Psi(x^*, y^*)} \geq \varphi(x_1) = \max_{y \in I} \frac{\Phi(x_1, y)}{\Psi(x_1, y)} \geq \frac{\Phi(x_1, y_0)}{\Psi(x_1, y_0)}$$

or

$$D(x_1) = \Phi(x_1, y_0)\Psi(x^*, y_0) - \Phi(x^*, y_0)\Psi(x_1, y_0) \leq 0 = D(x^*),$$

that is,

$$D(x_1) \leq D(x^*).$$

Since $c(.,.) > 0$, the above inequality yields

$$c(x_1, x^*)\phi[D(x_1) - D(x^*)] \leq 0. \tag{6.89}$$

Since $D$ is strictly $(\phi, c, \beta, \gamma, \zeta, \eta)$- pseudo-univex, (89) implies

$$\beta(x_1, x^*)\frac{1}{2}[D'(x^*)\eta(x_1, x^*)] + \frac{1}{2}\gamma(x_1, x^*)[D'(x^*)\zeta(x_1, x^*)] < 0,$$

$$\beta(x_1, x^*)\frac{1}{2}[[\Psi(x^*, y_0)\Phi_1'(x^*, y_0) - \Phi(x^*, y_0)\Psi_1'(x^*, y_0)\eta(x_1, x^*)]$$

$$+ \quad \frac{1}{2}\gamma(x_1, x^*)$$

$$[[\Psi(x^*, y_0)\Phi_1'(x^*, y_0) - \Phi(x^*, y_0)\Psi_1'(x^*, y_0)]\zeta(x_1, x^*)] < 0. \tag{6.90}$$

By (63) and (90), we get

$$-\beta(x_1, x^*)[\frac{1}{2}\Psi(x^*, y_0)\Omega_1'(x^*, z_0)\eta(x_1, x^*)]$$

$$+\frac{1}{2}\gamma(x_1, x^*)[\Psi(x^*, y_0)\Omega_1'(x^*, z_0)\zeta(x_1, x^*)] < 0,$$

i.e.,

$$\beta(x_1,x^*)[\frac{1}{2}\Omega_1'(x^*,z_0)\eta(x_1,x^*)]$$

$$+\frac{1}{2}\gamma(x_1,x^*)[\Omega_1'(x^*,z_0)\zeta(x_1,x^*)] > 0. \tag{6.91}$$

Since $\Psi(x^*,y_0) > 0$ by the ($\phi$, $c$, $\beta$, $\gamma$, $\eta$, $\zeta$) - quasi-univexity of $\Omega$, (91) implies

$$c(x_1,x^*)\phi\,[\Omega(x_1,z_0) - \Omega(x^*,z_0)] > 0.$$

Since $c(.,.) > 0$, the above inequality yields

$$\Omega(x_1,z_0) > \Omega(x^*,z_0). \tag{6.92}$$

On the other hand, $x_1 \in F_p, z_0 \in R_+^m$ and $\Omega(x_1,z_0) \leq 0$. By (66), $\Omega\ (x^*,z_0) \geq 0$. Thus $\Omega(x_1,z_0) \leq \Omega\ (x^*,z_0)$, this contradicts (92). Hence (88) holds under $x_1 \neq x^*$. Therefore $x_1 = x^*$.

**Theorem 6.29.** (Weak duality). Let $x \in F_p, (u,y,z) \in K_2$ and define

$$D(x) = \Psi(u,y)\Phi(x,y) - \Psi(x,y)\Phi(u,y)$$

and

$$E(x) = D(x) + \Psi(u,y)\Omega(x,z).$$

Suppose, for each $y \in I, z \in R_+^m$, that any one of the following conditions holds:

**(a)** $\Phi(.,y), -\Psi(.,y)$ and $\Omega(.,z)$ are ($\phi$, $c$, $\eta$)- univex with respect to $c$, $\phi$ and $\eta$, where $c(.,.) > 0$, and $\phi$ is superlinear.

**(b)** $D$ is ($\phi$, $c$, $\eta$)- pseudo-univex and $\Omega(.,z)$ is ($\phi$, $c$, $\eta$)- quasi-univex with respect to $c$, $\phi$, and $\eta$, where $c(.,.) > 0$, and $\phi$ is strictly increasing.

**(c)** $D$ is ($\phi$, $c$, $\eta$)- quasi-univex and $\Omega(.,z)$ is strictly ($\phi$, $c$, $\eta$)- pseudo-univex with respect to $c$, $\phi$, and $\eta$, where $c(.,.) > 0$, and $\phi$ is increasing.

**(d)** $E$ is ($\phi$, $c$, $\eta$)- pseudo-univex with respect to $c$, $\phi$, and $\eta$, where $c(.,.) > 0$, and $\phi$ is increasing.

Then

$$\varphi(x) \geq \frac{\Phi(u,y)}{\Psi(u,y)}. \tag{6.93}$$

**Proof.** Suppose to contrary. Then we have

$$\frac{\Phi(u,y)}{\Psi(u,y)} > \varphi(x) = \max_{y_1 \in I}\frac{\Phi(x,y_1)}{\Psi(x,y_1)} \geq \frac{\Phi(x,y)}{\Psi(x,y)}.$$

This implies that

$$D(x) = \Phi(x,y)\Psi(u,y) - \Phi(u,y)\Psi(x,y) < 0 = D(u). \tag{6.94}$$

As $x \in F_p, z \in R_+^m$ and $\Omega(x,z) \leq 0$, the constraints inequality (6.64) of $(MWD)$ yields

$$\Omega(x,z) \leq\ 0 \leq \Omega(u,z). \tag{6.95}$$

Consequently, (94) and (95) yield

$$E(x) < E(u). \tag{6.96}$$

If condition (a) holds, then by the ($\phi$, $c$, $\eta$)- univexity of $\Phi$, $-\Psi$ and $\Omega$, we have

$$c(x,u)\phi\left[\Phi(x,y) - \Phi(u,y)\right] \geq \Phi_1^{'}(u,y)\eta(x,u), \tag{6.97}$$

$$-c(x,u)\phi\left[\Psi(x,y) - \Psi(u,y)\right] \geq -\Psi_1^{'}(u,y)\eta(x,u), \tag{6.98}$$

$$c(x,u)\phi\left[\Omega(x,z) - \Omega(u,z)\right] \geq \Omega_1^{'}(u,z)\eta(x,u). \tag{6.99}$$

Since $\Psi(u,y)$ and $\Phi(u,y)$ are nonnegative, multiplying (97) by $\Psi(u,y)$, (98) by $\Phi(u,y)$ and (99) by $\Psi(u,y)$, and summing up the resulting inequalities in light of superlinearity of $\phi$, we obtain from (94) and (95)

$$0 > \left[[\Psi(u,y)\Phi_1^{'}(u,y) - \Phi(u,y)\Psi_1^{'}(u,y) + \Psi(u,y)\Omega_1^{'}(u,z)]\eta(x,u)\right),$$

which contradicts (6.63). Hence (93) holds.
In condition (b), $D$ is ($\phi$, $c$, $\eta$)- pseudo-univex. Since $c(.,.) > 0$, (94) implies $D'(u)\eta(x,u) < 0$, that is,

$$\left[[\Psi(u,y)\Phi_1^{'}(u,y) - \Phi(u,y)\Psi_1^{'}(u,y)]\eta(x,u)\right] < 0. \tag{6.100}$$

It follows from (63) and (100) that

$$-\Psi(u,y)\Omega_1^{'}(u,z)\eta(x,u) < 0$$

or

$$\Omega^{'}(u,z)\eta(x,u) > 0,$$

since $\Psi(u,y) > 0$. But $\Omega$ is ($\phi$, $c$, $\eta$)- quasi-univex, so the above inequality implies

$$\Omega(x,z) > \Omega(u,z),$$

which contradicts (95). Hence, (79) holds.

Similarly, one can prove for the rest of two cases (c) and (d).

**Theorem 6.30.** (Strong duality). Let $x^*$ be an optimal solution of $(P)$. Then there exist $y^* \in I, z^* \in R_+^m$ such that $(x^*, y^*, z^*)$ is feasible for $(MWD)$. If any of the conditions of Theorem 6.24 holds, then $(x^*, y^*, z^*)$ is an optimal solution of $(MWD)$. Furthermore,

$(MWD)$ and $(P)$ have the same optimal values.

**Proof.** If $x^*$ is an optimal solution of $(P)$, then by Theorem 6.24, there exist $y^* \in I$ and $z^* \in R^m_+$ such that $(x,y,z) \in K_2$, the feasible solution of $(MWD)$ and

$$\varphi(x^*) = \frac{\Phi(x^*,y^*)}{\Psi(x^*,y^*)} \leq \max_{y\in I} \frac{\Phi(x^*,y)}{\Psi(x^*,y)}.$$

The optimality of $(MWD)$ follows from Theorem 6.23.

**Theorem 6.31.** (Strict converse duality). Let $x_1$ and $(x^*,y_0,z_0)$ be optimal solutions of $(P)$ and $(MWD)$, respectively. Assume that the hypothesis of Theorem 6.25 is fulfilled. Assume that $D(.) = \Psi(x^*,y_0)\Phi(.,y_0) - \Psi(.,y_0)\Phi(x^*,y_0)$ is strictly $(\phi,\, c,\, \eta)$- pseudo-univex and $\Omega(.,z_0)$ is $(\phi,\, c,\, \eta)$- quasi-univex with respect to $c$, $\phi$ and $\eta$. Then $x_1 = x^*$ is an optimal solution of $(P)$, and $(P)$ and $(MWD)$ have the same optimal values

$$\varphi(x_1) = \Phi(x^*,y_0)/\Psi(x^*,y_0).$$

**Proof.** Since $x_1$ is an optimal solution of $(P)$, by Theorem 6.27, there exist $y_1 \in I$ and $z_1 \in R^m_+$ such that $(x_1,y_1,z_1)$ is an optimal solution of $(MWD)$ and

$$\varphi(x_1) = \frac{\Phi(x_1,y_1)}{\Psi(x_1,y_1)}.$$

On the other hand, $(x^*,y_0,z_0)$ is an optimal solution of $(MWD)$ with optimal values equal to $\varphi(x_1)$, that is,

$$\left[\varphi(x_1) = \frac{\Phi(x_1,y_1)}{\Psi(x_1,y_1)} = \frac{\Phi(x^*,y_0)}{\Psi(x^*,y_0)}\right]. \tag{6.101}$$

If $x_1 \neq x^*$, we could deduce

$$\left[\varphi(x_1) > \frac{\Phi(x^*,y_0)}{\Psi(x^*,y_0)}\right], \tag{6.102}$$

which contradicts (101).

Assume (102) does not hold. Then $\varphi(x_1) = \Phi(x^*,y_0)/\Psi(x^*,y_0)$, that is,

$$\frac{\Phi(x^*,y^*)}{\Psi(x^*,y^*)} \geq \varphi(x_1) = \max_{y\in I} \frac{\Phi(x_1,y)}{\Psi(x_1,y)} \geq \frac{\Phi(x_1,y_0)}{\Psi(x_1,y_0)}$$

or

$$D(x_1) = \Phi(x_1,y_0)\Psi(x^*,y_0) - \Phi(x^*,y_0)\Psi(x_1,y_0) \leq 0 = D(x^*),$$

that is,

$$D(x_1) \leq D(x^*).$$

Since $c(.,.) > 0$, the above inequality yields

$$c(x_1,x^*)\phi[D(x_1) - D(x^*)] \leq 0. \tag{6.103}$$

Since $D$ is strictly $(\phi,\, c,\, \eta)$- pseudo-univex, (103) implies

$$\left[[D^{'}(x^{*})\eta(x_{1},x^{*})\right] < 0,$$

$$[[\Psi(x^{*},y_{0})\Phi_{1}^{'}(x^{*},y_{0}) - \Phi(x^{*},y_{0})\Psi_{1}^{'}(x^{*},y_{0})]\eta(x_{1},x^{*}) < 0. \tag{6.104}$$

By (63) and (104), we get

$$-]\Psi(x^{*},y_{0})\Omega_{1}^{'}(x^{*},z_{0})]\eta(x_{1},x^{*}) < 0,$$

i.e.,

$$[\Omega_{1}^{'}(x^{*},z_{0})]\eta(x_{1},x^{*})] > 0. \tag{6.105}$$

Since $\Psi(x^{*},y_{0}) > 0$ by the $(\phi,\, c,\, \eta)$ - quasi-univexity of $\Omega$, (43) implies

$$c(x_{1},x^{*})\phi\,[\Omega(x_{1},z_{0}) - \Omega(x^{*},z_{0})] > 0.$$

Since $c(.,.) > 0$, the above inequality yields

$$\Omega(x_{1},z_{0}) > \Omega(x^{*},z_{0}). \tag{6.106}$$

On the other hand, $x_{1} \in F_{p}, z_{0} \in R_{+}^{m}$ and $\Omega(x_{1},z_{0}) \leq 0$. By (64), $\Omega\ (x^{*},z_{0}) \geq 0$. Thus $\Omega(x_{1},z_{0}) \leq \Omega\ (x^{*},z_{0})$, this contradicts (106). Hence (6.102) holds under $x_{1} \neq x^{*}$. Therefore $x_{1} = x^{*}$.

## 6. General Remarks

The results established in this chapter can be generalized, based on the work of Pitea and Postolache [7], to applications in mechanical engineering, where curvilinear integral objectives are used largely driven by their physical meaning as mechanical work. These kinds of objectives have also significant roles in mathematical modeling of certain processes relating robotics, tribology, engines, and others.

# Bibliography


[1] S. Chandra, B. D. Craven and B. Mond, Generalized Fractional programming duality: a ratio game approach, *J. Austral. Math. Soc. Ser. B*, **28** (1986) 170 - 180.

[2] J.P. Crouzeix, J. A. Ferland and S. Schaible, Duality in generalized fractional programming, *Math. Prog.* **27**, (1983) 342 - 354.

[3] H. C. Lai and J. C. Liu, Minimax fractional programming involving generalized invex functions, *ANZIAM J.* **44**, (2003) 339 - 354.

[4] S. K. Mishra, K. Shukla and R. U. Verma, Minmax fractional integral programming problems involving univexity, *DCDIS, Series A: Mathematical Analysis* **18** (2012), 335 - 355.

[5] B. Mond, S. Chandra and I. Husain, Duality for variational problems with invexity, *J. Math. Anal. Appl.* **134** (1988) 322 - 328.

[6] B. Mond and I. Husain, Sufficient optimality criteria and duality for variational problems with generalized invexity, *J. Austral. Math. Soc. Ser. B* **31** (1) (1989) 108 - 121.

[7] A. Pitea and M. Postolache, Duality theorems for a new class of multitime multiobjective variational problems, *J. Global Optimization* **54** (2012), 47 - 58.

[8] I.M. Stancu-Minasian, *Fractional Programming Theory, Methods and Applications*, Kluwer, Dordrecht, 1997.

[9] R. U. Verma, *Semi-Infinite Fractional Programming*, Springer Nature/Springer 2017. https://link.springer.com/book/10.1007/10-978-981-10-6256-8

[10] R. U. Vema and G. J. Zalmai, Parameter-free duality models and applications to semiinfinite minmax fractional programming based on higher-order $(\phi, \eta, \rho, \theta, m)$-sonvexities, *OPSEARCH* **55**(2) (2018), 381 - 410, Springer Nature DOI: 10.1007/s 12597-017-0323-8

[11] G. J. Zalmai, Generalized $(\mathscr{F}, b, \phi, \rho, \theta)$-univex n-set functions and semiparametric duality models in multiobjective fractional subset programming, *Int. J. Math. Math. Sci.* **2005** (7) (2005), 1105 - 1133.

# General References


[1] Aghezzaf, B. and M. Hachimi (1999), Second-order optimality conditions in multiobjective optimization problems, *J. Optim. Theory Appl.*, **102**, 37-50.

[2] Ahmad, I., S. K. Gupta, K. Kailey and R.P. Agarwal (2011), Duality in nondifferentiable minmax fractional programming with $B-(p,r)$invexity, *Journal of Inequalities and Applications*, **2011**:75 (2011).

[3] Ahmad, I. and Z. Husain (2006), Optimality conditions and duality in nondifferentiable minmax fractional programming with generalized convexity, *Journal of Optimization Theory and Applications*, **129** (2), 255-275.

[4] Ahmad, I. (2003), Optimality conditions and duality in fractional minmax programming involving generalized $\rho-$invexity, *International Journal of Management and System*, **19**, 165-180.

[5] Al-Khayyal, F. A. and J. Kyparisis (1991), Finite convergence of algorithms for nonlinear programs and variational inequalities, *Journal of Optimization Theory and Applications*, **70** (2), 319-332.

[6] Antczak, T. (2005), Modified ratio objective approach in mathematical programming, *Journal of Optimization Theory and Applications*, **126**(1), 23-40.

[7] Antczak, T. (2001), On $(p,r)$-invexity-type nonlinear programming problems, *J. Math. Anal. Appl.*, **264** (2001), 382-397.

[8] Antczak, T. (2001), $(p,r)$-Invex sets and functions, *J. Math. Anal. Appl.*, **263**, 355-379.

[9] Antczak, T, (2002), Lipschitz $r$-invex functions and nonsmooth programming, *Numer. Funct. Anal. Optim.*, **23**, 265-283.

[10] Antczak, T. (2003). Generalized $(p,r)$-invexity in mathematical programming, *Numer. Funct. Anal. Optim.*, **24**, 437-453.

[11] Antczak, T. (2003), A class of B-$(p,r)$-invex functions and mathematical programming, *J. Math. Anal. Appl.*, **268**, 187-206.

[12] Antczak, T. (2004), B-$(p,r)$-pre-invex functions, *Folia Math.* Acta Univ. Lodziensis, **11**, 3-15.

[13] Antczak, T. (2004), Minimax programming under $(p,r)$-invexity, *European J. Oper. Res.* **158**, 1-19.

[14] Antczak, T. (2005), $r$-Pre-invexity and $r$-invexity in mathematical programming, *Comput. Math. Appl.*, **50**, 551-566.

[15] Antczak, T. (2005), The notion of V-$r$-invexity in differentiable multiobjective programming, *J. Appl. Anal.*, **11**, 63-79.

[16] Antczak, T. (2009), Optimality and duality for nonsmooth multiobjective programming problems with V-$r$-invexity, *J. Global Optim.*, **45**, 319-334.

[17] Antczak, T. (2009), Generalized B-$(p,r)$-invexity functions and nonlinear mathematical programming, *Numer. Funct. Anal. Optim.*, **30**, 1-22.

[18] Antczak, T. (2010), The $\ell_1$ penalty function method for nonconvex differentiable optimization problems with inequality constraints, *Asia-Pac. J. Oper. Res.*, **27**, 559-576.

[19] Barros, A. I. (1998), *Discrete and Fractional Programming Techniques for Location Models*, Kluwer Academic Publishers, Dordrecht-Boston-London. .

[20] Bigi, G. and M. Castellani (2000), Second order optimality conditions for differentiable multiobjective problems, *RAIRO Oper. Res.* **34**, 411-426.

[21] Beck, A. and M. Teboulle (2000), Global optimality conditions for quadratic optimization problems with binary constraints, *SIAM Journal of Optimization* **11** (1), 179-188.

[22] Bector, C. R. and S. Chandra (1986), Second-order duality for generalized fractional programming, *Methods Oper. Res.*, **56**, 11-28.

[23] Ben-Israel, A. and B. Mond (1986), *What is the invexity? Journal of Australian Mathematical Society Ser. B* **28**, 1-9.

[24] Bolintinéanu, S. and M. El Maghri (1998), Second-order efficiency conditions and sensitivity of efficient points, *J. Optim. Theory Appl.*, **98**, 569-592.

[25] Bonnans, J. F., R. Cominetti, and A. Shapiro (1999), Second-order optimality conditions based on parabolic second-order tangent sets, *SIAM J. Optim.*, **9**, 466-492.

[26] Burke, J. V. (1987), Second-order necessary and sufficient conditions for convex NDO, *Math. Prog.* **38**, 287-302.

[27] Burke, J. V. and M. C. Ferris (1991), Characterization of solution sets of convex programs, *Operations Research Letters* **10** (1), 57-60.

[28] Burke, J. V., M. C. Ferris and M. Qian (1992), On the Clarke subdifferential of the distance function to a closed set, *Journal of Mathematical Analysis and Applications* **166**, 199-213.

[29] Burke, J. V. and S.P. Han (1986), A Gauss-Newton approach to solving generalized inequalities, *Mathematics of Operations Research* **11**, 632-643.

[30] Burke, J. V. and J.J. More' (1988), On the identification of active constraints, *SIAM Journal of Numerical Analysis*, **25**, 1197-1211.

[31] Cambini, A. and L. Martein (2002), First and second order optimality conditions in vector optimization, *J. Stat. Manag. Syst.*, **5**, 295-319.

[32] Cambini, A., L. Martein, and M. Vlach (1999), Second order tangent sets and optimality conditions, *Math. Japonica*, **49**, 451-461.

[33] Cambini, R. (1998), Second order optimality conditions in multiobjective programming, *Optimization*, **44**, 139-160.

[34] Cambini, A. and L. Martein (2002), First and second order optimality conditions in vector optimization, *J. Stat. Manag. Syst.*, **5**, 292-319.

[35] Chaney, R. (1982), Second-order sufficiency conditions for nondifferentiable programming problems, *SIAM J. Control Optim.*, **20**, 20-23.

[36] Chaney, R. (1987), Second order necessary conditions in constrained semismooth optimization, *SIAM J. Control Optim.*, **25**, 1072-1081.

[37] Chaney, R. (1988), Second-order necessary conditions in semismooth optimization, *Math. Prog.*, **40**, 95-109.

[38] Chaney, R. (1988), Second-order sufficient conditions in nonsmooth optimization, *Math. Oper. Res.*, **13**, 660-673.

[39] Charnes, A. and W. W. Cooper (1977), Goal programming and multiobjective optimization, part 1, *European J. Oper. Res.*, **1**, 39-54.

[40] Chen, H. and C. F. Hu (2009), On the resolution of the Vasicek-type interest rate model, *Optimization* **58**, 809-822.

[41] Chinchuluum, A. and P. M. Pardalos (2007), A survey of recent developments in multiobjective optimization, *Annals of Operational Research*, **154** 29-50.

[42] Clarke, F. H. (1975), Generalized gradients and applications, *Transactions of the American Mathematical Society*, **205**, 247-262.

[43] Clarke, F. H. (1983), *Optimization and Nonsmooth Analysis*, Wiley, New York.

[44] Cominetti, R. (1990), Metric regularity, tangent sets, and second-order optimality conditions, *Appl. Math. Optim.*, **21**, 265-287.

[45] Craven, B. D. (1981), Invex functions and constrained local minima, *Bull. Austral. Math. Soc.* **24** (1981), 357 - 366.

[46] Crespi, G., D. La Torre, and M. Rocca (2003), Mollified derivatives and second-order optimality conditions, *J. Nonlin. Convex Anal.*, **4**, 437-454.

[47] Daum, S. and R. Werner (2011), A novel feasible discretization method for linear semi-infinite programming applied to basket option pricing, *Optimization*, **60**, 1379-1398.

[48] Dinkelbach, W. (1967), On nonlinear fractional programming, *Manag. Sci.* **13** (1967), 492 - 498.

[49] Dien, P. H. and P. H. Sach (1989), Second-order optimality conditions for the extremal problem under inclusion constraints, *Appl. Math. Optim.*, **20**, 71-80.

[50] Egudo, R. R. and M. A. Hanson (1993), Second order duality in multiobjective programming, *Opsearch*, **30**, 223-230.

[51] Floudas, C. A. and V. Visweswaran, Quadratic optimization, In *Handbook of Global Optimization*, R. Horsi and P.M. Pardalos, eds., Kluwer Academic Publishers, Dordrecht, p. 217-269.

[52] Ferris, M. C. and O. L. Mangasarian (1992), Minmum principle sufficiency, *Mathematical Programming* **57** (1), 1-14.

[53] Garey, M. R. and D. S. Johnson (1979), *Computers and Intractability: A Guide to Theory of NP-Completeness*, W. H. Freeman, San Francisco.

[54] Georgiev, P. G. and N. Zlateva (1996), Second-order subdifferentials of $C^{1,1}$ functions and optimality conditions, *Set-Valued Anal.*, **4**, 101-117.

[55] Gfrerer, H. (2006) Second-order optimality conditions for scalar and vector optimization problems in Banach spaces, *SIAM J. Optim.*, **45**, 972-997.

[56] Ginchev, I. and A. Guerraggio (1998), Second-order optimality conditions in nonsmooth unconstrained optimization, *Pliska Stud. Math. Bulg.*, **12**, 39-50.

[57] Ginchev, I., A. Guerraggio and M. Rocca (2005), Second-order conditions for $C^{1,1}$ constrained vector optimization, *Math. Prog.*, **104**, 389-405.

[58] Ginchev, I., A. Guerraggio, and M. Rocca (2006), Second order conditions in $C^{1,1}$ vector optimization with inequality and equality constraints, *Lecture Notes in Economics and Mathematical Systems*, No. 563, Springer-Verlag, Berlin, pp. 29-44.

[59] Goberna, M. A. López (1998), *Linear Semi-Infinite Optimization*, Wiley, New York.

[60] Goberna, M. A. López (2001), *Semi-infinite Programming - Recent Advances*, Kluwer, Dordrecht.

[61] Goemans, M. X. and D. P. Williamson (1995), Improved approximation algorithms for maximum cut and satisfiability problems using semidefinite programming, *J. ACM* **42**, 1115-1145.

[62] Guerraggio, A., D. T. Luc, and N. B. Minh (2001), Second-order optimality conditions for $C^1$ multiobjective programming problems, *Acta Math. Vietnam.*, **26**, 257-268.

[63] Gutierrez, C., B. Jiménez, and V. Novo (2009), New second-order derivatives and optimality conditions for scalar and vector optimization, *J. Optim. Theory Appl.*, **142**, 85-106.

[64] Guerraggio, A., D. T. Luc, and N. B. Minh (2001), Second-order optimality conditions for $C^1$ multiobjective programming problems, *Acta Math. Vietnam.*, **26**, 257-268.

[65] Hanson, M. A. (1981), On sufficiency of the Kuhn-Tucker conditions, *J. Math. Anal. Appl.*, **80**, 545-550.

[66] Hanson, M. A. (1993), Second order invexity and duality in mathematical programming, *Opsearch*, **30**, 313-320.

[67] Hanson, M. A. and B. Mond (1982), Further generalization of convexity in mathematical programming, *Journal of Information and Optimization Science*, **3** , 25-32.

[68] Hettich, R. (1976), Semi-infinite Programming, *Lecture Notes in Control and Information Sciences*, Vol. 7, Springer, Berlin.

[69] Hettich, R. and G. Still (1995), Second order optimality conditions for generalized semi-infinite programming problems, *Optimization*, **34**, 195-211.

[70] Hettich, R. (1986), An implementation of a discretization method for semi-infinite programming, *Mathematical Programming* **34**, 354-361.

[71] Hettich, R. (1986), An implementation of a discretization method for semi-infinite programming, *Mathematical Programming* **34**, 354-361.

[72] Husain, Z., I. Ahmad and S. Sharma (2009), Second order duality for minmax fractional programming, *Optim. Lett.*, 277-286.

[73] Ito, S. A. (2000), A dual parametrization method for convex semi-infinite programming, *Annals of Operations Research*, **98**, 189-213.

[74] Ivanov, V. I. (2007), Second-order optimality conditions in nonsmooth optimization, *C. R. Acad. Bulgare Sci.*, **60**, 1053-1058.

[75] Jagannathan, R. and S. Schaible (1983), Duality in generalized fractional programming via Farkas' lemma, *Journal of Optimization Theory and Applications*, **41** (3), 417-424.

[76] Jeyakumar, V. (1985), Strong and weak invexity in mathematical programming, *Methods Oper. Res.* **55**, 109-125.

[77] Jeyakumar, V. (1985), First and second order fractional programming duality, *Opsearch*, **22**, 24-41.

[78] Jiménez, B. and V. Novo (2002), First and second order sufficient conditions for strict minimality in multiobjective programming, *Numer. Funct. Anal. Optim.*, **23**, 303-322.

[79] Jiménez, B. and V. Novo (2003), First and second order sufficient conditions for strict minimality in nonsmooth vector optimization, *J. Math. Anal. Appl.*, **284**, 496-510.

[80] Jiménez, B. and V. Novo (2000), Second-order necessary conditions in set constrained differentiable vector optimization, *Math. Methods Oper. Res.*, **58**, 299-317.

[81] Jiménez, B. and V. Novo (2004), Optimality conditions in differentiable optimization via second-order tangent sets, *Appl. Math. Optim.*, **49**, 123-144.

[82] Jongen, H. Th. and O. Stein (1997), On generic one-parametric semi-infinite optimization, *SIAM Journal of Optimization* **7** (4), 1103-1137.

[83] Jongen, H. Th., P. Jonker and F. Twilt (2000), *Nonlinear Optimization in Finite Dimension*, Kluwer, Dordrecht.

[84] Jongen, H. Th., W. Wetterling and G. Zwier, On sufficient conditions for local optimality in semi-infinite programming, *Optimization* **18**, 165-178.

[85] Kanniappan, P. and P. Pandian (1996), On generalized convex functions in optimization theory - A survey, *Opsearch*, **33**, 174-185.

[86] Karney, D. F. (1981), Clarke's theorem for semiinfinite convex programs, *Advances in Applied Mathematics*, **2**, 7-12.

[87] Karney, D. F. (1983), A duality theorem for semi-infinite convex programs and their finite subprograms, *Mathematical Programming*, **27**, 75-82.

[88] Kawasaki, H. (1988), Second-order necessary conditions of the Kuhn-Tucker type under new constraint qualifications, *J. Optim. Theory Appl.* **57**, 253-264.

[89] Kawasaki, H. (1988), An envelope-like effect of infinitely many inequality constraints on second order necessary conditions for minimization problems, *Math. Prog.* **41**, 73-96.

[90] Kawasaki, H. (1992), Second-order necessary and sufficient optimality conditions for minimizing a sup-type function, *J. Appl. Math. Optim.*, **26**, 195-220.

[91] Kelly, T. K. and M. Kupperschmid (1998), Numerical verification of second-order sufficiency conditions for nonlinear programming, *SIAM Review*, **40**, 310-314.

[92] Khanh, P. Q. and N. D. Tuan (2006), First and second-order optimality conditions using approximations for nonsmooth vector optimization in Banach spaces, *J. Optim. Theory Appl.*, **130**, 289-308.

[93] Khanh, P. Q. and N. D. Tuan (2007), Optimality conditions for nonsmooth multiobjective optimization using Hadamard directional derivatives, *J. Optim. Theory Appl.*, **133**, 341-357.

[94] J. S. H. Kornbluth, J. S. H. (1973), *A survey of goal programming*, **1**, 193-205.

[95] Li, Z. (1993), Duality theorems for a class of generalized convex multuiobjective programming, *Acta Sci. Natur. Univ. Neimenggu* **24** (2), 113-118.

[96] Kuk, H. and H-J Kim (2003), Optimality conditions and duality for nonsmooth generalized fractional programming problems, *Journal of Statistics and Management* **6**, 207-215.

[97] Lee, G. M. (1994), Nonsmooth invexity in multiobjective programming, *Journal of Information and Optimization Sciences* **15**, 127-136.

[98] Li, X. F., J. L. Dong and Q. H. Liu (1997), Lipschitz $B-vex$ functions and nonsmooth programming *JOTA* **93**, 557-574.

[99] Liang, Z. A. and Z. W. Shi (2003), Optimality conditions and duality for minmax fractional programming with generalized convexity, *JMAA* **277**, 474-488.

[100] Liu, J. C. (1999), Second order duality for minimax programming, *Utilitas Math.*, **56**, 53-63

[101] Liu, J. C. (1996), Optimality and duality for generalized fractional programming involving nonsmooth $(F,\rho)-$convex functions, *Computers and Mathematics with Applications* **32**, 91-102.

[102] Liu, J. C. (1998), Optimality and duality for generalized fractional programming involving nonsmooth pseudoinvexity, *JMAA* **202**, 36-51.

[103] Liu, J. C. and C. S. Wu (1998), On minmax fractional optimality conditions with $(f,\rho)-$convex functions, *JMAA* **219**, 36-51.

[104] Lopez, M. and G. Still (2007), Semi-infinite programming, *European Journal of Operational Research*, **180**, 491-518.

[105] C. Malivert, C. (1990), First and second order optimality conditions in vector optimization, *Ann. Sci. Math. Québec*, **14**, 65-79.

[106] Maruyama, Y. (1994), Second-order necessary conditions for nonlinear optimization problems in Banach spaces by the use of Neustadt derivative, *Math. Japonica*, **40**, 509-522.

[107] H. Maurer, H. (1981), First- and second-order sufficient optimality conditions in mathematical programming and optimal control, *Math. Prog. Study*, **14**, 163-177.

[108] Maurer, H. and J. Zowe (1979), First and second order necessary and sufficient optimality conditions for infinite-dimensional programming problems, *Math. Prog.*, **16**, 98-110.

[109] Mangasarian, O.L. (1969), *Nonlinear Programming*, McGraw-Hill, New York.

[110] Mangasarian, O. L. (1975), Second- and higher-order duality theorems in nonlinear programming, *J. Math. Anal. Appl.*, **51**, 607-620.

[111] Mangasarian, O. L. ((1990), Error bounds of nondegenerate monotone linear complementarity problems, *Mathematical programming*, **48**,437-446.

[112] Mangasarian, O. L. (1988), A simple characterization of solution sets of convex programs, *Operations Research Letters*, **7** (1), 21-26.

[113] Mangasarian, O. L. and R. R. Meyer (1979), Nonlinear perturbation of linear programs, *SIAM Journal on Control and Optimization*, **17** (6), 745-752.

[114] Martin, D. H. (1985), The essence of invexity, *J. Optim. Theory Appl.*, **47**, 65-76.

[115] Miettinen, K. M. (1999), *Nonlinear Multiobjective Optimization*, Kluwer Academic Publishers, Boston.

[116] Mishra, S. K. (1996), Generalized proper efficiency and duality for a class of nondifferentiable multiobjective variational problems with V-invexity, *Journal of Mathematical Analysis and Applications* **2**, 53-71.

[117] Mishra, S. K. and R. N. Mukherjee (1994), On efficiency and duality for multiobjective variational problems, *Journal of Mathematical Analysis and Applications* **187**, 40-54.

[118] Mishra, S. K. and R. N. Mukherjee (1994), Duality for multiobjective fractional variational problems, *Journal of Mathematical Analysis and Applications* **186**, 711-725.

[119] Mishra, S. K. and N. G. Rueda (2000), Higher-order generalized invexity and duality in mathematical programming, *J. Math. Anal. Appl.*, **247**, 173-182.

[120] Mishra, S. K. and N. G. Rueda (2006), Second-order duality for nondifferential minimax programming involving generalized type I functions, *J. Optim. Theory Appl.*, **130**, 477-486.

[121] Mishra, S. K., M. Jaiswal and Pankaj (2010), Optimality conditions for multiple objective fractional subset programming with invex and related non-convex functions, *Communications on Applied Nonlinear Analysis*, **17 (3)**, 89-101.

[122] Mishra, S. K. (1997), Second order generalized invexity and duality in mathematical programming, *Optimization*, **42**, 51-69.

[123] Mishra, S. K. (2000), Second order symmetric duality in mathematical programming with F-convexity, *European J. Oper. Res.*, **127**00), 507-518.

[124] Mititelu, Şt. (2004), Invex functions, *Rev. Roumaine Math. Pures Appl.*, **49**, 529-544.

[125] Mititelu, Şt. (2007), Invex sets and nonsmooth invex functions, *Rev. Roumaine Math. Pures Appl.*, **52**, 665-672.

[126] Mititelu, Şt. and M. Postolachi (2011), Nonsmooth invex functions via upper directional derivative of Dini, *J. Adv. Math. Stud.*, **4**, 57-76.

[127] Mititelu, Şt. and I. M. Stancu-Minasian (1993), Invexity at a point: Generalizations and classification, *Bull. Austral. Math. Soc.*, **48**, 117-126.

[128] Mond, B. (1974), Second order duality for nonlinear programs, *Opsearch*, **11**, 90-99.

[129] Mond, B. and I. Smart (1988), Duality and sufficiency in control problems with invexity, *JMAA* **136**, 325-333.

[130] Mond, B. and T. Weir (1981-1983), Generalized convexity and higher-order duality, *J. Math. Sci.*, **16-18**, 74-94.

[131] Mordukhovich, B. S. (1988), *Approximation Methods in Problems of Optimization and Control*, Nauka, Moscow, Russian.

[132] Mukherjee, R. N. and S. K. Mishra (1994), Duality for multiobjective fractional variational problems, *Journal of Mathematical Analysis and Applications*, **186**, 711-725.

[133] Nahak, C. and S. Nanda (1997), Duality for multiobjective variational problems with pseudoinvexity, *Optimization*, **41**, 361-382.

[134] Nahak, C. and S. Nanda (1997), On efficiency and duality for multiobjective variational control problems with convexity, *JMAA* **209**, 415-434.

[135] Nahak, C. and S. Nanda (1996), Duality for multiobjective variational control problems with invexity, *Optimization* **36**, 235-258.

[136] Pandian, P. (2002), On sufficiency and duality for (b,F,$\rho$)-convex multiobjective programs, *Indian Journal of Pure & Applied Mathematics*, **33**, 463-473.

[137] Patel, R. B. (1997), Second order duality in multiobjective fractional programming, *Indian J. Math.*, **38**, 39-46.

[138] Pitea, A. and M. Postolache (2012), Duality theorems for a new class of multitime multiobjective variational problems, *Journal of Global Optimization*, **54**, 47-58.

[139] Preda, V. (1992), On efficiency and duality for multiobjective programs, *Journal of Mathematical Analysis and Applications*, **166**, 365-377.

[140] Reemtsen, R. and J. -J. Rückmann (1998), *Semi-Infinite Programming*, Kluwer, Boston.

[141] Robinson, S. M. (1984), Local structure of feasible sets in nonlinear programming, Part II, *Nondegeneracy, Mathematical Programming*, **22** (1), 217-230.

[142] Rockafellar, R. T. (1970), *Convex Analysis*, Princeton University Press.

[143] Rockafellar, R. T. and R. J-B. Wets (2004), *Variational Analysis*, Springer-Verlag, New York.

[144] Rückman, J. J. and A. Shapiro (2001), Second-order optimality conditions in generalized semi-infinite programming, *Set-Valued Anal.*, **9**, 169-186.

[145] Sawaragi, Y., H. Nakayama and T. Tanino (1986), *Theory of Multiobjective Optimization*, Academic Press, New York.

[146] Schaible, S. (1974), Parameter-free convex equivalent and dual programs of fractional programming problems, *Zeitschrift fur Operations Research*, **18**, 187-196.

[147] Schaible, S. (1976), Minimization of ratios, *Journal of Optimization Theory and Applications*, **19** (2), 347-352.

[148] Schaible, S. (1976), Fractional Programming I, duality, *Management Science*, **22** (8), 858-867.

[149] Schaible, S. (1981), Fractional Programming: Applications and Algorithms, *European Journal of Operational Research*, **7** (2), 111-120.

[150] Schaible, S. and T. Ibaraki (1983), Fractional Programming, *European Journal of Operational Research*, **12** (3), 325-338.

[151] Schaible, S. and J. Shi (2003), Fractional Programming: The Sum of ratios case, *Optimization Methods and Software*, **18** (2), 219-229.

[152] Schaible, S. (2002), Fractional programming: A recent survey, *J. Stat. Manag. Syst.*, **5**, 63-86.

[153] Schaible, S. and J. Shi (2004), Recent developments in fractional programming: single ratio and max-min case, *Nonlinear Analysis and Convex Analysis*, Yokohama Publ., Yokohama, pp. 493-506.

[154] Sagan, H. (1969), *Introduction to the Calculus of Variations*, McGraw-Hill, New York.

[155] Shapiro, A. (2005), On duality theory of convex semi-infinite programming, *Optimization*, **54**, 535-543.

[156] Shapiro, A. (2009), Semi-infinite programming, duality, discretization and optimality conditions, *Optimization*, **58**, 133-161.

[157] Shapiro, A. (1990), On concepts of directional differentiability, *Journal of Mathematical Analysis and Applications*, **66**, 477-487.

[158] Shi, A. (2001), A combined algorithm for fractional programming, *Annals of Operations Research*, **103**, 135-147.

[159] Shor, N. Z. (1985), On a bounding method for quadratic extremal problems with 0-1 variables, *Kibernetika* **2**, 48-50.

[160] Srivastava, M. K. and M. Bhatia (2006), Symmetric duality for multiobjective programming using second order $(F,\rho)$-convexity, *Opsearch*, **43**, 274-295.

[161] Srivastava, K. K. and M. G. Govil (2000), Second order duality for multiobjective programming involving $(F,\rho,\sigma)$-type I functions, *Opsearch* **37**, 316-326.

[162] Suneja, S. K., C. S. Lalitha and S. Khurana (2003), Second order symmetric duality in multiobjective programming, *European J. Oper. Res.*, **144**, 492-500.

[163] Stancu-Minasian, I. M. (1997), *Fractional Programming: Theory, Models and Applications*, Kluwer, Dordrecht.

[164] Stein, O. (2003), *Bilevel Strategies in Semi-infinite Programming*, Kluwer, Boston.

[165] Stein, O. and G. Still (2000), On optimality conditions for generalized semi-infinite programming problems, *Journal of Optimization Theory and Applications*, **104**, 443-458.

[166] Stein, O. and G. Still (2002), On generalized semi-infinite optimization and bilevel optimization, European *Journal of Operational Research*, **142** (3), 444-462.

[167] Stein, O. and G. Still (2003), Solving semi-infinite optimization problems with interior point techniques, *SIAM Journal on Control and Optimization*, **42** (3), 769-788.

[168] Still, G. (1999), Generalized semi-infinite programming: Theory and Methods, *European Journal of Operational Research*, **119**, 301-313.

[169] Still, G. (2001), Discretization in semi-infinite programming: The rate of convergence, *Mathematical Programming Series A*, **91** (1), 53-69.

[170] Vartak, M. N. and I. Gupta (1987), Duality theory for fractional programming problems under $\eta$-convexity, *Opsearch*, **24**, 163-174.

[171] Verma, R. U. (2016), Generalized higher order invexities in parametric optimality conditions for discrete minmax fractional programming, *Journal of Basic & Applied Sciences*, **12**, 283-300.

[172] Verma, R. U. (2016), Parametric duality models for discrete minmax fractional programming problems on higher order invex functions, *Annals of TPS*, **14**, 3-19.

[173] Verma, R. U. (2015), Generalized hybrid invexities and efficiency conditions for multiobjective fractional programming, *Tbilisi Mathematical Journal*, **8** (2), 159-180.

[174] Verma, R. U. (2013), New $\varepsilon-$optimality conditions for multiobjective fractional subset programming problems, *Transactions on Mathematical Programming and Applications*, **1 (1)**, 69-89.

[175] Verma, R. U. (2012), Second order $(\Phi,\Psi,\rho,\eta,\theta)-$invexity frameworks and $\varepsilon-$efficiency conditions for multiobjective fractional programming, *Theory and Applications of Mathematics & Computer Science*, **2 (1)**, 31-47.

[176] Verma, R. U. (2013), Role of second order $B-(\rho,\eta,\theta,\tilde{p},\tilde{r})-$invexities and parametric sufficient conditions in semiinfinite minimax fractional programming, *Transactions on Mathematical Programming and Applications*, **1 (2)**, 13-45.

[177] Verma, R. U. (2012), A generalization to Zalmai type second order univexities and applications to parametric duality models to discrete minimax fractional programming, *Advances in Nonlinear Variational Inequalities*, **15 (2)**, 113-123.

[178] Verma, R. U. (2012), A generalization to Zalmai type second order univexities and applications to parametric duality models to discrete minimax fractional programming, *Advances in Nonlinear Variational Inequalities*, **15** (2), 113-123.

[179] Verma, R. U. (2013), Generalized $(\mathscr{G}, b, \beta, \phi, h, \rho, \theta)$-univexities with applications to parametric duality models for Discrete Minimax Fractional Programming, *Transactions on Mathematical Programming and Applications*, **1**(1), 1-14.

[180] Verma, R. U. (2013), Advanced higher order univexities with applications to parametric duality models in discrete minimax fractional programming, *Transactions on Mathematical Programming and Applications*, **1 (3)**, 77-99.

[181] Verma, R. U. (2013), Generalized $(\mathscr{G}, b, \beta, \phi, h, \rho, \theta)$-univexities with applications to parametric duality models for discrete minimax fractional programming, *Transactions on Mathematical Programming and Applications*, **1 (1)**, 1-14.

[182] Verma, R. U. (2012), Second order $(\Phi, \Psi, \rho, \eta, \theta)-$invexity frameworks and $\varepsilon-$efficiency conditions for multiobjective fractional programming, *Theory and Applications of Mathematics & Computer Science* **2 (1)**, 31-47.

[183] Verma, R. U. (2013), Role of second order $B-(\rho, \eta, \theta, \tilde{p}, \tilde{r})-$invexities and parametric sufficient conditions in semiinfinite minimax fractional programming, *Transactions on Mathematical Programming and Applications* **1 (2)**, 13-45.

[184] Verma, R. U. (2013), Hybrid $(\mathscr{G}, \beta, \phi, \rho, \theta, \tilde{p}, \tilde{r})$-univexities of higher-orders and applications to minimax fractional programming, *Transactions on Mathematical Programming and Applications*, **1** (5), 63-86.

[185] Verma, R. U. (2013), Hybrid $(\mathscr{G}, \beta, \phi, \rho, \theta, \tilde{p}, \tilde{r})$-univexities of higher-orders and applications to minimax fractional programming, *Transactions on Mathematical Programming and Applications*, **1** (5), 63-86.

[186] Verma, R. U. and J. Liu (2017), Higher-order parametric optimality conditions for discrete minmax fractional programming based on univexity, *PanAmerican Mathematical Journal*, **27** (1), 98-117.

[187] Verma, R. U. and R. N. Mohapatra (2015), Generalized higher order hybrid invexities with second order parametric optimality criteria for discrete minmax fractional programming, *Communications on Applied Nonlinear Analysis*, **22** (4), 79-94.

[188] Verma, R. U. and Y. Seol (2015), Some sufficient efficient conditions in semiinfinite multiobjective fractional programming based on exponential type invexities, *Journal of Inequalities and Applications*, 2015-252 DOI 10.1186/s 13660-0775-4

[189] Verma, R. U. and G. J. Zalmai (2016), Second-order parametric optimality conditions in discrete minmax fractional programming, *Communications on Applied Nonlinear Analysis*, **23** (3), 1-32.

[190] Verma, R. U. and G. J. Zalmai (2016), Generalized second-order parametric optimality conditions in discrete minmax fractional programming and second order univexity, *Statistics, Optimization & Information Computing*, **4**, 15-29.

[191] Verma, R. U. and G. J. Zalmai (2015), Generalized parametric duality models in discrete minmax fractional programming based on second-order optimality conditions, *Communications on Applied Nonlinear Analysis*, **22** (2), 17-36.

[192] Verma, R. U. and G. J. Zalmai (2015), Generalized second-order parameter-free optimality conditions in discrete minmax fractional programming, *Communications on Applied Nonlinear Analysis*, **22** (2), 57-78.

[193] Verma, R. U.and G. J. Zalmai (2014), Parameter-free duality models in discrete minmax fractional programming based on second-order optimality conditions, *Transactions on Mathematical Programming and Applications*, **2** (11), 1-37.

[194] Verma, R. U. and G. J. Zalmai (2016), Second-order parametric optimality conditions in discrete minmax fractional programming, *Communications on Applied Nonlinear Analysis* **23** (3), 1-32.

[195] Verma, R. U. and G. J. Zalmai (2016), Generalized parametric duality models in discrete minmax fractional programming based on second-order optimality conditions, *PanAmerican Mathematical Journal*, **26** (3), 25-60.

[196] Verma, R. U. and G.J. Zalmai (2017), New generation parametric optimality conditions for semiinfinite discrete minmax fractional programming problems, *Transactions on Mathematical Programming and Applications* **5** (1), 25-50.

[197] von Neumann, J. (1945), A model of general economic equilibrium, *Review Econ. Stud.*, **13**, 1-9.

[198] Wang, S. Y. (1991), Second order necessary and sufficient conditions in multiobjective programming, *Numer. Funct. Anal. Optim.*, **12**, 237-252.

[199] Weber, G. -W., S. Z. Alparslan-Gök and A. Söyler (2009), A new mathematical approach in environmental and life sciences: gene-environment networks and their dynamics, *Environ. Model. Assess.*, **14**, 267-288.

[200] Weber, G. -W. and A. Tezel (2007), On generalized semi-infinite optimization of genetic networks, *TOP*, **15**, 65-77.

[201] Werner, J. (1988), Duality in generalized fractional programming, *Internat. Ser. Numer. Anal.*, **84**, 341-351.

[202] Xu, Z. (1996), Mixed type duality in multiobjective programming problems, *JMAA*, **198**, 621-653.

[203] Yang, X. M., X. Q. Yang, K. L. Teo and S. H. Hou (2004), Second order duality for nonlinear programming, *Indian J. Pure Appl. Math.*, **35** , 699-708.

[204] Yang, X. (1994), Generalized convex duality for multiobjective fractional programs, *Opsearch*, **31**, 155-163.

[205] Yang, X. Q. (1993), Second-order conditions in $C^{1,1}$ optimization with applications, *Numer. Funct. Anal. Optim.*, **14**, 621-632.

[206] Yang, X. Q. (1998), Second-order global optimality conditions of convex composite optimization, *Math. Prog.*, **81**, 327-347.

[207] Yang, X. Q. (2004), Second-order global optimality conditions for optimization problems, *J. Global Optim.*, **30**, 271-284.

[208] Yokoyama, K. (1996), Epsilon approximate solutions for multiobjective programming problems, *Journal of Mathematical Analysis and Applications*, **203 (1)**, 142-149.

[209] Yang, X. M., K. L. Teo and X. Q. Yang (2004), Higher-order generalized convexity and duality in nondifferentiable multiobjective mathematical programming, *J. Math. Anal. Appl.*, **29**, 48-55.

[210] Yang, X. M., X. Q. Yang and K. L. Teo (2003), Nondifferentiable second order symmetric duality in mathematical programming with F-convexity, *European J. Oper. Res.*, **144**, 554-559.

[211] Yang, X. M., X. Q. Yang and K. L. Teo (2005), Huard type second-order converse duality for nonlinear programming, *Appl. Math. Lett.*, **18**, 205-208.

[212] Yang, X., X. Q. Yang and K. L. Teo (2008), Higher-order symmetric duality in multiobjective programming with invexity, *J. Ind. Manag. Optim.*, **4**, 385-391.

[213] Zalmai, G. J. (1989), Optimality conditions and duality for constrained measurable subset selection problems with minmax objective functions, *Optimization*, **20**, 377-395.

[214] Zalmai, G. J. (2012), Generalized second-order $\mathscr{F},\beta,\phi,\rho,\theta)$-univex functions and parametric duality models in semiinfinite discrete minmax fractional programming, *Advances in Nonlinear Variational Inequalities*, **15**, 63-91.

[215] Zalmai, G. J. (2007), General parametric sufficient optimality conditions for discrete minmax fractional programming problems containing generalized $(\theta,\eta,\rho)$-V-invex functions and arbitrary norms, *Journal of Applied Mathematics & Computing*, **23 (1-2)**, 1-23.

[216] Zalmai, G. J. (2013), Semiinfinite multiobjective fractional programming problems involving Hadamard directionally differentiable functions, part I: Sufficient efficiency conditions, *Trans. Math. Prog. Appl.*, **1**(9), 129-147.

[217] Zalmai, G. J. (2013), Semiinfinite multiobjective fractional programming problems involving Hadamard directionally differentiable functions, part II: First-order parametric duality models, *Trans. Math. Prog. Appl.*, **1**(10), 129-147.

[218] Zalmai, G. J. and Qinghong Zhang (2013), Necessary efficiency conditions for semiinfinite multiobjective optimization problems involving Hadamard directionally differentiable functions, *Trans. Math. Prog. Appl.*, **1**(1), 129-147.

[219] Zalmai, G. J. (2012), Hanson-Antczak-type generalized invex functions in semiinfinite minmax fractional programming, Part I: Sufficient optimality conditions, *Communications on Applied Nonlinear Analysis*, **19(4)**, 1-36.

[220] Zalmai, G. J. (2013), Hanson - Antczak - type generalized $(\alpha,\beta,\gamma,\xi,\eta,\rho,\theta)$-V- invex functions in semiinfinite multiobjective fractional programming Part I, Sufficient efficiency conditions, *Advances in Nonlinear Variational Inequalities*, **16** (1), 91-114.

[221] Zalmai, G. J. (2012), Generalized second-order $(\mathscr{F},\beta,\phi,\rho,\theta)$-univex functions and parametric duality models in semiinfinite discrete minmax fractional programming, *Advances in Nonlinear Variational Inequalities*, **15** (2), 63-91.

[222] Zalmai, G. J. (2007), Parametric duality models for discrete minmax fractional programming problems containing generalized $(\theta,\eta,\rho)$-V-invex functions and arbitrary norms, *Journal of Applied Mathematics & Computing*, **24** (1-2), 105-126.

[223] Zalmai, G. J. and Q. Zhang (2007), Generalized $(\mathscr{F},\beta,\phi,\rho,\theta)$-univex functions and parametric duality models in semiinfinite discrete minmax fractional programming, *Advances in Nonlinear Variational Inequalities*, **10** (2), 21-42.

[224] Zalmai, G. J. and Q. Zhang (2008), Global semiparametric sufficient effficient conditions for desiinfinite multiobjective fractional programming problems involving $(\alpha,\eta,\rho)-V-$ invex functions, *Southeast Asian Bulletin of Mathematics*, **32**, 573-599.

[225] Zalmai, G. J. and Q. Zhang (2011), Semiinfinite multiobjective fractional programming, Part II, Duality models, *Journal of Applied Analysis*, **17**, 1-35.

[226] G. J. Zalmai, G. J. and Q. Zhang (2010), Semiinfinite multiobjective fractional programming, Part I, Sufficient efficiency conditions, *Journal of Applied Analysis*, **16**, 199-224.

[227] Zalmai, G. J. and Qinghong Zhang (2007), Generalized $(F,\beta,\phi,\rho,\theta)$-univex functions and global parametric sufficient optimality conditions in semiinfinite discrete minmax fractional programming, *PanAmerican Math. J.*, **17**, 1-26.

[228] Zhang, J. and B. Mond (1996), Second order b-invexity and duality in mathematical programming, *Utilitas Math.*, **50**, 19-31.

[229] Zhang, J. and B. Mond (1997), Second order duality for multiobjective nonlinear programming involving generalized convexity, in *Proceedings of the Optimization*

*Miniconference III* (B. M. Glover, B. D. Craven, and D. Ralph, eds.), University of Ballarat, pp. 79-95.

[230] Zalmai, G. J. (2013), Hanson-Antczak-type generalized $(\alpha, \beta, \gamma, \xi, \eta, \rho, \theta)$-V-invex functions in semiinfinite multiobjective fractional programming, Part I: Sufficient efficiency conditions, *Adv. Nonlin. Var. Ineq.*, **16**, 91-114.

[231] Zalmai, G. J. and Qinghong Zhang (2013), Necessary efficiency conditions for semiinfinite multiobjective optimization problems involving Hadamard directionally differentiable functions, *Transactions on Mathematical Programming and Applications*, **1** (1), 129-147.

[232] Zalmai, G. J. and R. U. Verma (2016), Generalized parameter-free duality models in discrete minmax fractional programming based on second-order optimality conditions, *Mathematical Sciences*, **10** (4),185-199.

[233] Zeidler, E. (1985), *Nonlinear Functional Analysis and its Applications III: Variational Methods and Optimization*, Springer-Verlag, New York.

[234] Zhang, J. and B. Mond (1997), Duality for a nondifferentiable programming problem, *Bull. Australian Math. Soc.* **55**, 29-44.

[235] Zhian, L. and Y. Qingkai (2001), Duality for a class of multiobjective control problems with generalized invexity, *Journal of Mathematical Analysis and Applications* **256**, 446-461.

# About the Author

**Ram U. Verma, PhD**
President
International Publications USA
Mathematical Sciences Division
Denton, TX, US
Email: ramverma114@yahoo.com

Ram U. Verma is President of International Publications USA, and prior to this, he held faculty positions (lecturer, Assistant Professor, Associate Professor and Professor) at the University of Cape Coast through University of Tripoli, New York University, University of Orient, University of Puerto Rico, University of Toledo, Texas State University, and beyond. He is the Founding Editor-in-Chief of four journals from International Publications USA as follows:

Editor-in-Chief, *Advances in Nonlinear Variational Inequalities*
Editor-in-Chief, *Communications on Applied Nonlinear Analysis*
Editor-in-Chief, *Pan American Mathematical Journal*
Editor-in-Chief, *Transactions on Mathematical Programming and Applications*
Editor, *Involve* (University of California at Berkeley)
Editor, *International Journal of Math & Math Sciences (Hindawi)*
Editor, *Journal of Operators (Hindawi)*
Associate Editor, *Applied Mathematics and Computation* (Elsevier)

His research interests encompass mathematical programming, fractional programming, semi infinite fractional programming, multi objective fractional programming, numerical analysis, generalized Newton's methods, next generation Newton type methods, nonlinear functional analysis, applied analysis, evolution equations and semigroup, stochastic analysis, mathematics education, and determinant theory for singular integral equations. His overall research publications are 725 in peer-reviewed international journals (including *Applied Mathematics and Computation*, *Applicable Analysis*, *Archivum Math*, *Communications in Nonlinear Science and Numerical Simulations*, *Czechoslovak Mathematical Journal*, *Electron. Journal of Differential Equations*, *Journal of Computational Analysis and Applications*, *Journal of Mathematical Analysis and Applications*, *Journal of Optimization Theory and Applications*, *Nonlinear Analysis: TMA*, *Numerical Functional Analysis and Optimization*, *Proceedings of the American Mathematical Society*, *Proceedings of the Royal Irish Academy*, and *ZAMM: Z. Angew. Math. Mech.*).

# Index

## A



## B



## C



## D



## E



## F

## G



## H



## I



## L



## M



## N



## O



## P



## Q



## R



## S

## T



## U



## V



## W



## Y